Transformative Concepts for Drug Design:
Target Wrapping

Ariel Fernández

Transformative Concepts for Drug Design: Target Wrapping

 Springer

Prof. Dr. Ariel Fernández
Karl F. Hasselmann Chair in Engineering
Rice University
Department of Bioengineering
E200K George R. Brown Hall
Houston TX 77005
USA

ISBN 978-3-642-11791-6   e-ISBN 978-3-642-11792-3
DOI 10.1007/978-3-642-11792-3
Springer Heidelberg Dordrecht London New York

Library of Congress Control Number: 2010922998

Cover design: WMXDesign GmbH, Heidelberg

Printed on acid-free paper

Springer is part of Springer Science+Business Media (www.springer.com)

Preface

To a man with a hammer everything looks like a nail.
Mark Twain

Notwithstanding the enticing promises of the post-genomic era, the pharmaceutical world appears to be in a state of disarray. Drug discovery seems riskier and more uncertain than ever as projects get routinely terminated in mid-stage clinical trials, as the dearth of new targets becomes apparent, and as successful therapeutic agents are often recalled whenever an idiosyncratic side effect is detected. Exploiting the huge output of genomic data to make more efficacious and safer drugs has proven to be much more difficult than anticipated. More than ever, the lead in the pharmaceutical industry depends on the ability to harness innovative research, and this type of innovation can only come from one source: *fundamental knowledge*. This book has a place in this scenario, as it introduces fundamental discoveries in basic biomolecular research that hold potential to become transformative and broaden the technological base of the pharmaceutical industry.

The book takes a fresh and fundamental look at the problem of how to design an effective drug with controlled specificity. Within the pharmaceutical industry, it is of course superfluous to recall that the principal bottleneck in developing new drugs is the clinical uncertainty stemming from the lack of control of specificity. Chemists know how to increase affinity, but when they do this, the affinity of the drug to structurally similar molecules also increases, target discrimination becomes very difficult, and adverse side-effects due to unwanted binding are usually sufficiently severe to render the drug unusable.

The secret of how nature manages to design molecules with extraordinarily high and specific affinities lies in cooperativity. In medicine, we are nearly always working in aqueous media and therefore cooperativity needs to be looked at in the specific context of aqueous systems.

Recognizing that these concepts are unfamiliar to most practitioners, the first part of this book (Chaps. 1, 2, 3, 4, 5, and 6) explains these matters very carefully starting from a fairly elementary physico-chemical level. The second part of the book (Chaps. 7, 8, 9, 10, 11, 12, 13, and 14) is devoted to practical applications. We are aiming at nothing less than a paradigm shift in drug design.

Thus, cooperativity emerges as a molecular design principle in Chaps. 7, 8, 9, 10, 11, 12, 13, and 14, but this incarnation is only possible after the concept is explored from architectural, biophysical, bioinformatics and evolutionary perspectives in the preparatory Chaps. 1, 2, 3, 4, 5, and 6.

This book is above all addressed to scientists working at the cutting edge of research in the pharmaceutical industry, but the material is at the same time fully accessible to senior undergraduates or graduate students interested in fundamental concepts on drug discovery. It essentially covers my lectures on systems biology and molecular design, an elective undergraduate and graduate level course for bioengineering majors at Rice University.

It has been a pleasure to work with the talented staff at Springer. I am especially grateful to Marion Hertel (executive editor), and to Cornelia Kinsky, Beate Siek and Sam Roobesh for their helpful cooperation and enduring patience.

Houston, USA   Ariel Fernández

Contents

Chapter 1
Protein Cooperativity and Wrapping: Two Themes in the Transformative Platform of Molecular Targeted Therapy

In spite of the enticing promises of the post-genomic era, the pharmaceutical world appears to be in a state of disarray. Projects get routinely terminated in mid-stage clinical trials, the scarcity of new targets is ever so apparent, and successful therapeutic agents are often recalled as idiosyncratic side effects are detected in patient subpopulations. The vast and seemingly endemic problems of the pharmaceutical industry are not confined to the scientific realm but the latter has much to do with the current stagnation. Properly harvesting and ultimately exploiting the output of genomic forays to make more efficacious and safer drugs has proven to be much more difficult than originally thought. In spite of the huge output of integrative post-genomic studies, drug discovery and development remain essentially a serendipitous endeavor where high-throughput screening and toxicological studies are favored over rational molecular design. Thus, more than ever, the lead in the pharmaceutical industry depends pivotally on our ability to harness innovative high-risk research. This chapter and ultimately this book may have a place in this scenario, as we introduce fundamental discoveries in basic biomolecular-level research that hold potential to become transformative and broaden the technological base of the pharmaceutical industry.

This chapter sets the tone for the entire book as it delineates the molecular basis of cooperativity, a crucial biomolecular concept largely overlooked in drug design. Cooperativity is shown to be tightly related to a particular molecular attribute of target proteins known as "wrapping." This structure-based feature and its exploitation in the contexts of drug safety, specificity, and personalized therapy will become the leitmotiv of the book.

1.1 Many-Body Problems for the Drug Designer

As we examine the current state of molecular targeted therapy, the first thing that strikes our attention is the fact that the design of therapeutic drugs is hardly ever rational. While the biophysical principles governing the affinity of a drug for a target biomolecule are believed to be understood, the control of specificity, the safety, and the idiosyncratic efficacy of the therapeutic agents remain very opaque subjects. They are typically dealt with through painstaking trial and error and at an

A. Fernández, *Transformative Concepts for Drug Design: Target Wrapping*,
DOI 10.1007/978-3-642-11792-3_1, © Springer-Verlag Berlin Heidelberg 2010

enormous cost because our a priori understanding of the therapeutic context is sketchy at best. Thus, clinical uncertainty and unpredictable adverse effects often hamper or impede drug development and this situation is unlikely to change unless a higher level of conceptual innovation is effectively incorporated in the discovery pipeline.

We shall narrow down our treatment of these vast problems to small-molecule drugs purposely engineered to target human proteins and thereby inhibit their biological function. Aiming at a paradigm shift in the field, we advocate a translational top-down approach that takes us back to the very fundamentals of protein associations as we introduce a foundational platform for a next generation of safer and more effective drugs.

In molecular therapy, we often deal with water-soluble proteins that are targeted by man-made ligands; therefore the efficacy and target specificity of a molecular design depends pivotally on our understanding of protein–ligand associations. In this regard, there is a crucial property that seems to have been missed altogether in rational drug design: *cooperativity*. We somewhat narrowly define this property as the concurrent participation of different regions of the biomolecule to promote and sustain intramolecular or intermolecular interactions. In plain terms, "cooperativity is the nonadditive contribution to protein interactions," a peculiar property often illustrated by the phrase "the whole is more than the sum of the parts." Cooperativity is thus an attribute of natural proteins that plays a decisive role in determining how the peptide chain folds into its native 3D structure and form associations or complexes with other molecules.

In our context of interest, the nonadditive nature of protein associations implies that *the rational drug designer faces a many-body problem*: the interactions between the protein target and the drug/ligand involve more than groups matched up in a pairwise fashion at the target–ligand interface. Because protein–ligand interactions take place in an aqueous medium, this many-body problem is a very special one. As we shall advocate throughout this book, matching groups with complementary biochemical properties across the target/ligand interface is only one aspect of what rational design is about and by no means the decisive one, as the evidence attests. The next generation of molecular designs must take into account modes of association or binding above and beyond pairwise intermolecular interactions involving groups in the ligand and their purported matched groups in the target.

Be as it may, the current design paradigm is unlikely to change anytime soon unless a clear case can be made for cooperativity, and the right computational tools are brought to fruition to operationally incorporate this concept in drug design. Cooperativity will broaden the technological base of drug design only if it is introduced in a manner that is suitable to address the manifold practical problems that the pharmaceutical industry is currently addressing. Thus, as we deal with cooperativity, perhaps the first core question that needs to be dealt with is: *What sort of many-body problem is the drug designer facing and how can this knowledge play advantageously to address the major therapeutic imperatives of today and tomorrow?* The answer to this pressing question will unravel as we labor through the pages of this book and will be hinted at in this chapter.

1.2 Cooperative Protein Interactions: The Need for the Wrapping Concept

Protein structure in solution is assumed to arise and be sustained by forces that are essentially electrostatic [1–4]. Even the hydrophobic attraction between two nonpolar groups, an entropic effect arising from the minimization of unfavorable interfaces with water, includes a major electrostatic contribution as it increases the extent of hydrogen bonding among surrounding water molecules [3]. The forces that drive protein folding and protein associations are actually modulated by an important factor often neglected: *the shaping of the solvent microenvironment wherein the forces become operational* [2]. Since the shaping of the microenvironment around a pairwise intramolecular interaction requires the participation of other regions of the molecule, we may state that cooperativity is inherent to the folding of a protein chain [4–9].

To illustrate the importance of cooperativity, we may recall that an electrostatic interaction occurring in bulk water is 78 times weaker than the same interaction in an anhydrous medium [2, 10]. Thus, the stability and strength of pairwise interactions between different parts of the peptide chain is determined not only by the atomic groups directly engaged in the interaction but also by the groups involved in shaping their microenvironment by promoting the expulsion of surrounding water [10]. The latter contributors are just as important, as they determine either the persistence or the ephemeral nature of the interactions and, ultimately, the integrity of the protein structure [2]. In fact, low-permittivity microenvironments around the backbone hydrogen bonds of a self-interacting polypeptide chain are essential to promote and sustain its structure and have been the focus of much attention as we attempt to underpin the physical basis of cooperativity [2, 4, 11].

The backbone of a protein or peptide chain is highly polar, comprising an amide and carbonyl group per residue. This chemical feature introduces constraints on the nature of the hydrophobic collapse [9] and on the chain composition of foldable proteins, i.e., those capable of sustaining such a collapse [12, 13]. Thus, the hydrophobic collapse entails the dehydration of backbone amides and carbonyls and such a process would be thermodynamically unfavorable unless amides and carbonyls engage in hydrogen bonding with each other [9]. Only a hydrophobic collapse that ensures the formation and protection of backbone hydrogen bonds is likely to be conducive to sustainable folding [11].

The hydration of amides and carbonyls competes with the formation of the intramolecular hydrogen bonds. Thus, the structural integrity of proteins is compromised by a "deficiently wrapped" backbone [11, 14]. Wrapping refers to a cluster of nonpolar groups around a pre-formed Coulombic interaction [13]. The need for protection of intramolecular hydrogen bonds from water attack is an important factor in determining the chain composition of a *foldable* protein, that is, of a chain capable of sustaining a soluble structure and folding expeditiously and reproducibly [11].

As noted above, the strength and stability of backbone hydrogen bonds clearly depend on the microenvironment where they occur: The proximity of nonpolar groups to a hydrogen bond enhances the electrostatic interaction by de-screening the partial charges or lowering the local environment permittivity [10, 12]. These nonpolar groups also stabilize the hydrogen bond by destabilizing the nonbonded state, i.e., by hindering the hydration of the polar groups in the nonbonded state [12, 13]. Thus, to guarantee the integrity of soluble protein structure, most intramolecular hydrogen bonds must be surrounded or "*wrapped*" by nonpolar groups fairly thoroughly as to become significantly dehydrated [11–14].

To make the wrapping concept more precise, we need a definition that enables a direct assessment of the extent of hydrogen bond protection from structure coordinates. This parameter, denoted ρ, is given by the number of side chain carbonaceous nonpolar groups (CH_n, $n = 0$, 1, 2, 3) contained within a desolvation domain that represents the hydrogen bond microenvironment. This domain is defined as the reunion of two intersecting spheres of fixed radius ($\sim$thickness of three water layers) centered at the α-carbons of the residues paired by the hydrogen bond. In structures of PDB-reported soluble proteins, backbone hydrogen bonds are protected on average by $\rho = 26.6 \pm 7.5$ side chain nonpolar groups for a desolvation sphere of radius $r = 6\,\text{Å}$. The desolvation domain adopted for a hydrogen bond is a residue-based feature, incorporating a descriptor of the local environment of each of the paired residues [13]. It fully subsumes the local environment of the hydrogen bond itself since the heavy atoms N and O are invariably within 6 Å of the α-carbons of the paired residues and hence fully contained in the intersection of the desolvation spheres.

Accordingly, if we treat the solvent implicitly, we may identify a *wrapping interaction* as a three-body *ijk*-interaction, where nonpolar group k contributes to the removal of water from the microenvironment of an amide–carbonyl hydrogen bond that pairs groups i and j. In other words, the "wrapping" nonpolar group k lies within the desolvation domain of the *ij*-interaction. This type of three-body contribution is cooperative by definition and needed to maintain the structural integrity of the protein [2, 13]. An illustration of wrapping interactions in the native structure of human *ubiquitin* (PDB accession code PDB.1UBI) is given in Fig. 1.1. Thus, the wrapping of the Lys27–Gln31 backbone hydrogen bond by residues Ile36 and Lys29 is represented by thin blue lines. These residues contribute four and three nonpolar groups, respectively, to the desolvation domain ($r = 6\,\text{Å}$) of the backbone hydrogen bond.

Taken together, the hydration propensity of amide and carbonyl and the dehydration-induced strengthening of their electrostatic association represent two conflictive tendencies, suggesting that there must be a crossover point in the dehydration propensity of a backbone hydrogen bond [15]. If the bond is poorly wrapped by a few nonpolar groups that cluster around it, then hydration of the paired amide

---

Fig. 1.1 (a) Cooperative "wrapping" interactions involving residues Lys29 and Ile36 that contribute to stabilize and enhance the backbone hydrogen bond Lys27–Gln31 (*thin white line*) in human ubiquitin (PDB.1UBI). Each interaction generates a three-body energy term, and a residue contributing with nonpolar side chain groups to the desolvation domain of the hydrogen bond is represented by a *thin blue line* joining the α-carbon (*gray sphere*) of the residue with the center of

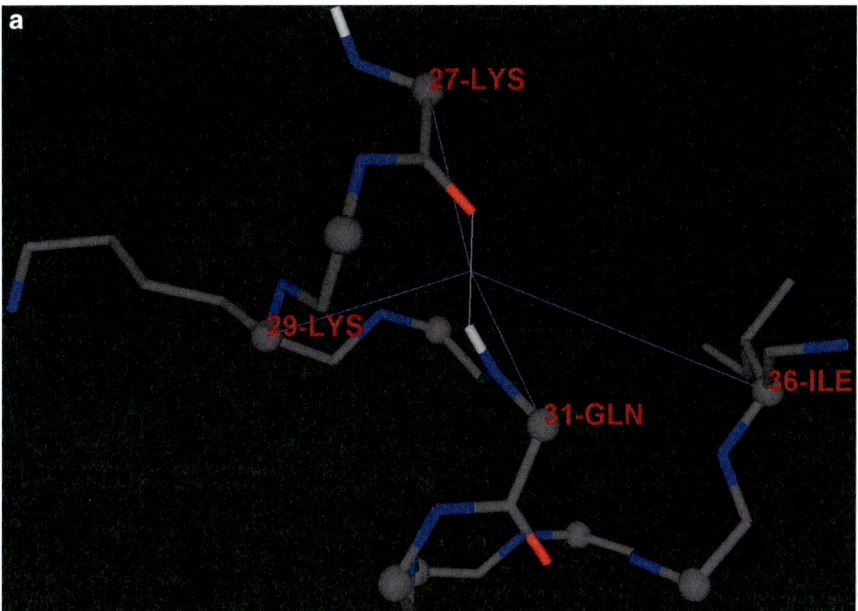

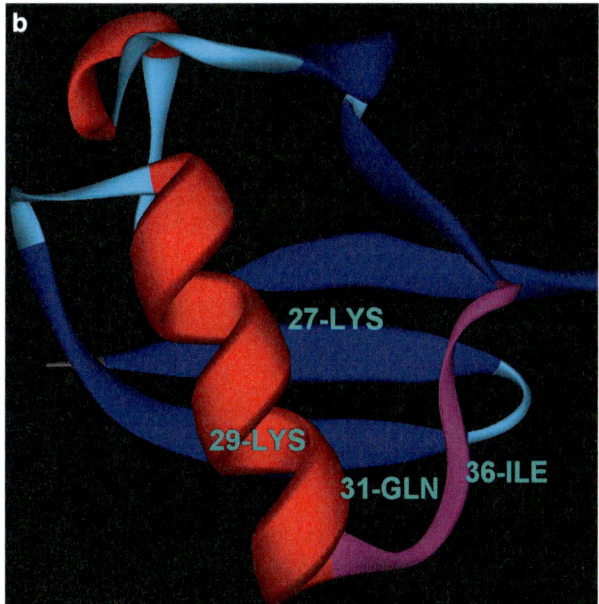

Fig. 1.1 (continued) the amide–carbonyl hydrogen bond. Conventional colors are used for atom representation and the protein backbone is represented schematically, except for the two residues paired by the backbone hydrogen bond that are displayed in full backbone detail. Only the side chains of the wrapping residues 29Lys and 36Ile are shown. (**b**) Location of the residues in a ribbon rendering the native structure of human ubiquitin

and carbonyl is favored and prevails, but as the hydrogen bond becomes better wrapped, the surrounding water loses too many hydrogen bonding partnerships and thus may be favorably removed [13]. This observation is essential to rationalize the cooperative two-state nature of the folding of single-domain proteins [16–18], as shown in Chap. 3: we may say that the state of hydration of a protein hydrogen bond is in a statistical sense a local reflection of the degree of progress of the folding process.

1.3 Poorly Wrapped Hydrogen Bonds are Promoters of Protein Associations

The structural integrity of a soluble protein is contingent on its ability to exclude water from its amide–carbonyl hydrogen bonds [11, 12]. Thus, water-exposed intramolecular hydrogen bonds, the so-called *dehydrons*, constitute structural weaknesses taking the particular form of wrapping deficiencies [12, 19, 20]. On the other hand, these defects favor the removal of surrounding water as a means to strengthen and stabilize the underlying electrostatic interaction [13, 20, 21], and thus are predictably implicated in protein associations [21], aberrant aggregation [22], and macromolecular recognition [23, 24]. By exogenously contributing to the wrapping of pre-formed hydrogen bonds, these associations in effect remove the wrapping defects, thereby stabilizing the structure.

Intramolecular hydrogen bonds that are not "wrapped" by a sufficient number of nonpolar groups in the protein itself may become stabilized and strengthened by the attachment of a ligand, i.e., a potential drug, or a binding partner that further contributes to their dehydration (Fig. 1.2) [19, 21]. Ample bioinformatics evidence on the distribution of dehydrons at the interface of protein complexes support this physical picture [13, 21]. Thus, dehydrons are decisive factors driving association in 38% of the PDB complexes and constitute significant factors (interfacial dehydron density larger than average on individual partners) in about 95% of all complexes reported in the PDB [13].

Dehydrons may be identified from atomic coordinates of proteins with reported structure, as illustrated in Fig. 1.2. Thus, we need to introduce an auxiliary quantity, the extent of hydrogen bond wrapping, ρ, indicating the number of nonpolar groups contained within a "desolvation domain" around the bond. In structures of soluble proteins at least two-thirds of the backbone hydrogen bonds are wrapped on average by $\rho = 26.6 \pm 7.5$ nonpolar groups for a desolvation ball radius 6 Å. Dehydrons lie in the tails of the distribution, i.e., their microenvironment contains 19 or fewer nonpolar groups, so their ρ-value is below the mean ($\rho = 26.6$) minus 1 standard deviation ($\sigma = 7.5$) [12, 13].

Thus, the overall under-wrapping of a protein may be assessed by determining the percentage of intramolecular hydrogen bonds satisfying the inequality $\rho \leq 19$, that is, the percentage of dehydrons in its structure. An example of the under-wrapping of a protein is given in Fig. 1.3, where the dehydron pattern of human ubiquitin is displayed.

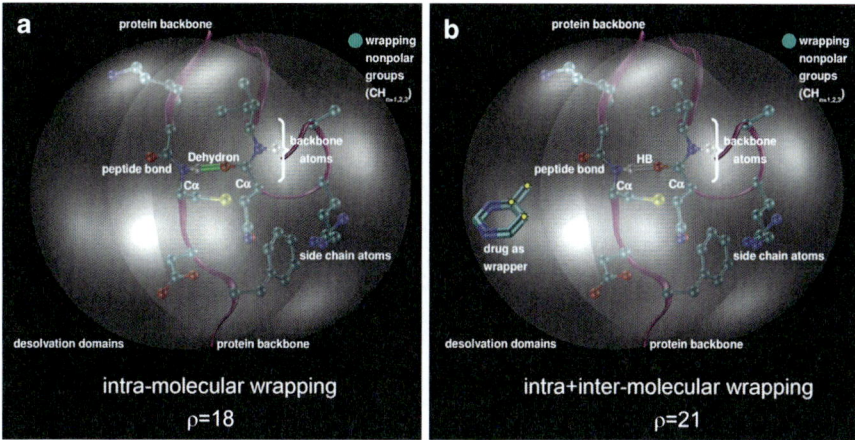

Fig. 1.2 (**a**) Dehydron in a soluble protein. The dehydron ($\rho = 18$), marked in *green*, pairs two backbone groups (amide and carbonyl, conventional colors for atoms). The microenvironment is indicated by two intersecting *gray* spheres centered at the α-carbons of the paired residues. Wrapping side chain groups are shown in *light blue* and only side chains contributing (fully or partially) to the dehydration of the hydrogen bond are indicated. (**b**) The drug depicted in the figure acts as an exogenous wrapper of the hydrogen bond (*gray bond*, $\rho = 21$) turning the dehydron into a well-protected bond (the three atoms marked with $*$ complete the desolvation of the dehydron)

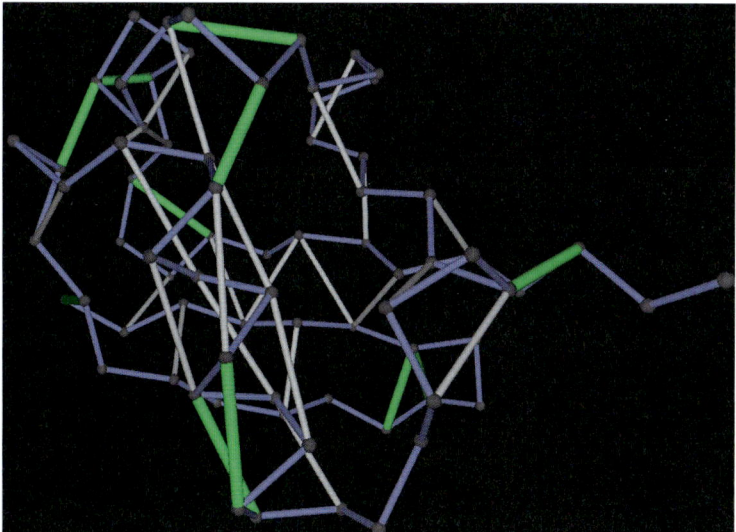

Fig. 1.3 Illustration of the under-wrapping of protein structure. Dehydron pattern of human ubiquitin (ribbon display in Fig. 1.1b). Dehydrons are indicated as *green* segments joining the α-carbons of the paired units, well-wrapped hydrogen bonds ($\rho > 19$) are shown in *light gray*, and the protein backbone is conventionally shown as *blue* virtual bonds joining the α-carbons of consecutive amino acid units. The displayed structure has 33 backbone hydrogen bonds, of which 11 are dehydrons. Thus, the extent of under-wrapping for this protein is 33%

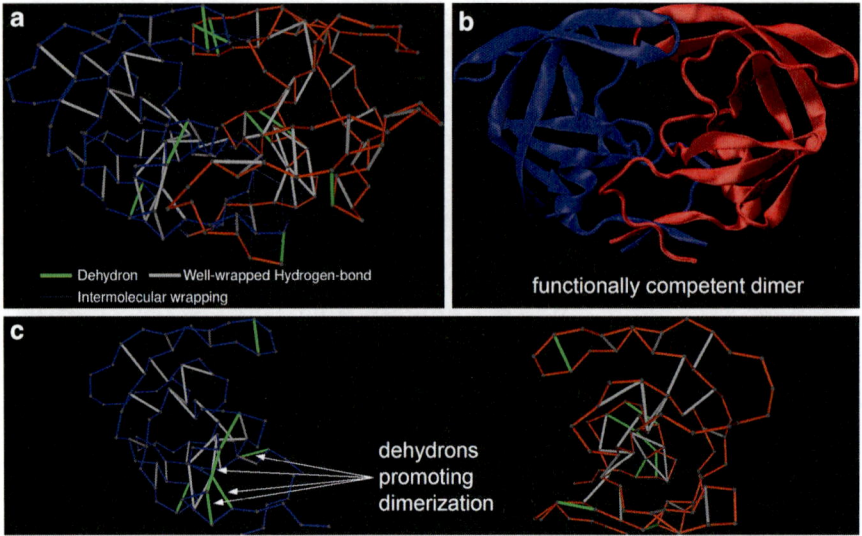

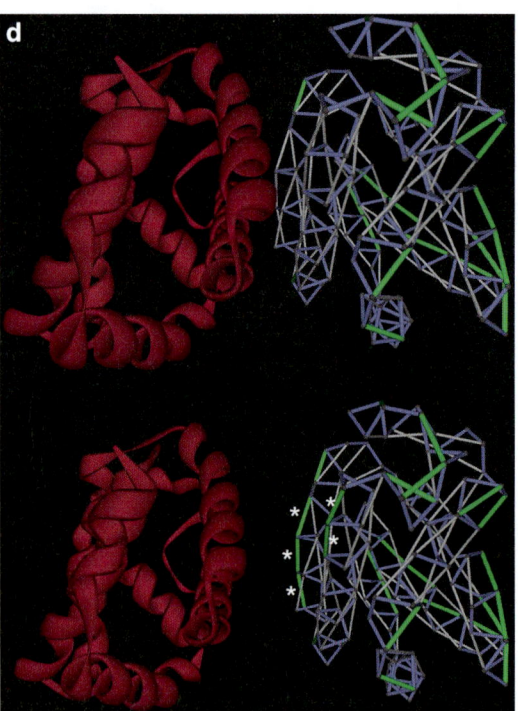

Fig. 1.4 (**a**) Intermolecular wrapping in the human HIV-1 protease dimer (PDB.1A30) as a means of protecting the enzyme structure from water attack. Dehydrons are indicated as *green* segments joining the α-carbons of the paired units, well-wrapped hydrogen bonds are shown in *light gray*, and the protein backbone is conventionally shown as virtual bonds joining the α-carbons of

Dehydron-rich regions in soluble proteins are typical hot spots for protein associations because of their propensity toward further dehydration [13, 21]. A functional perspective reinforces this view, since dehydrons constitute vulnerabilities that need to be "corrected" to maintain the integrity of the protein structure and its functional competence. Thus, specific residues of the binding partner contribute to the desolvation of dehydrons, as they penetrate the desolvation domain of intramolecular hydrogen bonds upon association. This intermolecular wrapping is illustrated in Fig. 1.4a–c, displaying the functionally competent human HIV-1 protease dimer. The purported interfacial region of the homodimer contains seven dehydrons when the monomeric domains are considered in isolation and separated from each other. Upon association, specific residues of the binding partner contribute to the desolvation of some of the dehydrons, as they enter the desolvation domain of the intramolecular hydrogen bonds. This intermolecular wrapping reduces the vulnerability of the protease, which only has three dehydrons at the interface upon complexation, instead of the original seven in the monomeric form. Similarly, the allosteric clam hemoglobin, a functionally competent homodimer, utilizes five dehydrons of the monomeric state to accommodate its quaternary structure. These five dehydrons become well wrapped in the dimer (Fig. 1.4d).

Throughout the book, dehydrons will be referred to in different ways depending on the context. Thus, the terms *packing defect, wrapping defect, dehydron, structural deficiency, structural vulnerability* will be used synonymously. Far from introducing a notational chaos, this name multiplicity bespeaks of the richness of the concept.

1.4 Wrapping Defects Are Sticky

As indicated above, dehydrons have unique physico-chemical properties: They represent structural vulnerabilities of the protein, but they also constitute sticky spots promoting the removal of surrounding water [12–14, 19, 21]. This latter property could only be established by addressing the following questions: How do we effectively demonstrate that a dehydron attracts nonpolar test groups? Can we measure the mechanical equivalent of its dehydration propensity?

Fig. 1.4 (continued) consecutive amino acid units. Complexation reduces the structural vulnerability of the protease by reducing the number of dehydrons at the interface from 7 to 3. Intermolecular wrapping is depicted by *thin blue lines* from the α-carbon of the wrapping residue to the middle of the hydrogen bond that is intermolecularly protected. Thus, an intermolecular wrapping residue contributes with nonpolar groups to the dehydration of the pre-formed hydrogen bond from the binding partner. (**b**) Ribbon display of the dimer. (**c**) Wrapping of separate domains of the human HIV-1 protease. (**d**) Allosteric clam hemoglobin, a functionally competent homodimer that utilizes five dehydrons (marked by *asterisks* in *lower panel*) of the monomeric state to create a favorable interface for its quaternary structure. The five dehydrons become well wrapped in the dimeric state (*upper panel*). The wrapping pattern of one monomer is displayed while the other monomer is shown in ribbon representation

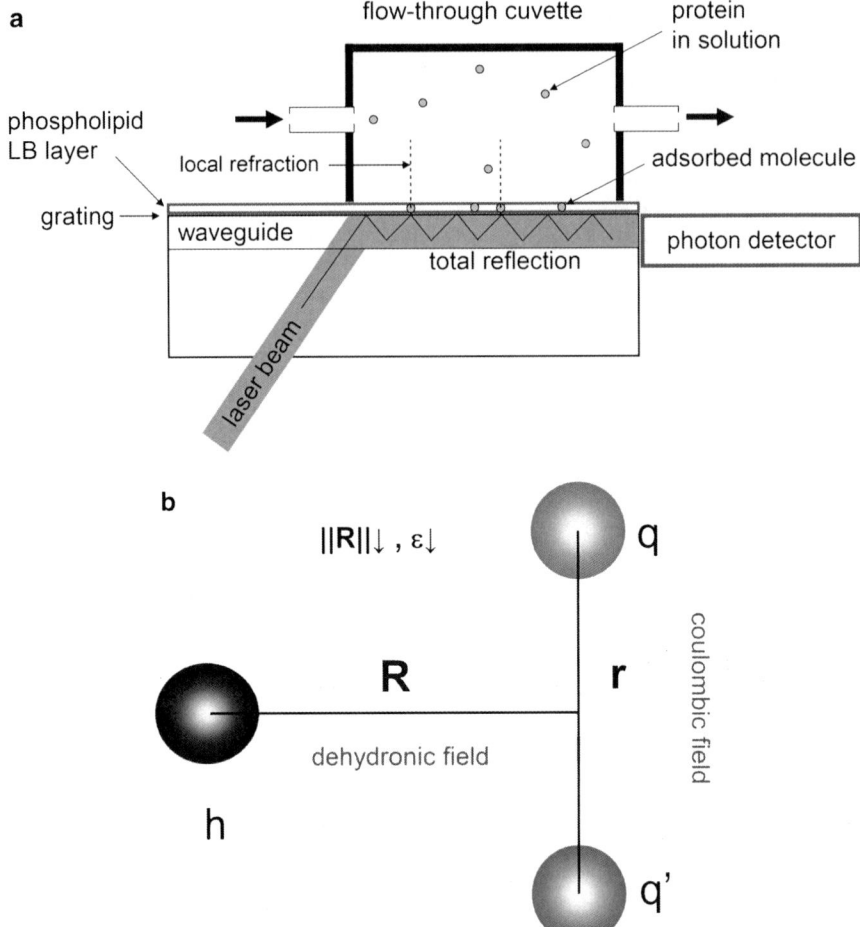

Fig. 1.5 **(a)** High-precision total reflection setup to measure the adsorption uptake of proteins onto a hydrophobic surface under controlled hydrodynamic conditions [19]. The adsorption uptake is proportional to the photon loss due to local alterations in the refractive index of a Langmuir–Blodgett (LB) layer that constitutes the wrapping medium for the protein. **(b)** Orthogonality between dehydronic field exerted on the test hydrophobe (h) along coordinate **R** and the Coulomb field exerted along coordinate **r** between two spherical charges q, $q\prime$

Reported experimental work addressed these questions by measuring the adsorption of proteins with wrapping defects and equivalent surface hydrophobicity (area of solvent-exposed nonpolar surface) onto a "wrapping" layer [19]. This wrapping medium consists of a Langmuir–Blodgett phospholipid film coating a waveguide, as shown in Fig. 1.5a. These high-precision experiments made use of evanescent-field spectroscopic interrogation of the wrapping medium enabling a direct measurement of the protein adsorption uptake. This observable is determined by detecting local changes in refractive index of the phase within which total reflection of the incident light occurred [19]. Thus, a beam from a He–Ne laser travels through a waveguide at an incidence angle suitable for total reflection within the medium. The adsorbed molecules alter the refractive properties of the hydrophobic layer and consequently alter the critical angle for total reflection. Thus, protein adsorption is commensurate with photon loss resulting from the extent of local refraction or "evanescent field." Hence adsorption uptake can be determined by the loss of photons due to refractive photon leakage from the total reflection pattern.

For proteins with comparable surface hydrophobicity, the adsorption uptake correlates strongly with the extent of protein under-wrapping [19]. As an adequate control, only proteins with the same extent of surface hydrophobicity or solvent-exposed nonpolar area were included in the comparative analysis. Hence, the attractive drag exerted by dehydrons on test hydrophobes became accessible. The net gain in Coulomb energy associated with wrapping a dehydron has been experimentally determined to be ~ 4 kJ/mol [19]. The adhesive force exerted by a dehydron on a hydrophobe at 6 Å distance is $\sim$7.8 pN, a magnitude comparable to the hydrophobic attraction between two nonpolar moieties that frame unfavorable interfaces with water.

This study was motivated by the earlier observations that dehydrons play a pivotal role in driving protein associations, as such associations contribute intermolecularly to the wrapping of pre-formed structure [12,14], as discussed in the previous section. In consistency with current terminology, the force stemming from the dehydration propensity of the partially wrapped hydrogen bond is hithertofore termed *dehydronic*. The dehydronic force arises as a nonpolar group approaches a dehydron with a net effect of immobilizing and ultimately removing surrounding water molecules. This displacement lowers the polarizability of the microenvironment which, in turn, de-shields the paired charges [12, 19]. Thus, a net attractive force is exerted by the dehydron on a nonpolar group and this force represents the mechanical equivalent of the dehydration propensity of an unburied pre-formed hydrogen bond. Since the water molecules solvating an amide and carbonyl paired by a dehydron are necessarily depleted of some hydrogen bonding partners, the work required for their ultimate removal from the bond surroundings is minimal [12, 22]. The dehydronic field, denoted $\mathbf{\Phi}(\mathbf{R})$, is necessarily orthogonal to the Coulomb field generated by the polar (amide–carbonyl) pair and may be described within a quasi-continuous treatment of the solvent by the equation

$$\mathbf{\Phi}(\mathbf{R}) = -\nabla_{\mathbf{R}} \left[4\pi \varepsilon(\mathbf{R}) \right]^{-1} qq'/r, \qquad (1.1)$$

where $\mathbf{R}$ represents the position vector of the hydrophobe or nonpolar group with respect to the center of mass of the hydrogen-bonded polar pair, $\nabla_{\mathbf{R}}$ is the gradient taken with respect to this vector, r is the distance between the charges of magnitude q and $q\prime$ paired by the hydrogen bond (Fig. 1.5b), and the local permittivity coefficient $\varepsilon = \varepsilon(\mathbf{R})$ subsumes the polarizability of the microenvironment, which is generically dependent on the position of the test hydrophobe [2, 14]. An appropriate expression for $\varepsilon(\mathbf{R})$ valid at nanoscales is unavailable at present [25], because of the discreteness of the dielectric medium and the need to include individual solvent dipole correlations [26]. Although a mean-field dielectric description is unsatisfactory, it is still possible to assert that $\Phi(\mathbf{R})$ is an attractive force since a decrease in $\|\mathbf{R}\|$ entails a decrease in local polarization which, in turn, enhances the Coulomb attraction.

Building on this analysis, we may quantify the net hydrophobicity η of a hydrogen bond by taking into account the surface flux of the dehydronic field generated by the hydrogen bond. This field is given by $\omega^{-1}\Phi(\mathbf{R})$ (ω = volume of test hydrophobe). Thus, in accord with Gauss theorem we obtain

$$\eta = \int\int_{\Sigma} \Phi(\mathbf{R}).d\,\sigma(\mathbf{R}), \tag{1.2}$$

where Σ is the closed surface of the dehydration domain of the hydrogen bond (cf. Fig. 1.2) and $d\sigma(\mathbf{R})$ is the differential surface area vector at position $\mathbf{R}$.

1.5 Cooperative Drug–Target Associations: A Window into Molecular Engineering Possibilities

Like in any protein–ligand complex, the wrapping intermolecular interactions between a drug and its target protein are expected to play a pivotal role in determining the affinity. It is worth noting here that such cooperative interactions between the drug and pre-formed hydrogen bonds in the target protein actually entail a net gain to the stability of the protein of ∼4 kJ/mol per wrapping interaction. Thus, the drug contributes with nonpolar groups to the desolvation of pre-existing dehydrons in the target and each wrapping contribution translates into a ∼4 kJ/mol decrease in the free energy change associated with drug binding. As an illustration, the wrapping contributions between the powerful anticancer drug imatinib (Gleevec®) and one of its primary therapeutic targets, the KIT kinase [27, 28], are depicted in Fig. 1.6. The wrapping interactions are promoted by the dehydronic fields generated by the deficiently wrapped hydrogen bonds 673Cys–676Gly, 595Leu–603Val, 640Glu–644Leu, and 811Phe–814Ala.

The kinase inhibitor imatinib was not *purposely* designed as a wrapper of its purported protein targets, yet, as shown in Fig. 1.6, five wrapping interactions contribute to its affinity for the primary target KIT kinase. A crude estimate would indicate that this contribution to the association free energy is actually significant (∼5 × 4 kJ/mol = 20 kJ/mol). Notice that while imatinib was selected for its pharmacological properties that conferred anticancer activity [27], it is unlikely to have emerged from a conventional structure-based design. Matching a nonpolar group in the drug against a carbonyl or amide in the target protein is hardly conceivable in conventional structure-based design as it would entail a hindrance to the hydration of the polar groups. Yet, as a three-body contribution, the proximity of a nonpolar group to an unburied polar pair is justified given the dehydronic field generated by the latter. Furthermore, given the extent to which wrapping interactions are likely to affect drug affinity, one wonders whether the wrapping concept can be fruitfully exploited to guide drug design.

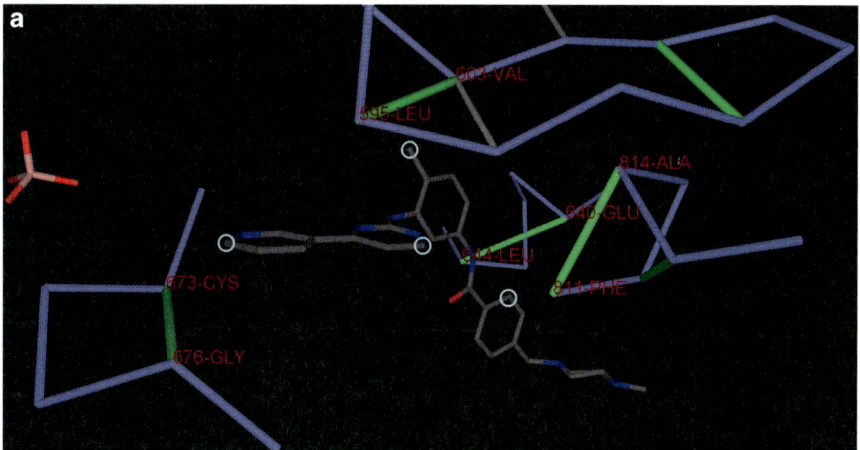

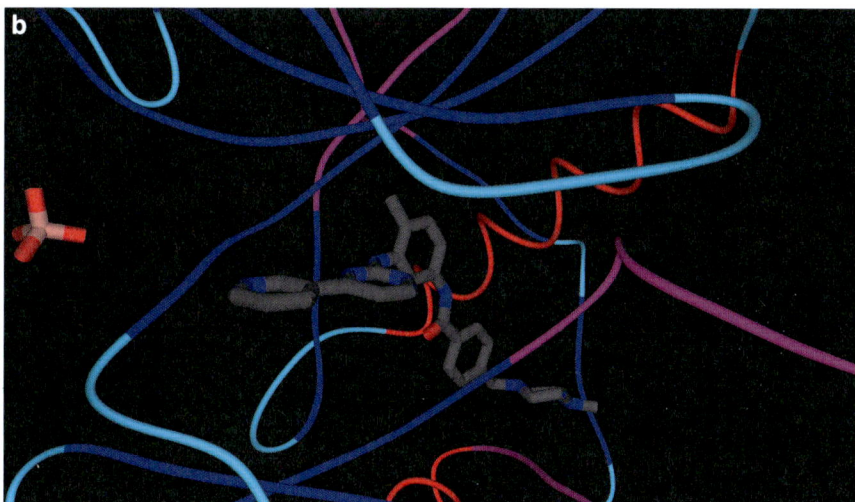

Fig. 1.6 (**a**) Dehydrons in the KIT kinase wrapped intermolecularly by the kinase inhibitor imatinib in the crystallized drug/target complex (PDB.1T46). The drug nonpolar groups contributing to the wrapping upon association are marked by *circles*. The dehydrons (*green*) wrapped by the drug involve residue pairs 673Cys–676Gly, 595Leu–603Val, 640Glu–644Leu, and 811Phe–814Ala. (**b**) Simplified tube rendering of the protein backbone provided as visual aid. (**c**) Detail of intermolecular wrapping interaction between imatinib and the KIT kinase. The drug penetrates the desolvation domain (intersecting *pink spheres*) of KIT dehydron 640Glu–644Leu upon binding, contributing with two nonpolar groups to the desolvation of the pre-formed hydrogen bond

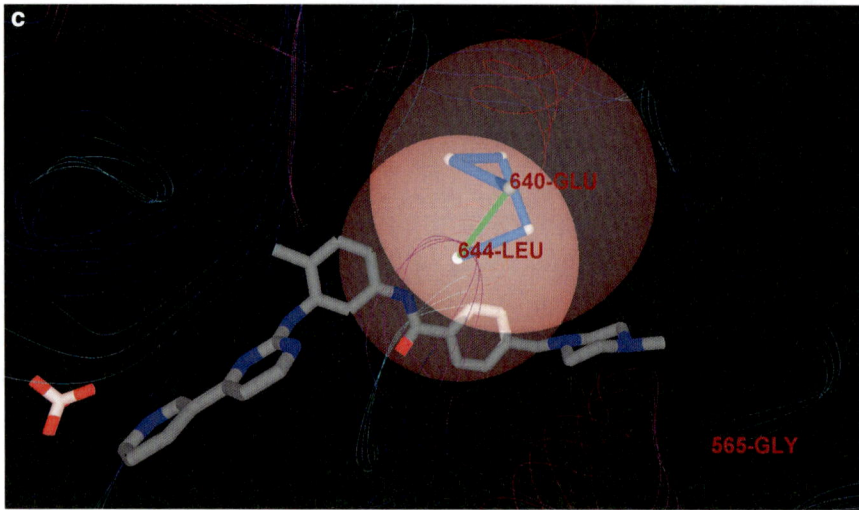

Fig. 1.6 (continued)

This possibility naturally elicits some basic questions:

- Can we purposely design "cooperative drugs" that wrap dehydrons in the protein target?
- What is the potential impact of such a concept on the technological base of drug discovery and what are the advantages it is likely to provide when compared with conventional structure-based design?
- Can this type of cooperative-based design lead to a better exploitation of genomic information to yield safer and more efficacious therapeutic agents?
- Can we harness idiosyncratic differences in the wrapping patterns across patients to develop a personalized treatment?

These are all crucial problems in molecular targeted therapy that will be addressed in the subsequent chapters as we assess the transformative power of the wrapping concept.

References

1. Anfinsen CB (1973) Principles that govern the folding of protein chains. Science 181:223–230
2. Fernández A, Sosnick TR, Colubri A (2002) Dynamics of hydrogen-bond desolvation in folding proteins. J Mol Biol 321:659–675
3. Chandler D (2005) Interfaces and the driving force of hydrophobic assembly. Nature 437: 640–647
4. Jewett A, Pande VS, Plaxco KW (2003) Cooperativity, smooth energy landscapes and the origins of topology-dependent protein folding rates. J Mol Biol 326:247–253
5. Scalley-Kim M, Baker D (2004) Characterization of the folding energy landscapes of computer generated proteins suggests high folding free energy barriers and cooperativity may be consequences of natural selection. J Mol Biol 338:573–583

6. Fernández A, Colubri A, Berry RS (2002) Three-body correlations in protein folding: The origin of cooperativity. Physica A 307:235–259

7. Fernández A, Kostov K, Berry RS (1999) From residue matching patterns to protein folding topographies: General model and bovine pancreatic trypsin inhibitor. Proc Natl Acad Sci USA 96:12991–12996

8. Fernández A, Colubri A, Berry RS (2000) Topology to Geometry in protein folding: Beta-lactoglobulin. Proc Natl Acad Sci USA 97:14062–14066

9. Fernández A, Kardos J, Goto J (2003) Protein folding: Could hydrophobic collapse be coupled with hydrogen-bond formation? FEBS Lett 536:187–192

10. Fernández A (2001) Conformation-dependent environments in folding proteins. J Chem Phys 114:2489–2502

11. Fernández A, Kardos J, Scott R, Goto Y, Berry RS (2003) Structural defects and the diagnosis of amyloidogenic propensity. Proc Natl Acad Sci USA 100:6446–6451

12. Fernández A (2004) Keeping dry and crossing membranes. Nat Biotechnol 22: 1081–1084

13. Pietrosemoli N, Crespo A, Fernández A (2007) Dehydration propensity of order-disorder intermediate regions in soluble proteins. J Proteome Res 6:3519–3526

14. Fernández A, Scott R (2003) Dehydron: A structure-encoded signal for protein interactions. Biophys J 85:1914–1928

15. Avbelj F, Baldwin RL (2003) Role of backbone solvation and electrostatics in generating preferred peptide backbone conformations: distributions of phi. Proc Natl Acad Sci USA 100:5742–5747

16. Krantz BA, Moran LB, Kentsis A, Sosnick TR (2000) D/H amide kinetic isotope effects reveal when hydrogen bonds form during protein folding. Nat Struct Biol 7:62–71

17. Fersht A (2000) Transition-state structure as a unifying basis in protein-folding mechanisms: Contact order, chain topology, stability, and the extended nucleus mechanism. Proc Natl Acad Sci USA 97:1525–1929

18. Plaxco KW, Simmons KT, Baker D (1998) Contact order, transition state placement and the refolding rates of single domain proteins. J Mol Biol 277:985–994

19. Fernández A, Scott LR (2003). Adherence of packing defects in soluble proteins. Phys Rev Lett 91:018102

20. Fernández A, Zhang X, Chen J (2008) Folding and wrapping soluble proteins: Exploring the molecular basis of cooperativity and aggregation. Prog Nucleic Acids Res Transl Sci 83:57–87

21. Fernández A, Scheraga HA (2003) Insufficiently dehydrated hydrogen bonds as determinants of protein interactions. Proc Natl Acad Sci USA 100:113–118

22. Fernández A, Berry RS (2003) Proteins with H-bond packing defects are highly interactive with lipid bilayers: Implications for amyloidogenesis. Proc Natl Acad Sci USA 100: 2391–2396

23. Deremble C, Lavery R (2005) Macromolecular recognition. Curr Opin Struct Biol 15: 171–175

24. Ma B, Elkayam T, Wolfson H, Nussinov R (2003) Protein-protein interactions: Structurally conserved residues distinguish between binding sites and exposed protein surfaces. Proc Natl Acad Sci USA 100:5772–5777

25. Fernández A (2003) What caliber pore is like a pipe? Nanotubes as modulators of ion gradients. J Chem Phys 119:5315–5319

26. Despa F, Fernández A, Berry RS (2004) Dielectric modulation of biological water. Phys Rev Lett 93:228104

27. Demetri G (2002) Efficacy and safety of imatinib mesyalte in advanced gastrointestinal stromal tumors. N Engl J Med 347:472–480

28. Fernández A, Sanguino A, Peng Z et al (2007) An anticancer C-kit kinase inhibitor is reengineered to make it more active and less cardiotoxic. J Clin Invest 117:4044–4054

Chapter 2
Wrapping Defects and the Architecture of Soluble Proteins

Wrapping defects in soluble proteins represent local weaknesses of the native structures and have received little attention, especially by the drug design community. The protein structure may be inherently weak at sites where hydration of the backbone is locally hampered by formation of an intramolecular hydrogen bond which in turn is not stabilized through complete burial within a hydrophobic environment. This chapter explores the architectural implications stemming from the existence of these vulnerabilities. Thus, the unburied backbone hydrogen bonds or *dehydrons* are shown to be compensated by disulfide bridges that are needed to maintain the structural integrity in extracellular environments. Examination of all reported soluble structures reveals that the number of disulfide bonds correlates tightly with the number of dehydrons in a 1:5 ratio. The results have implications for biomolecular design as they introduce universal constraints in the architecture of water-soluble proteins.

2.1 How Do Soluble Proteins Compensate for Their Wrapping Defects?

Backbone hydration, prevalent in the unfolded state of a polypeptide chain, is often hindered in a soluble folded state as backbone amides and carbonyls are paired through hydrogen bonds [1, 2]. Yet, the thermodynamic cost of dehydration is not always compensated, especially if the backbone hydrogen bond is not completely sequestered from solvent. In soluble proteins, such bonds may be readily identified from the structural coordinates by determining the number of nonpolar groups within the bond microenvironment [3–5]. These unburied backbone hydrogen bonds constitute structural deficiencies and represent markers for protein associations [5]. In turn, these associations are required to maintain the structural integrity of the protein through intermolecular protection of the pre-formed hydrogen bonds [5].

A thorough examination of the protein data bank (PDB) singles out toxin peptides with picomolar affinity for the Kv1.3 potassium channel [6], such as HsTX1 (PDB.1QUZ), as members of the protein family with the highest extent of structural

A. Fernández, *Transformative Concepts for Drug Design: Target Wrapping*, DOI 10.1007/978-3-642-11792-3_2, © Springer-Verlag Berlin Heidelberg 2010

deficiency. The unburied hydrogen bonds in such proteins can make up to 100% of the backbone hydrogen bonds. This observation immediately suggests a reason for the extremely high target affinity of neurotoxins: according to Chap. 1, we expect a huge dehydronic field for such biomolecules.

A separate analysis reveals that such proteins contain an inordinately large number of disulfide bonds, with an average of 11 when normalized to 100 amino acids. These observations prompt us to investigate the relation between structural deficiency and disulfide bonds in search for a balance equation that reflects a statistical correlation between structural strengths and vulnerabilities of soluble proteins and polypeptides. The balance equation unraveled in this chapter is likely to impact the design of soluble proteins and enable a better control of their functional modulation in relation to environmental redox conditions.

A comprehensive wrapping analysis of the PDB yields the structural deficiencies of an exhaustive nonredundant set of 2,989 monomeric uncomplexed soluble proteins or peptides with disulfide links and 8,975 proteins without disulfide links [7]. These data are compiled as illustrated in Table 2.1 for some selected PDB entries. Peptide chains were excluded from the analysis if their structural integrity required prosthetic groups or cation coordination. In order to compare protein or peptide chains of different lengths, two normalized parameters were used to characterize a protein structure: $Y =$ number of structural deficiencies per 100 backbone hydrogen bonds and $X =$ number of disulfide bonds per 100 amino acids. Proteins were binned according to their X-value in integer groups with $n = 0, 1, \ldots, 18$, where proteins with no disulfide bonds ($X = 0$) belong to group $n = 0$ and proteins with X in the range $n < X \leq n + 1$ belonged to group $n + 1$. The mean Y-value and standard deviation were computed for each n-group and the results are shown in Fig. 2.1a. A tight X–Y linear correlation ($R^2 = 0.96$) results and is further corroborated by the raw X–Y-trendline generated by linear regression on all (X, Y) data points (Fig. 2.1b). Taken together, the results from Fig. 2.1 unambiguously reveal a simple balance relation $Y = 5X + 20$. *This statistical relation introduces a 1:5 ratio to buttress vulnerable proteins and allows for a 20%-baseline in structural deficiency.*

Table 2.1 Structural parameters for few selected monomeric uncomplexed soluble proteins with disulfide bridges lacking prosthetic groups and scaffolding cation coordination

PDB entry	Disulfide bonds	Chain length	Structural deficiencies	Backbone HBs	X	Y	Non-α/β residues (%)
2PNE	2	81	37	37	2.47	100.00	100.00
1M4F	4	25	8	8	16.00	100.00	68.00
1EZG	8	81	33	48	9.87	68.75	77.78
1HOE	2	74	11	34	2.70	32.35	54.05
2Z9T	1	100	10	40	1.00	25.00	50.00
135L	4	129	35	109	3.10	32.11	63.57
153L	2	185	38	185	1.08	20.54	48.65
154L	2	185	39	194	1.08	20.10	48.65
172L	1	164	47	188	0.61	25.00	30.49
1A2J	1	189	57	196	0.53	29.08	39.15
1A39	9	402	69	250	2.24	27.60	59.70

Table 2.1 (continued)

PDB entry	Disulfide bonds	Chain length	Structural deficiencies	Backbone HBs	X	Y	Non-α/β residues (%)
1A3P	2	45	9	18	4.44	50.00	82.22
1A43	1	87	22	67	1.15	32.84	49.43
1A67	2	108	20	62	1.85	32.26	48.15
1A7M	3	180	37	155	1.67	23.87	41.67
1AC5	3	483	91	370	0.62	24.59	51.76
1ACJ	3	537	77	404	0.56	19.06	50.09
1ACW	3	29	18	18	10.34	100.00	41.38
1ACX	2	108	20	44	1.85	45.45	56.48
1ADX	3	40	5	7	7.50	71.43	100.00
1ADZ	3	71	11	35	4.23	31.43	76.06
1AE5	4	225	33	123	1.78	26.83	62.22
1AEC	3	218	33	155	1.38	21.29	55.96
1AFH	4	93	39	74	4.30	52.70	46.24
1AG2	1	103	15	80	0.97	18.75	43.69
1AGG	4	48	11	14	8.33	78.57	87.50
1AGI	3	125	18	82	2.40	21.95	48.80
1AGY	2	200	43	158	1.00	27.22	51.50
1AH1	2	129	15	51	1.55	29.41	60.47
1AHK	3	129	13	43	2.33	30.23	80.62
1AHL	3	49	6	14	6.12	42.86	87.76
3DHM	1	100	10	50	1.00	20.00	52.00
3DIH	7	122	35	97	5.74	36.08	52.46
3EGP	1	108	8	37	0.93	21.62	58.33
3EHS	3	476	93	359	0.63	25.91	40.34
3EMY	1	329	44	198	0.30	22.22	49.85
3ENG	7	213	34	119	3.29	28.57	66.67
3EO5	1	171	47	125	0.58	37.60	48.54
3EOW	2	221	21	75	0.90	28.00	70.14
3ETP	2	187	22	89	1.07	24.72	51.87
3EXD	4	129	35	113	3.10	30.97	63.57
3EZM	2	101	31	57	1.98	54.39	42.57
3GF1	3	70	25	53	4.29	47.17	77.14
3LYM	4	129	36	109	3.10	33.03	63.57
3LZ2	4	129	32	91	3.10	35.16	64.34
3MAN	1	302	58	252	0.33	23.02	50.99
3PTE	1	349	70	297	0.29	23.57	50.43
3RAT	4	124	28	81	3.23	34.57	49.19
3RSD	4	124	25	78	3.23	32.05	49.19
3SEB	1	238	27	153	0.42	17.65	50.00
3SSI	2	113	26	69	1.77	37.68	59.29
3TGF	3	50	13	34	6.00	38.24	68.00
3TGL	3	269	42	221	1.12	19.00	49.81
4AIT	2	74	13	35	2.70	37.14	54.05
4APE	1	330	49	195	0.30	25.13	47.88
4CMS	3	323	53	203	0.93	26.11	42.72
4ENG	7	210	34	119	3.33	28.57	67.14
4RAT	4	124	29	82	3.23	35.37	49.19
4TGL	3	269	36	204	1.12	17.65	50.19

Table 2.1 (continued)

PDB entry	Disulfide bonds	Chain length	Structural deficiencies	Backbone HBs	X	Y	Non-α/β residues (%)
4TSV	1	150	15	70	0.67	21.43	50.67
5LYT	4	129	32	102	3.10	31.37	62.79
5LYZ	4	129	29	90	3.10	32.22	64.34
5PEP	3	326	51	200	0.92	25.50	46.01
5RAT	4	124	28	79	3.23	35.44	49.19
5RNT	2	104	21	62	1.92	33.87	64.42
6HIR	3	65	10	13	4.62	76.92	90.77
6LYT	4	129	32	103	3.10	31.07	62.79
6LYZ	4	129	27	95	3.10	28.42	62.79
7LYZ	4	129	24	88	3.10	27.27	62.79
7RAT	4	124	28	81	3.23	34.57	49.19
8PTI	3	58	10	29	5.17	34.48	65.52
8RAT	4	124	25	76	3.23	32.89	49.19
9RAT	4	124	28	76	3.23	36.84	49.19

To illustrate the role of this relation in defining protein structure we focus on specific examples of proteins with widely diverse levels of structural deficiency. Figure 2.2a, b shows the structural deficiencies of the α-amylase inhibitor HOE-467A (PDB.1HOE; length $N = 74$; 2 disulfide bonds). This protein contains 23 fully buried backbone hydrogen bonds (5 are double bonds, with each paired residue contributing both proton donor and acceptor, and 13 are single bonds) and 11 structural deficiencies (2 double bonds and 7 single bonds). Hence, its parameters are $Y = 11 \times 100/(11 + 23) = 32.35$ and $X = 2 \times 100/74 = 2.70$. Applying the balance relation, we get the estimate $Y = 5 \times 2.7 + 20 = 33.5$, differing in less than 1% from the real Y-value. β-2-Microglobulin (PDB.2Z9T, $N = 100$, $X = 1$) has 30 fully buried hydrogen bonds (9 double bonds and 12 single bonds) and 10 structural deficiencies (1 double bond), as shown in Fig. 2.2c, d. Hence, following the balance equation, Y is estimated at $5 \times 1 + 20 = 25$, which agrees exactly with the actual value $Y = 10 \times 100/(10 + 30)$. Finally, the antimicrobial hormone hepcidin (PDB.1M4F, $N = 25$, 4 disulfide bonds or $X = 16.00$) has a fully defective structure with $Y = 100$ (Fig. 2.2e, f), fitting exactly the balance equation $100 = 5 \times 16 + 20$.

Anomalously large Y-values may be explained as we incorporate the polyproline II (PPII)-conformation content [8–10] to the structural analysis. The hydrogen-bonded amide and carbonyls in the backbone of residues in PPII conformation maximize their solvent exposure and hence hydrogen bond burial is not required as a provider of thermodynamic compensation. Hence, proteins with large PPII content in their folded native state maximize their backbone exposure to solvent to a level where backbone hydrogen bonds no longer hinder the hydration of amides and carbonyls [9]. At this anomalously high level of backbone hydration, the structure-destabilizing contribution of structural deficiencies represented by unburied backbone hydrogen bonds (dehydrons) is reduced significantly, and therefore, so is the need for disulfide-bond compensation.

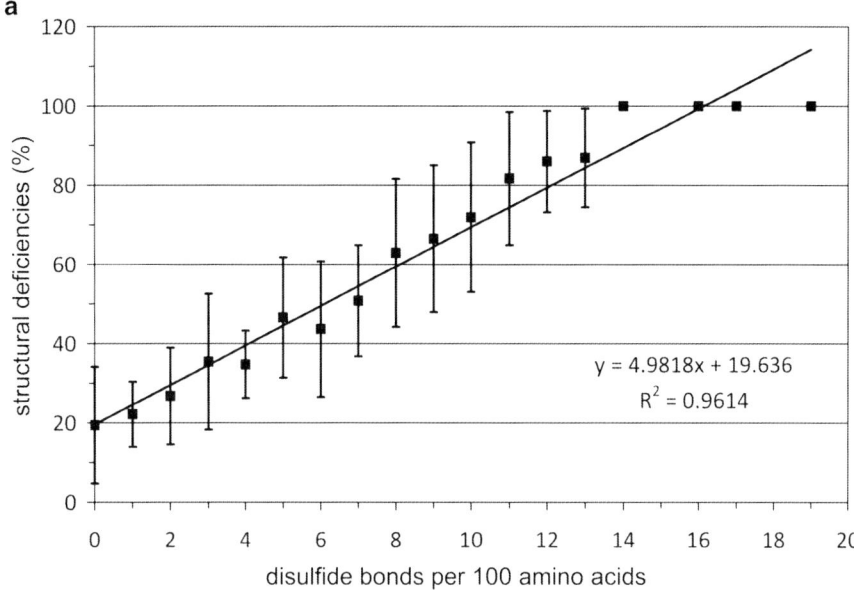

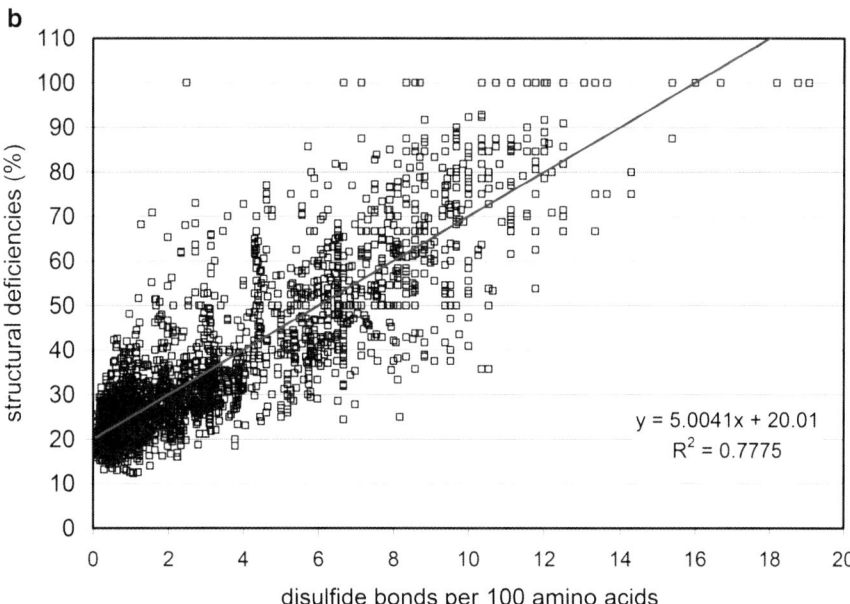

Fig. 2.1 Number of structural deficiencies (unburied backbone hydrogen bonds, dehydrons) normalized to 100 backbone hydrogen bonds (Y) plotted against number of disulfide bonds normalized to 100 amino acids (X) for PDB-reported soluble proteins. **a** Mean Y-value (*square*) and standard deviation (error bar) for proteins grouped according to their number of disulfide bonds. Proteins were binned according to their X-value in integer groups with $n = 0, 1, \ldots, 18$, where proteins with no disulfide bonds ($X = 0$) belong to group $n = 0$ and proteins with $n < X \le n + 1$ belong to group $n + 1$. **b** All (X, Y)-data points from the nonredundant exhaustive set of PDB entries for uncomplexed soluble proteins

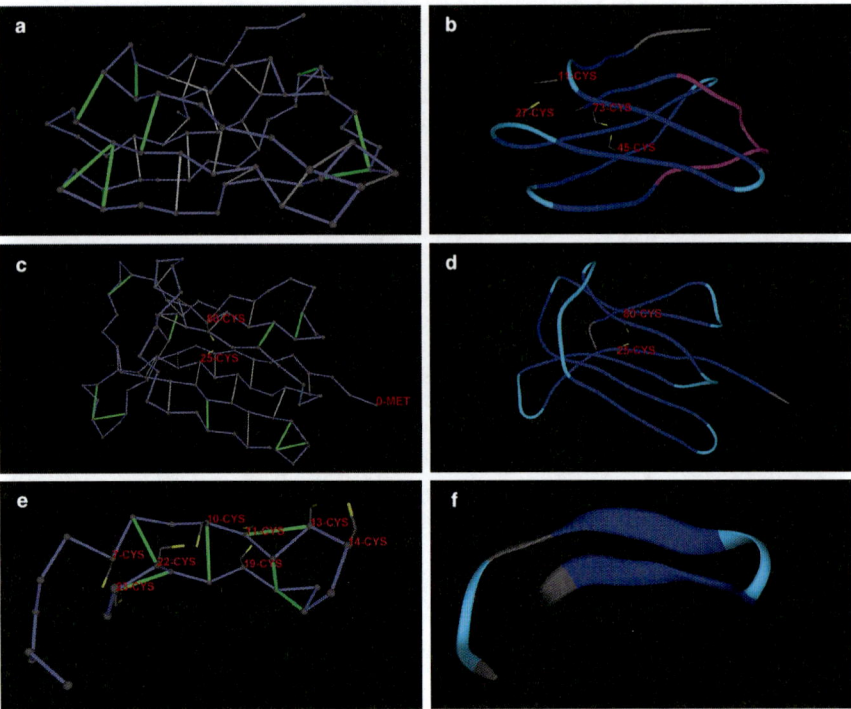

Fig. 2.2 Structural deficiencies in soluble proteins. The protein backbone is shown as virtual bonds (*blue*) joining consecutive α-carbons in the peptide chain. Light-gray segments joining α-carbons represent completely buried backbone hydrogen bonds and *green* segments represent structural deficiencies (unburied backbone hydrogen bonds). A tube/ribbon representation is added for visual aid. Cysteines involved in disulfide bonds are identified by side chain display. Structural deficiencies (**a, c, e**) and tube/ribbon representation (**b, d, f**), respectively, of α-amylase inhibitor HOE-467A (PDB.1HOE) (**a, b**), β-2 microglobulin (PDB.2Z9T) (**c, d**), and antimicrobial hormone hepcidin (PDB.1M4F) (**e, f**)

2.2 Thermodynamic Support for the Dehydron/Disulfide Balance Equation

The dehydron/disulfide balance relation clearly identifies proteins with excess ($Y > 5X + 20$) or lack ($Y < 5X + 20$) of structural deficiencies, with the former likely to be more favorably denatured than the latter under equivalent redox and denaturation conditions. To test this prediction, thermodynamic data on thermal denaturation (Table 2.2) were obtained for an exhaustive set of proteins for which structural information was also available [7]. Thus, the thermal denaturation free energy change, ΔG, under reducing conditions and comparable temperatures [11], was obtained for monomeric uncomplexed PDB-reported proteins with disulfide bonds and lacking prosthetic groups or ion coordination. A significant anticorrelation was found

Table 2.2 *Thermodynamic and structural parameters of soluble proteins.* Thermal denaturation free energy change, ΔG, under reducing conditions and comparable temperatures for an exhaustive set of monomeric uncomplexed proteins with disulfide bonds and without prosthetic groups or ion coordination [11]. Deviations from the balance relation are measured by $Y–(5X + 20)$ and shown to anticorrelate tightly ($R^2 = 0.72$, Fig. 2.3) with the denaturation free energies

PDB entry	$Y–(5X+20)$	ΔG (kcal/mol)	T (C)	pH	Reference
1BSQ	−9.83	11.10	40.00	7.00	Int J Biol Macromol 38, 9–17 (2006)
1RTB	−5.00	10.10	25.00	8.40	Biophys Chem 127, 51–63 (2007)
4LYZ	−3.30	9.02	26.85	7.00	Biopolymers 85, 264–273 (2007)
1CX1	−3.26	5.38	24.85	7.09	Biochemistry 37, 3529–3537 (1998)
1QG5	1.13	8.80	40.00	7.00	Int J Biol Macromol 38, 9–17 (2006)
2AIT	2.39	6.70	25.00	5.00	J Mol Biol 223, 769–779 (1992)
3SSI	8.82	4.07	20.00	7.00	J Mol Biol 249, 625–635 (1995)
1HIC	25.58	5.02	25.00	7.00	Eur J Biochem 202, 67–73 (1991)
1PMC	38.33	1.10	20.00	3.00	Nat Struct Biol 3, 45–53 (1996)

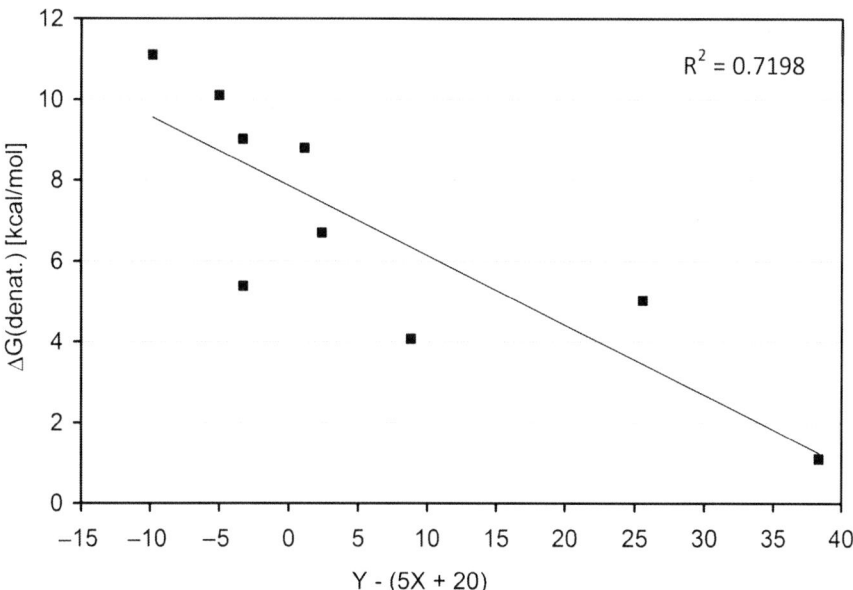

Fig. 2.3 Anticorrelation between denaturation free energy (ΔG) and excess structural defects with respect to the balance relation, measured by $Y–(5X + 20)$. The exhaustive set of monomeric uncomplexed proteins with disulfide links and the respective denaturation conditions are given in Table 2.2. The coefficient $R^2 = 0.72$ for the linear fit was obtained by linear regression

($R^2 =$ 0.72, Fig. 2.3) between the deviation from the balance equation, mea-sured as $Y-(5X + 20)$, and the thermal denaturation free energy (ΔG). This tight anticorrelation provides a thermodynamic validation of the balance equation.

The 5:1 rule may be justified on thermodynamic grounds. Thus, Doig and Williams [12] addressed the inconsistencies in Flory's treatment of the entropic contribution to protein denaturation, calculating $\Delta \Delta G$ for denaturation for a cross-linked protein versus its non-cross-linked counterpart. At physiological temperature of 300 K, they estimated $\Delta \Delta G \approx 4.4$ kcal/mol. This value is essentially independent of protein length and loop size and best represents the insensitivity of experimental values to loop size-dependent configurational entropies [13].

This 4.4 kcal/mol constant agrees reasonably well with the free energy contribu-tion associated with the native-state destabilization brought about by five structural deficiencies. If we take into account that –0.93 kcal/mol is the free energy change associated with complete dehydration of an unburied backbone hydrogen bond [4, 14], we may estimate the net destabilization effect promoted by five structural deficiencies at 0.93 kcal/mol $\times$ 5 = 4.65 kcal/mol. This value is in close proxim-ity to the Doig–Williams constant. The thermodynamic agreement supports the 5:1 golden ratio for protein buttressing arising from structural analysis.

2.3   Evolutionary Support for the Balance Equation

Since the evolutionary axis is germane to any biological analysis, the following question naturally arises: Is the architectural constraint defined by the dehy-dron/disulfide balance equation respected by evolution? Bioinformatics evidence on orthologous proteins (homologs across species) supports the tenet of evolutionary conservation. Thus, we may compare the structural deficiency and normalized num-ber of disulfide bonds across 1105 homolog pairs that differ in at least one disulfide bond (one homolog may have no disulfide bond). The changes in X and Y (ΔX, ΔY) were obtained for homolog pairs identified by their respective PDB accessions and the deviation from ideality was measured as $\Gamma = \Delta Y- 5\Delta X$. For all homolog pairs we obtained $\Gamma/Y < 11\%$, with Y associated with either homolog. Table 2.3 illustrates the tightest evolutionary conservation of the architectural constraint across homolog pairs with nontrivial buttressing differences.

2.4   Wrapping Translates into Protein Architecture

This chapter introduced a basic design principle that can be rationalized through an analogy. Just like defiance of gravity in building engineering requires buttress-ing to preserve the integrity of the building, protein design allowing for backbone hydration (the force counteracting structural cohesion) requires disulfide bridges to maintain the structural integrity of the protein. In this regard, this work unravels two constants that define a fundamental architectural constraint in soluble proteins: After suitable normalization, a single disulfide bond stabilizes *five* structural deficiencies

Table 2.3 Evolutionary conservation of the balance relation $Y = 5X + 20$.

SCOP family [a]	Sequence identity	$\Delta Y - 5\Delta X$	ΔX	ΔY	Homolog 1	Homolog 2
Fibronectin type II module	63.125	−0.1	0.21	0.95	1QO6	1E8B
β-Glycanases	40	−0.1	0.31	1.45	1FH8	1I1X
β-Glycanases	42.13836478	−0.1	0.31	1.45	1EOV	1B30
C-Type lysozyme	57.69230769	−0.09	0.79	3.86	1KXY	1LHM
C-Type lysozyme	58.46153846	−0.09	0.79	3.86	1JIS	1LHM
Ribonuclease A-like	30.23255814	−0.09	0.79	3.86	1RUV	1B1E
Ribonuclease A-like	31.00775194	−0.09	0.79	3.86	1RUV	1H52
Acetylcholinesterase-like	30.82437276	−0.07	0.18	0.83	2CKM	1K4Y
C-Type lysozyme	67.69230769	−0.07	0.79	3.88	1LMP	2BQM
C-Type lysozyme	56.92307692	−0.07	0.79	3.88	1LSM	2BQK
Eukaryotic proteases	31.47410359	−0.07	0.29	1.38	1EX3	1DST
Snake venom toxins	32.89473684	−0.07	0.4	1.93	1TXA	1ERA
Pepsin-like	40	−0.06	0.31	1.49	1PSN	1FQ6
Ribonuclease A-like	31.74603175	−0.06	0.73	3.59	1RTB	1K5B
C-Type lysozyme	59.23076923	−0.05	0.79	3.9	1IR7	1HNL
C-Type lysozyme	59.23076923	−0.05	0.79	3.9	1XEK	1HNL
C-Type lysozyme	59.23076923	−0.05	0.79	3.9	1UIH	1HNL
Ribonuclease A-like	31.00775194	−0.05	0.79	3.9	4RAT	1K5A
C-Type lysozyme	58.46153846	−0.05	0.79	3.9	2HS7	2BQK
Plant proteinase inhibitors	47.74774775	−0.04	0.34	1.66	1TIH	1FYB
Papain-like	70.21943574	−0.03	0.84	4.17	1ITO	2PBH
β-Glycanases	36.36363636	−0.03	0.31	1.52	2XYL	1B3X
C-Type lysozyme	84.72222222	−0.03	1.02	5.07	3LZ2	1LSG
Ribonuclease A-like	31.00775194	−0.03	0.79	3.92	1RHB	1K5A

([a]) SCOP: structural classification of proteins (Murzin AG, Brenner SE, Hubbard T, Chothia C (1995) SCOP: a structural classification of proteins database for the investigation of sequences and structures. J Mol Biol 247:536–540)

and every soluble protein has a 20%-baseline level of structural deficiency. These constants define statistically a design principle.

The baseline structural deficiency $Y = 20$ represents the maximum of a tight distribution (standard deviation $\sigma = 2.25$) of Y-values for the structural deficiency of soluble proteins with no disulfide bridges. This baseline Y-value implies that soluble proteins are not perfectly packed and maintain at least 20% of unburied backbone hydrogen bonds. Since such structural deficiencies locally promote backbone hydration, they belong to an intermediate region between order and disorder and hence represent markers of structural flexibility. Thus, because of its universality, the $Y = 20$ constant may be interpreted as the baseline flexibility needed for protein function.

In this chapter, dehydrons were characterized as structural deficiencies. These deficiencies are of a special kind: They are promoters of backbone hydration [4] and hence destabilizers of the native structure. On the other hand, disulfide bonds pre-formed in the denatured state reduce the structure-destabilizing conformational entropy cost associated with the folding process [12], hence stabilizing the native structure. Thus, it should be generally expected that the two major and opposite contributors to native structure destabilization would be correlated, as revealed by the balance equation described in this chapter.

This relation is hence likely to assist the molecular engineering of soluble proteins. Furthermore, since disulfide bridges can be formed or dismantled in accord with redox environmental conditions, the relation presented is likely to enable the type of design fine-tuning that may be required for an environmental modulation of the protein function.

As the dehydron/disulfide organizational principle is established and shown to hold for soluble proteins, we cannot fail to notice that it also introduces a new set of problems stemming from a basic question *What is in physical terms the fate of a soluble protein whose structure significantly violates the architectural constraints defined by the balance equation?* This issue will be explored in Chap. 5.

References

1. Baldwin RL (2003) In search of the energetic role of peptide hydrogen bonds. J Biol Chem 278:17581–17588
2. Powers ET, Deechongkit S, Kelly JW (2006) Backbone-backbone H-bonds make context-dependent contributions to protein folding kinetics and thermodynamics: Lessons from amide-to-ester mutations. In: Peptide Solvation and H-bonds, Baldwin RL, Baker D, eds, Adv Protein Chem 72:40–79, Elsevier Academic Press, San Diego, California
3. Fernández A, Berry RS (2002) Extent of Hydrogen-bond protection in folded proteints: A constraint on packing architectures. Biophys J 83:2475–2481
4. Pietrosemoli N, Crespo A, Fernández A (2007) Dehydration propensity of order-disorder intermediate regions in soluble proteins. J Proteome Res 6:3519–3526
5. Fernández A, Scheraga HA (2003) Insufficiently dehydrated hydrogen bonds as determinants for protein interactions. Proc Nat Acad Sci USA 100:113–118
6. MacKinnon R, Reinhart PH, White MN (1988) Charybdotoxin block of Shaker K^+ channels suggests that different types of K^+channels share common features. Neuron 1:997–1001
7. Fernández A, Berry RS (2002) Extent of Hydrogen-bond protection in folded proteins: A constraint on packing architectures. Biophys J 83:2475–2481
8. Pentelute BL, Gates ZP, Tereshko V et al (2008) X-ray structure of snow flea antifreeze protein determined by racemic crystallization of synthetic protein enantiomers. J Am Chem Soc 130:9695–9701
9. Adzhubei AA, Sternberg MJE (1993) Left-handed polyproline II helices commonly occur in globular proteins. J Mol Biol 229:472–493
10. Shi Z, Woody RW, Kallenbach NR (2002) Is polyproline II a major backbone conformation in unfolded proteins? In: Unfolded Proteins, Rose G, ed, Adv Protein Chem 62:163–240
11. Kumar MD (2006) ProTherm and ProNIT: thermodynamic databases for proteins and protein-nucleic acid interactions. Nucleic Acids Res 34:D204–D206
12. Doig AJ, Williams DH (1991) Is the hydrophobic effect stabilizing or destabilizing in proteins: The contribution of disulphide bonds to protein stability. J Mol Biol 217:389–398
13. Betz SF (1993) Disulfide bonds and the stability of globular proteins. Protein Sci 2: 1551–1558
14. Fernández A, Scott R (2003) Adherence of packing defects in soluble proteins. Phys Rev Lett 91:018102

Chapter 3
Folding Cooperativity and the Wrapping of Intermediate States of Soluble Natural Proteins

This chapter focuses on the molecular basis of cooperativity as a means to understand the folding of soluble natural proteins. We explore the concept of protein wrapping, its intimate relation to cooperativity, and its bearing on the expediency of the folding process for natural proteins. As previously described, wrapping refers to the environmental modulation or protection of intramolecular electrostatic interactions through an exclusion of surrounding water that takes place as the chain folds onto itself. Thus, a special many-body picture of the folding process is shown to emerge where the folding chain not only interacts with itself but also shapes the microenvironments that stabilize or destabilize the interactions. This picture reflects a competition between chain folding and backbone hydration leading to the prevalence of backbone hydrogen bonds for natural foldable proteins. A constant of motion governing the folding process emerges from the analysis.

3.1 Many-Body Picture of Protein Folding: Cooperativity and Wrapping

The physical underpinnings to the protein folding process remain elusive or, rather, difficult to cast in a useful form that enables structure prediction [1–10]. Thus, the possibility of inferring the folding pathway of a soluble protein solely from physical principles continues to elude major research efforts.

A major difficulty arises as we attempt to tackle this problem: *as a peptide chain folds onto itself, it also shapes the microenvironments of the intramolecular interactions, and hence the strength and stability of such interactions need to be rescaled according to the extent to which they become "wrapped" or surrounded by other parts of the chain.* Thus, interactions between different parts of the peptide chain not only entail the units directly engaged in the interaction but also the units involved in shaping their microenvironment, and the latter are just as important as they determine either the persistence or the ephemeral nature of such interactions. This fact makes the folding problem essentially a many-body problem and points to the heart of cooperativity, a pivotal attribute of the folding process [4, 6]. Furthermore, it highlights the intimate link between cooperativity and wrapping: *intramolecular*

A. Fernández, *Transformative Concepts for Drug Design: Target Wrapping*,
DOI 10.1007/978-3-642-11792-3_3, © Springer-Verlag Berlin Heidelberg 2010

hydrogen bonds prevail only if properly wrapped and this requires a cooperative process.

To further explore the molecular basis of cooperativity, we need to examine the folding process from a physico-chemical perspective: With an amide and carbonyl group per residue, the backbone of the protein chain is highly polar and this molecular property imposes severe constraints on the nature of the hydrophobic collapse and on the chain composition of proteins capable of sustaining such a collapse [2, 9, 11]. Thus, the hydrophobic collapse entails the dehydration of backbone amides and carbonyls and such a process would be thermodynamically disfavored if it were not for the possibility of amides and carbonyls to engage in hydrogen bonding with each other. Hence, *not every hydrophobic collapse qualifies as being conducive to folding the protein chain: Only a collapse that ensures the formation and protection of backbone hydrogen bonds is likely to ensure an expedient folding of the chain* [2]. On the other hand, polar group hydration competes with intramolecular hydrogen bonds, compromising the structural integrity of proteins with a deficiently wrapped backbone [12]. Thus, the need for formation and protection of intramolecular hydrogen bonds from water attack imposes constraints on the chain composition of an efficient folder capable of sustaining a reproducible and expedient collapse.

In accord with this picture, it has been postulated that as water-soluble proteins fold, the hydrogen bond pairing of backbone amides and carbonyls is concurrent with the hydrophobic collapse of the chain [11, 13]. This fact has been rationalized taking into account that the thermodynamic cost associated with the dehydration of unpaired polar groups is relatively high and that the hydrophobic collapse hinders the backbone hydration by shielding it from water. On the other hand, the strength and stability of hydrogen bonds clearly depend on the microenvironment where they occur: The proximity of nonpolar groups to a hydrogen bond enhances the electrostatic interaction by de-screening the partial charges and stabilizes it by hindering the hydration of the polar groups in the nonbonded state. *Thus, to guarantee the integrity of soluble protein structure and the expediency of the folding process, most intramolecular hydrogen bonds must be surrounded or "wrapped" by nonpolar groups fairly thoroughly so as to become significantly dehydrated at all times during the folding process.* This observation has implications at an ensemble-average level accessible to experimentalists [13, 14]. As shown below, it may help understand the fact that single domain proteins are likely to be two-state folders, with a single kinetic barrier dominating the folding process at the ensemble-average level [13, 15].

The hydration propensity of amide and carbonyl and the dehydration-induced strengthening of their electrostatic association represent two conflictive tendencies, suggesting that there must be a crossover point in the dehydration propensity of a backbone hydrogen bond. If the bond is poorly wrapped by a few nonpolar groups that cluster around it, then hydration of the paired amide and carbonyl is likely to be favorable, but as the hydrogen bond becomes better wrapped intramolecularly, the surrounding water loses too many hydrogen bonding partnerships and thus further removal of surrounding water is promoted. *Thus, the dehydronic field per hydrogen bond must significantly increase as hydrogen bonds get sufficiently dehydrated and this enhancement promotes further compaction of the chain and commits its fold. This switch-over behavior from backbone hydration to dehydration of backbone hydrogen bonds translated into an enhanced dehydronic field reflects the commitment to fold into a compact structure in which most backbone hydrogen bonds will be thoroughly dehydrated. This general picture is schematized in Fig. 3.1. The rationalization of the two-state folding kinetics of single domain proteins is thus based on the cooperative nature of wrapping interactions.*

The crossover point in hydrogen bond dehydration propensity may be regarded as representing a local characterization of the folding transition state if we adopt the backbone hydrogen bond dehydration as a generic folding coordinate. Once the folding process has progressed beyond the crossover point, further dehydration of the backbone is favored in consonance with the downhill nature of the folding process beyond the transition state [15]. Thus, *a transition state conformation commits the chain to fold partly because the partially wrapped hydrogen bonds trigger their further desolvation, in turn fostering further chain compaction* (Fig. 3.1). This compaction is essential to augment the number of nonpolar groups within the hydrogen bond microenvironments, thus protecting the bonds from water attack.

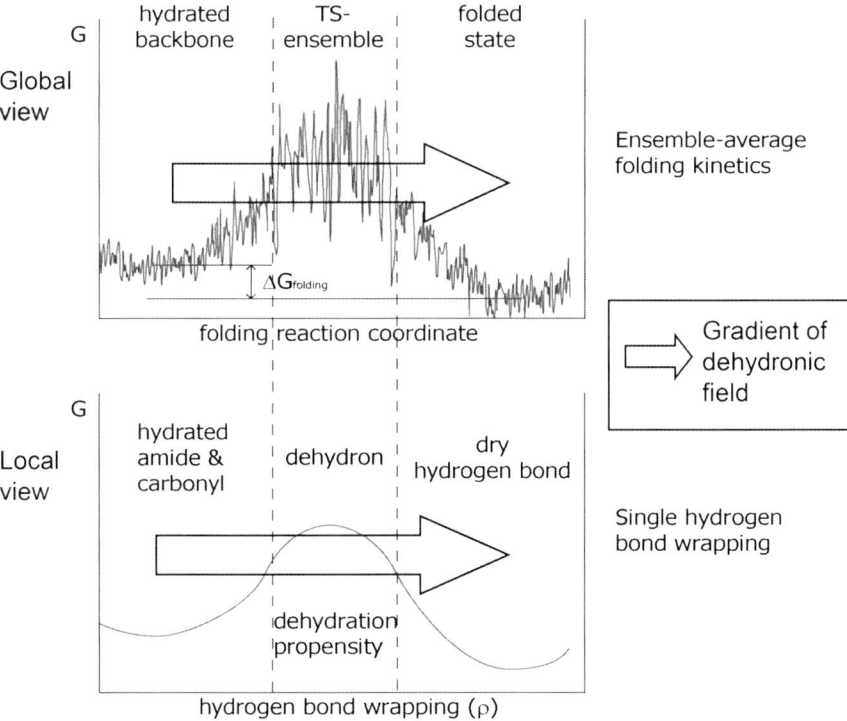

Fig. 3.1 The cross-over behavior from backbone hydration to dehydration of backbone hydrogen bonds triggered by hydrophobic collapse translates at the ensemble folding-kinetics level into a commitment of the chain to fold as the folding reaction progresses beyond the transition state

However compelling, the scenario described above requires a rigorous justification that entails an understanding of the crossover behavior at a local level. This understanding prompts us to focus on the solvent environment of individual hydrogen bonds during the conformational exploration performed by the protein chain as the hydrophobic collapse takes place [12, 16, 17].

3.2 Hydrogen Bond Wrapping Requires Cooperative Folding

The dynamics of hydrogen bond wrapping concurrent with the folding process cannot be probed directly at this time, much like the conformational exploration defining the folding process still remains inaccessible to experimentalists [13–15]. To make progress in our understanding of the wrapping dynamics, we need to capture this process through ab initio folding algorithms independently benchmarked and validated against PDB-reported structures and experimental data on the folding kinetics [2, 13, 15].

To properly describe folding cooperativity, we exploit a program that algorithmically reproduces folding dynamics by stochastically and coarsely representing transitions in the conformation space of individual residues and effectively incorporating wrapping three-body interactions. To access realistic folding timescales beyond the microsecond range, the program builds on an implicit solvent model of cooperativity [6] and exploits a coarse-grained representation of the torsional Ramachandran (Φ, Ψ)-state of each residue [10]. The implicit treatment of the solvent dictates that the program assesses at each step the microenvironments of intramolecular hydrogen bonds that form as byproducts of the chain attempt at achieving a sustainable compaction.

In order to reduce computation time and access relevant timescales, the microenvironmental dielectric within which intramolecular hydrogen bonds form is computed in a coarse manner at each step of the simulations. The strength and stability of intramolecular hydrogen bonds are thus rescaled with each computational step in consonance with the degree of dryness or dehydration of the bond determined by the overall folding state of the chain. Thus, the extent of bond dehydration defines its net hydrophobicity (see (1.2)), which in turn promotes further wrapping begetting further compaction of the chain.

The ab initio folding algorithm reproduces the essential dynamical features of cooperativity while possessing predictive potential in regard to both endpoint structure and folding pathway [2]. Thus, the algorithm appears to reproduce a basic feature of the folding kinetics of single domain proteins: its purported two-stateness observed at an ensemble-average level of experimentation [13]. This dynamic feature is likely to arise as an overall consequence of the crossover behavior in hydrogen bond dehydration propensity, although a "hard proof" of this statement has not been produced to this day.

In spite of the difficulties inherent to a direct probing, the folding of a soluble protein is believed to be accompanied by a progressive structuring, immobilization, and eventual exclusion of water surrounding backbone hydrogen bonds. This process turns hydrogen bonds into major determinants of the folding pathway and compensates for the thermodynamic penalties associated with desolvation of the backbone polar groups [18]. Thus, the need to wrap hydrogen bonds as a means to ensure their integrity has been shown to determine a constraint and hence define a constant of motion in the long-time limit behavior of coarse-grained ab initio simulations. In this regard, an examination of one of the longest available (1 μs) all-atom simulations with explicit solvent [19] reveals that *the extent of wrapping is a constant of motion for the folding trajectory* [2]. It is well documented that such stabilization is best achieved by clustering five residues with nonpolar side chain groups around the backbone hydrogen bonds, an arrangement that yields an optimal stabilization of the intramolecular hydrogen bond with a minimal conformational entropy cost [2]. The results hereby reported thus support and clarify the view that hydrophobic surface burial should be commensurate with hydrogen bond formation [13, 20] and enable us to introduce a basic wrapping motif inherent to the folding dynamics of soluble proteins. *Ultimately, the crossover behavior outlined in Fig. 3.1 is reproduced, lending ample support to the two-state picture that emerged from ensemble-averaged kinetic studies and to the observed expediency of the folding process.*

The connection between hydrophobic interactions, water structuring, and the strength of hydrogen bonds was first delineated by Scheraga and co-workers [21]. Thus, the inaccessibility of hydrogen bonds to solvent takes place as the protein strategically places hydrophobes around its backbone polar groups. This process induces hydrogen bond formation as a means to compensate for the unfavorable burial of the backbone polar groups. In this regard, natural questions arise and will be addressed through the examination of wrapping dynamics along the folding process:

(a) Does the wrapping or protection of backbone hydrogen bonds promote an expedient folding and if so, how?
(b) How can we identify the conformation or conformational ensemble that commits the chain to fold expeditiously into its native structure?
(c) Can we establish whether hydrophobic collapse precedes or is concurrent with secondary structure formation?

We approach these problems by establishing a relationship between the modulation of electrostatic shielding and the wrapping of the protein conformation along generated folding trajectories. This link is delineated based on statistical information on hydrophobic clustering around native backbone hydrogen bonds, information that leads us to establish a pervasive wrapping motif in native structures [22].

The analysis is carried out exploiting both a coarse-grained ab initio folding algorithm based on an implicit solvent model [18] and a long (1 μs) all-atom molecular dynamics simulation [19] of the type pioneered by the Karplus [23], Levitt [24], and Brooks [25] groups. Both algorithmic approaches reveal a nearly constant average extent of hydrogen bond wrapping along the folding trajectories with relatively small dispersion, suggesting the existence of a constant of motion.

These results provide evidence supporting a dynamic picture of folding in which surface burial is commensurate with hydrogen bond formation or, phrased differently, hydrophobic collapse is concurrent with the formation of secondary structure [9, 13, 26]. Furthermore, the results add a new element to the picture: *hydrophobic collapse occurs productively in so far as hydrophobes can also exert a protective role* [26], *wrapping the backbone hydrogen bonds, and they do so to the same extent throughout the folding pathway, irrespective of the number of hydrogen bonds present at a given time.* This last assertion enables us to postulate a *constant of motion* underlying the folding process.

3.3 Generating Cooperative Folding Trajectories

To validate the previous scenario, the dynamics of backbone desolvation are studied using the so-called folding machine (FM), an ab initio wrapping-based algorithm designed to generate low-resolution folding pathways [2, 7, 8, 18], and contrasted against wrapping information extracted from all-atom explicit solvent simulations [19] as well as experimental kinetics data [13].

The trustworthiness of the FM-generated folding pathways is guaranteed by an independent corroboration of the predictive value of the FM. This algorithm has been successful at predicting crucial dynamic features of complex folders that do not follow the two-state kinetic picture, such as β-lactoglobulin [8]. The native-like and nonnative conformations occurring along the folding pathway in this presumed nonhierarchical folder as well as the productive role of nonnative interactions in preventing misfolding or incorrect structure condensation were predicted through the FM and subsequently validated experimentally [27]. This corroboration added significant leverage to the wrapping-based FM algorithm, making it a powerful tool to study folding cooperativity.

Rather than encompassing all the structural detail for each step, the FM focuses on the time evolution of backbone torsional constraints imposed by steric clashes with side chains and rescales the chain's intramolecular potential according to the wrapping microenvironments around pairwise dielectric-dependent interactions. Thus, each pairwise interaction of the peptide chain, whether hydrophobic or electrostatic, is effectively enhanced or weakened according to the extent of burial of the paired groups. The net decrease of free energy associated with the backbone hydrogen bond desolvation is typically only in the range 0.5–1.2 kcal/mol, due to the opposing increase in the solvation free energy of the polar amide and carbonyl groups. The implicit treatment of the solvent requires that we introduce three-body correlations to characterize the wrapping of pre-formed intramolecular interactions, and accordingly, rescale the internal energy terms with each iteration; that is, every time the pattern of three-body correlations around an interaction has changed. Both the reduction in structural resolution and the implicit solvent treatment are essential to make realistic folding timescales (> 1 μs) accessible to the FM computations. Thus, the FM ab initio approach is geared to generate folding pathways with a coarse structural resolution needed to make folding timescales computationally accessible.

The FM algorithm utilizes no a priori information on target folds (unlike the so-called Go models which use the native fold as input for the simulator [28–30]), nor does it incorporate any energetic biases. The model focuses on the torsional Φ, Ψ constraints that are applied to backbone torsional state due to the steric hindrances imposed by the side chains. Each residue is assigned to a region, or basin, in the Ramachandran map, and changes in configuration occur by hopping to a new basin.

By dealing with the evolution of constraints (i.e., Ramachandran basins) rather than the backbone torsional coordinates themselves, the dynamics are judiciously simplified [31]. The algorithm consists of a stochastic simulation of the coarsely resolved dynamics, simplified to the level of time-evolving Ramachandran basin assignments. An operational premise is that steric restrictions imposed by the side chains on the backbone may be subsumed into the basin-hopping dynamics. The side chain constraints define regions in the Ramachandran map that can be explored in order to obtain an optimized pattern of nonbonded interactions.

The basin location of each residue coarsely defines the topology of the protein conformation. This string of basin locations, termed the local topology matrix or LTM(t), reflects the inherent geometrical constraints of a real polypeptide chain. The precise coordinates of the chain (i.e., the physical realization of an LTM) are defined by explicit Φ, Ψ angles determined by an optimization process that is turned on every 10 hopping steps. To maintain structural continuity during a folding trajectory, the explicit dihedral angles are retained for each residue from one time step to the next until that residue Ramachandran basin is scheduled to change.

To make torsional moves in 3D space, translating the "*modulo-basin topology,*" a conformation is generated with a set of explicit Φ, Ψ angles compatible with the basin string. This explicit realization is used to identify the extent of structural involvement of each residue. As expected, the more structurally involved the residue, the less likely it may be engaged in a basin hopping, and the algorithmic rules do in fact reflect this fact. The degree of structural involvement is quantified energetically with a semiempirical potential. This potential is used to determine which residues change their Ramachandran basin in the next step. Upon a basin transition, the new structure is energetically minimized by changing Φ, Ψ angles within the chosen basins.

The basin-hopping probability is dependent on the extent of structural engagement of the residue, which is defined by the energetic cost associated with the virtual move of changing basin, higher the more structurally engaged the residue is. On the other hand, the probability of hopping to a target basin (given that a hopping move is scheduled to occur) depends on the target basin lake area or its microcanonical entropy. To fit experimental folding measurements [13, 15, 32], a free residue is assigned a basin-hopping rate fixed at 10^9 Hz. The basic tenets governing interbasin hopping in the FM algorithm are as follows: (a) interbasin hopping is slower than intrabasin exploration. This "adiabatic tenet" warrants a subordination of the backbone (Φ, Ψ) search to the LTM evolution or "*modulo-basin dynamics*"; (b) side chain torsional exploration occurs on a faster timescale than backbone LTM dynamics. The last premise introduces a second adiabatic approximation, justifying the averaging of side chain torsional motions in the stages of folding that precede a final side chain fine-tuning on the native backbone fold. This simplification is adequate to represent early stages of compaction and hydrogen bond wrapping.

The FM captures the molecular basis of folding cooperativity by introducing an effective enhancement of dielectric-dependent two-body interactions according to the extent of wrapping of the interaction (see Figs. 1.1 and 1.2). *The cooperative effect translates as a rescaling of the zero-order (in-bulk) pairwise contributions depending on the number of wrapping side chain groups.* The weakening of hydrophobic attractions depending on the extent of hydrophobic burial of the paired nonpolar groups is treated in a similar manner.

Three alternative and interchangeable representations of the folding state of a chain are simultaneously captured by the FM, as depicted in Fig. 3.2a–c: the modulo-basin Ramachandran torsional state (LTM), the 3D rendering, and the wrapping pattern. Figure 3.2 shows the endpoint conformation of a representative simulation for the thermophilic variant of protein G (PDB.1GB4) performed at 313 K, pH 7, and consisting of 10^6 steps [2]. The endpoint was largely reproduced in 66 of 91 runs and has RMSD ∼3.0 Å from PDB entry 1GB4.

The wrapping model subsumed into the FM algorithm reflects the fact that hydrogen bonds are extremely context-sensitive. The algorithm, however, treats the solvent implicitly. This simplification requires that we introduce three-body correlations involving the wrapping residues (Fig. 3.2) as an operational means to incorporate rescalings of the intramolecular potential according to the microenvironmental modulations that take place during the course of folding. Explicitly, these correlations rescale the "zeroth-order" pairwise interactions by determining their extent of desolvation. For consistency, the wrapping of a hydrogen bond is also introduced in a coarse-grained manner in this analysis. Thus, in contrast with the detailed wrapping assessment (cf. Fig. 1.2), here the wrapping parameter, denoted ρ_{coar}, indicates the number of residues contributing with nonpolar side chain groups to the dehydration of the hydrogen bond. Each residue contributing to the dehydration of a hydrogen bond determines a three-body correlation (cf. Fig. 1.1). Notice that we have introduced a notational distinction absent in Chap. 1: a three-body correlation denotes something coarser than a three-body interaction. The latter refers to a wrapping group while the former refers to residues contributing with wrapping groups (often more than one) to shield a hydrogen bond. The overall number of three-body correlations, denoted $C_3 = C_3(t)$, is invariably smaller than the number of three-body interactions $I_3 = I_3(t)$, and both are roughly proportional for most natural proteins ($I_3 \approx 2.5C_3$).

The wrapping effect may also be cast in thermodynamic terms: due to their destabilizing effect on the nonbonded state, the hydrophobes surrounding a dielectric-dependent interactive pair become enhancers of the interaction. As folding progresses, the effective hydrophobic energy contribution becomes progressively transferred into an effective energy of backbone desolvation in which the amide–carbonyl hydrogen bonds become determinants of protein structure and compensate for the rise in self-energy of the paired groups.

a

	1	2	3	4	5	6	7	8	9	10
Aminoacid	MET-M	THR-T	THR-T	PHE-F	LYS-K	LEU-L	ILE-I	ILE-I	ASN-N	GLY-G
R-basin										
Phi-angle	-60.48	-53.47	-111.35	-146.29	-77.71	-142.19	-60.03	-136.53	-146.28	-83.41
Psi-angle	-50.21	-46.87	114.70	166.19	134.23	132.85	117.82	139.68	137.50	0.57
Omega-angle	180.00	180.00	180.00	180.00	180.00	180.00	180.00	180.00	180.00	180.00

	11	12	13	14	15	16	17	18	19	20
Aminoacid	LYS-K	THR-T	LEU-L	LYS-K	GLY-G	GLU-E	ILE-I	THR-T	ILE-I	GLU-E
R-basin										
Phi-angle	-77.16	-130.78	-61.56	-110.69	-76.55	-128.50	-141.75	-129.74	-131.06	-57.29
Psi-angle	2.03	132.05	-25.00	104.99	86.44	106.57	175.45	137.67	129.69	-32.23
Omega-angle	180.00	180.00	180.00	180.00	180.00	180.00	180.00	180.00	180.00	180.00

	21	22	23	24	25	26	27	28	29	30
Aminoacid	ALA-A	VAL-V	ASP-D	ALA-A	ALA-A	GLU-E	ALA-A	GLU-E	LYS-K	ILE-I
R-basin										
Phi-angle	-63.64	-60.87	-62.93	-144.65	-61.10	-59.75	-61.47	-61.16	-54.21	-61.01
Psi-angle	-32.49	-47.50	-33.90	136.37	-43.25	-27.33	-50.01	-46.13	-43.25	-46.07
Omega-angle	180.00	180.00	180.00	180.00	180.00	180.00	180.00	180.00	180.00	180.00

	31	32	33	34	35	36	37	38	39	40
Aminoacid	PHE-F	LYS-K	GLN-Q	TYR-Y	ALA-A	ASN-N	ASP-D	ASN-N	GLY-G	ILE-I
R-basin										
Phi-angle	-63.08	-54.65	-63.38	-58.81	-118.93	-67.19	-88.71	-78.66	80.98	-61.82
Psi-angle	-45.96	-44.88	-49.83	-27.40	114.72	-53.04	0.80	138.34	-94.77	-33.78
Omega-angle	180.00	180.00	180.00	180.00	180.00	180.00	180.00	180.00	180.00	180.00

	41	42	43	44	45	46	47	48	49	50
Aminoacid	ASP-D	GLY-G	GLU-E	TRP-W	THR-T	TYR-Y	ASP-D	ASP-D	ALA-A	THR-T
R-basin										
Phi-angle	-65.78	84.02	-66.69	-133.65	-64.78	-145.61	-76.24	-132.24	-68.87	-127.87
Psi-angle	-48.66	-3.46	-52.64	174.31	124.11	171.01	137.33	143.91	-54.06	114.38
Omega-angle	180.00	180.00	180.00	180.00	180.00	180.00	180.00	180.00	180.00	180.00

	51	52	53	54	55	56	57
Aminoacid	LYS-K	THR-T	PHE-F	THR-T	VAL-V	THR-T	GLU-E
R-basin							
Phi-angle	62.05	-81.78	-73.09	-129.16	-141.35	-62.94	59.51
Psi-angle	58.19	131.81	136.63	137.30	132.79	120.79	55.61
Omega-angle	180.00	180.00	180.00	180.00	180.00	180.00	180.00

Fig. 3.2 Three representations of the conformational state of thermophilic variant of protein G (PDB code: 1GB4) obtained after 10^6 FM iterations. (**a**) LTM or backbone torsional state represented "modulo Ramachandran basin." Each basin is indicated by the quadrant in the (Φ, Ψ)-torus where it occurs. Thus, *blue* stands for the extended residue conformations including the β-strand states, *red* corresponds to a manifold of conformations containing the right-handed α-helix, *green* denotes the basin containing the left-handed helical conformation, while *gray* represents the basin in the lower right quadrant which is only accessible to Gly. (**b**) Three-dimensional ribbon representation of the endpoint chain conformation. (**c**) Wrapping state of the chain in the endpoint conformation. The chain backbone is shown as virtual bonds joining α-carbons depicted in *pink*, hydrogen bonds are shown as *gray* segments joining α-carbons, and three-body "wrapping" correlations (cf. Fig. 1.1) are shown as *thin blue lines* joining the α-carbon of the wrapping residue with the center of the wrapped hydrogen bond. A wrapping residue is defined as a contributor of nonpolar groups to the hydrogen-bond microenvironment. Thus, the FM keeps track of the pairwise interactions as well as of the evolving microenvironments of such interactions determined by the evolving chain conformation. Reprinted from [35], with permission from Elsevier

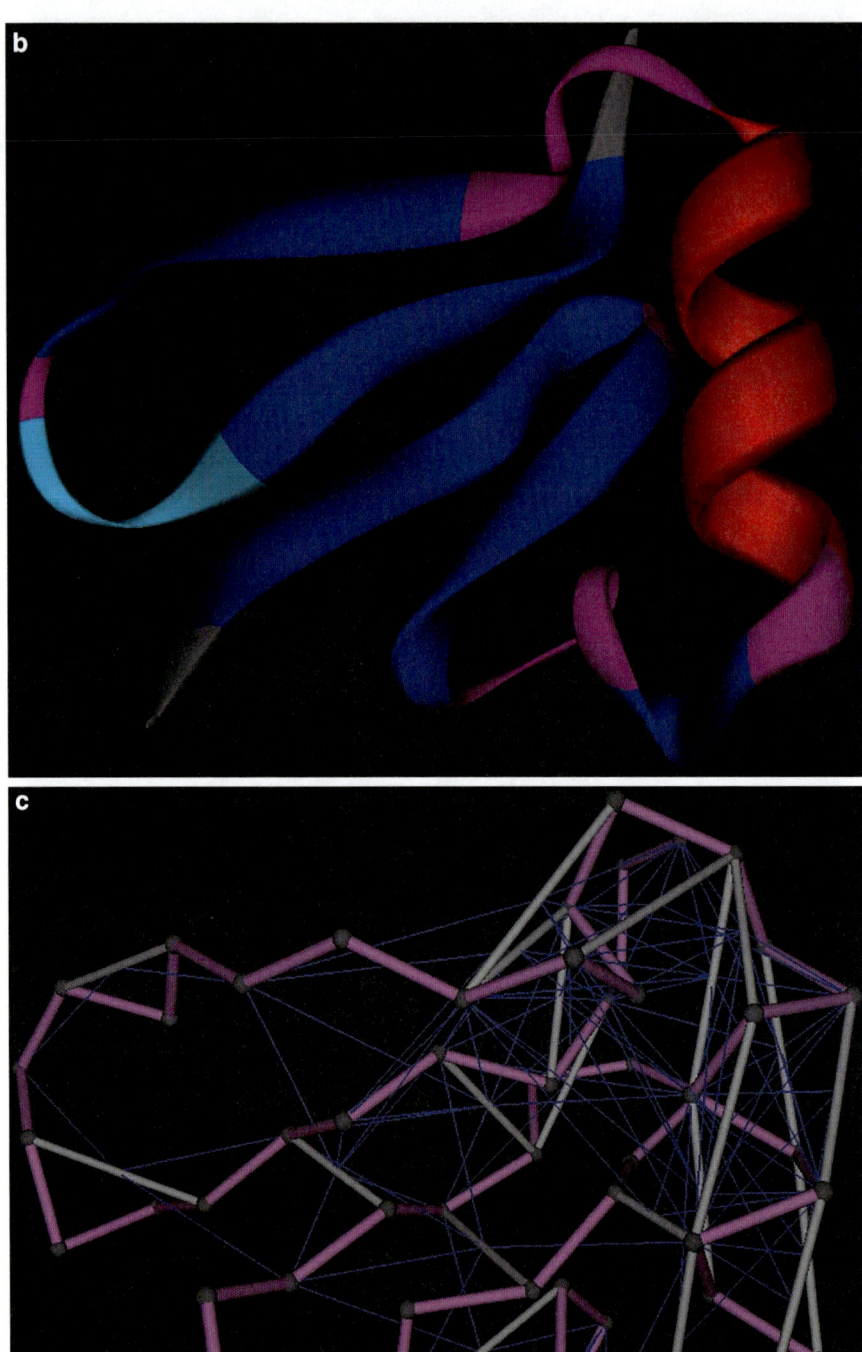

Fig. 3.2 (continued)

To take into account this effect, we incorporate a phenomenological approach to wrapping electrostatics [10, 33] as a means of incorporating changes in permittivity associated with changes in the conformation-dependent environments that affect pairwise interactions. This procedure entails a rescaling of the intramolecular potential terms as folding progresses, an operation requiring keeping track of vicinal hydrophobe positions with each FM iteration (Fig. 3.2c).

3.4 Wrapping Patterns Along Folding Trajectories

To take into account the effect of wrapping on dielectric-dependent pairwise interactions and assess its role in defining cooperativity, we have adopted a semiempirical procedure to algorithmically keep track of the conformation-dependent microenvironments. Thus, the in-bulk potential energy contributions are regarded as zero-order terms, while cooperative effects arise due to the wrapping of favorable interactions brought about by hydrophobic third-body participation (cf. Fig. 3.2c).

A basic question is addressed through this analysis: What is the dynamic relevance of optimal wrapping in regard to the protein's commitment to fold? To tackle this question, we analyze a representative simulation for the thermophilic variant of protein G. This simulation performed at 313 K, pH 7, consists of 10^6 steps and was essentially reproduced in 66 of 91 FM runs. All the runs generated a stationary fold within RMSD ~4 Å from the native structure (PDB entry 1GB44) and a dramatic decrease in potential energy around 0.6 ms (Fig. 3.3).

In accord with experimental tenets, direct examination of the time-dependent behavior of contact order suggests a *nucleation process*, whereby a sustainable large-scale organization is achieved only at 8×10^{-4}s after a relatively lengthy trial-and-error process (0–0.63 ms), followed by a critical regime (0.63–0.8 ms). Direct inspection of Fig. 3.4 reveals that the transition from local to large-scale organization is actually defined by a sudden burst in the number of three-body correlations starting at $6.0 \times 10^{-4} - 7 \times 10^{-4}$ s.

The nucleation picture revealed by Figs. 3.3 and 3.4 has been further confirmed by examination of total internal energy of the peptide chain and solvent-exposed area plots. The energy experiences a sudden decrease in the $6.3 \times 10^{-4} - 8.0 \times 10^{-4}$ s region concurrent with a dramatic decrease in the solvent-exposed area. The point at which the protein is actually committed to fold (Fig. 3.5) can be inferred by performing runs with different starting conformations extracted from the $6.3 \times 10^{-4} - 8.0 \times 10^{-4}$s time window. This commitment arises when a sustainable number of three-body correlations (native or nonnative!) equal to or larger than the final almost stationary number is reached (Fig. 3.4). In the case of protein G, the burst time window is $6.3 \times 10^{-4} - 8.0 \times 10^{-4}$ s and a sustainable population of three-body correlations is maintained in the region $7.0 (\pm 0.2) \times 10^{-4}$ s.

A similar FM computation was carried out for ubiquitin [32]. The exposed surface area at the transition state is estimated to be 7200 Å^2, while the random coil conformation exposes approximately 10,800 Å^2. Thus, we find that the transition

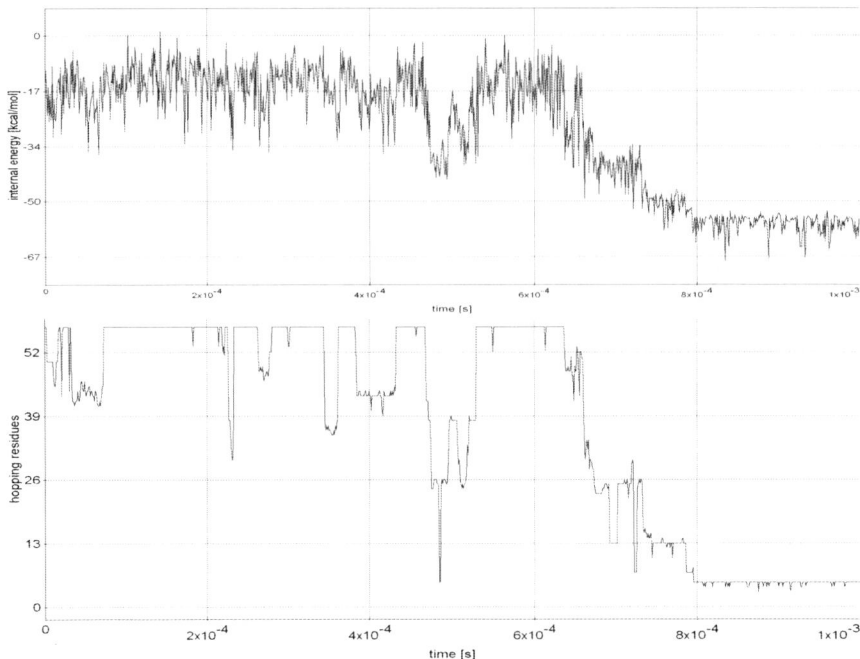

Fig. 3.3 Internal energy and number of hopping residues along a reproducible representative FM trajectory for the thermophilic variant of protein G. A hopping residue is defined as tagged for a Ramachandran basin transition in the coarse-grained stochastic process that underlies the FM torsional dynamics. Reprinted from [35], with permission from Elsevier

state buries 3,600 Å^2, approximately 60% of the total area buried in the native fold, in good agreement with the experimental results [13].

The dominant folding pathway for the variant of protein G is coarsely displayed in Fig. 3.4, where the abscissas denote residue numbers and the ordinates, the time axis. The Ramachandran basin assignment for each residue is given as a function of time and the topology of the entire chain is designated by a row in the histogram. The blue color (basin 1) designates the large basin containing the β strand and PP2 conformations, the red color (basin 2) designates the basin containing the right-handed helix conformation, green (basin 3) designates the basin containing the left-handed helix coordinates, and gray (basin 4) corresponds to the fourth basin present only in glycine. The figure clearly reveals the emergence of a stable large-scale organization that prevails after a critical period located at the $6.3 \times 10^{-4} - 8.0 \times 10^{-4}$ s interval.

We have also reported the extent of protection of backbone hydrogen bonds along the folding pathway. Figure 3.6 shows that the average extent of hydrogen bond protection, $\rho_{\text{coar}}(t)$, converges to the value $\rho_{\text{coar}} = 5$ in the long-time limit that starts right after the trial-and-error period; that is, at the sharp burst in $C_3 = C_3(t)$. This regime is associated with the region $C_3 > 60$. The stationary native-like population

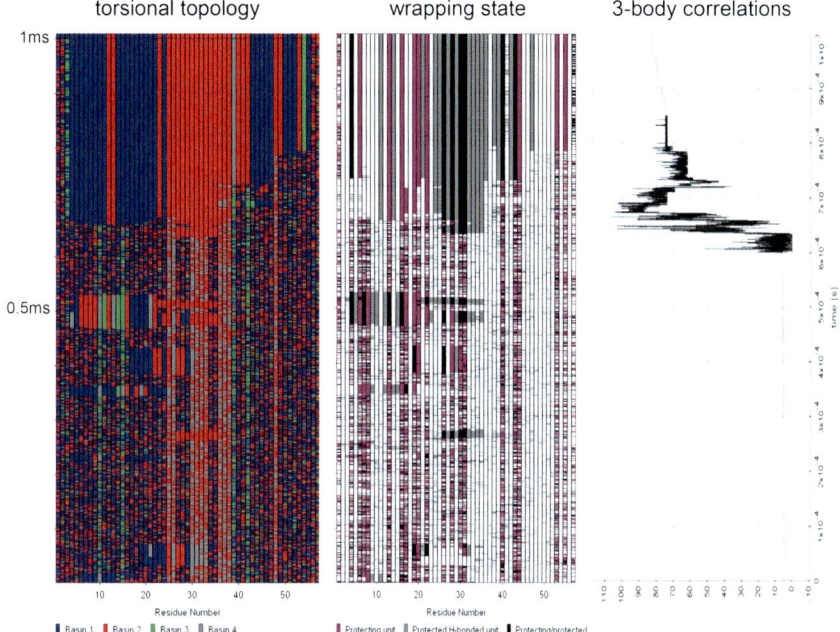

Fig. 3.4 Three views of the wrapping/folding dynamics for the thermophilic variant of protein G obtained from a representative expeditious FM trajectory. The *left panel* represents the time evolution of the local topology matrix (LTM) with the abscissas denoting residue numbers and the ordinates, the time axis. The Ramachandran basin assignment for each residue is given as a function of time and the topology of the entire chain is designated by a row in the histogram. Color convention is consistent with Fig. 3.2b. The *middle panel* represents the different roles exerted by residues along the chain at different times. Thus, a *pink* residue-time entry indicates a residue exerting a protecting or wrapping role at the particular time, a *gray entry* indicates a residue engaged in pairwise interaction which is being protected or wrapped by other hydrophobes, and a *black entry* indicates a dual role as protector or wrapper and also engagement in a hydrogen bond which is being exogenously protected. The *right panel* indicates the total number of three-body correlations representing the wrapping dynamics concurrent with the expeditious folding process. *Notice that the burst phase in three-body correlations coincides with the region of transition from a trial-and-error phase to a sustainable structure.* Reprinted from [35], with permission from Elsevier

of 15($\pm$ 1) backbone hydrogen bonds are protected by 75($\pm$ 5) three-body correlations. We see that the value $\rho_{coar} = 5$ becomes an approximate constant of motion in the critical region and beyond, that is for $C_3 > 60$. A similar result holds for ubiquitin [2]: The native-like stationary population of 28($\pm$ 3) backbone hydrogen bonds is now wrapped by 140($\pm$ 5) three-body correlations.

The wrapping results from Figs. 3.4 and 3.6 are more specific and informative than earlier attempts at establishing whether buried surface area is commensurate with hydrogen bond formation [13, 32]. It is difficult to infer from such studies whether hydrophobic collapse triggers hydrogen bond formation or whether the latter directs the former. However, Figs. 3.4 and 3.6 reveal that the productive

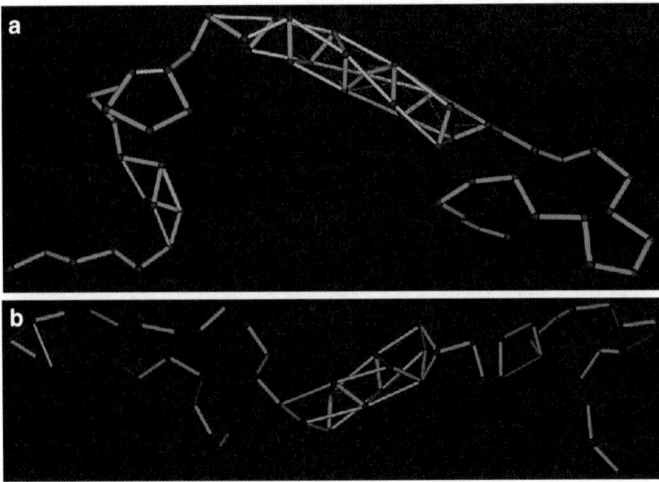

Fig. 3.5 Wrapping patterns for chain conformations occurring during the burst phase (cf. Fig. 3.4) of protein-G variant generated at 6.4×10^{-4} s (*upper panel*) and 6.5×10^{-4} s (*lower panel*). These conformations commit the chain to fold are part of the "transition state ensemble," and *do not contain exclusively native interactions*, as it becomes apparent in the *upper panel*. Reprinted from [35], with permission from Elsevier

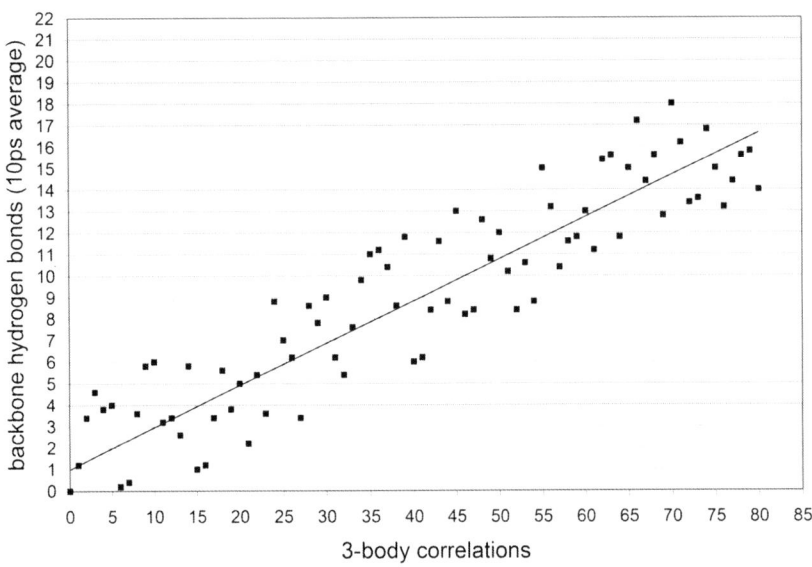

Fig. 3.6 Number of backbone hydrogen bonds plotted against the number of three-body correlations (C_3) extracted from representative FM trajectory capturing the wrapping/folding dynamics for the thermophilic variant of protein G

buildup in hydrogen bond formation beyond the "trial-and-error folding regime" is invariably assisted by the nearly constant $\rho_{coar} = 5$ wrapping value.

To further validate our conclusions by contrasting them against all-atom molecular dynamics simulations, we have analyzed one of the longest all-atom simulations with explicit solvent: the 1 μs simulation of the villin headpiece by Duan and Kollman [19]. Their trajectory was examined using the FM's representation and visualization tools and provides valuable information on the wrapping dynamics in the trial-and-error regime. *The analysis revealed an almost constant proportionality between C_3 and the number of backbone hydrogen bonds along the entire trajectory. The results reveal again that $\rho_{coar} = 5$ is a constant of motion for the folding trajectory.*

Testing the universal validity of this constant of motion may prove to be daunting, as the wrapping of most good folders would need to be investigated dynamically. Nevertheless, a theoretical analysis based on a nanoscale treatment of the solvent further supports this hypothesis. Thus, we now rigorously prove that *a hydrogen bond is embedded in the lowest dielectric when surrounded by five average-shaped hydrophobic residues, and this optimal wrapping arrangement represents a compromise between crowding and proximity to the hydrogen bond subject to the steric constraints determined by a minimum approach distance.*

This approximate law holds for both native structure [22] and folding dynamics [2]. In this regard, this wrapping motif may be regarded as a structural element that captures the basic component of energy transduction from hydrophobic association to structure formation. Furthermore, it implies that a fundamental constraint in protein architecture applicable to native structures applies also throughout the folding trajectory.

Kinetic amide isotope studies [13, 32] imply that helical hydrogen bond formation and surface area burial form to the same degree in the transition state for single domain proteins that fold in a two-state manner. Extensive helix formation does not occur prior to significant hydrophobic association at the limiting step. The surveyed results on individual folding trajectories indicate that commensurate burial occurs both on the way up to the limiting barrier (pre-nucleation) and afterwards, in consistency with the ensemble-average evidence stemming from the kinetic folding experiments.

3.5 Nanoscale Solvation Theory of Folding Cooperativity: Dynamic Benchmarks and Constant of Motion

The goal of this section is to describe a semiempirical model of nanoscale solvation that captures the dielectric modulation brought about by the approach of a hydrophobe to a protein hydrogen bond. In essence, the model captures the solvent-ordering effect promoted by the hydrophobe and quantifies the effect of this induced organization on the electrostatics of a pre-formed amide–carbonyl hydrogen bond. *This model reproduces the crossover point in hydrogen bond dehydration propensity that characterizes the folding transition state* (Fig. 3.1) *as we adopt the extent of backbone hydrogen bond dehydration as generic reaction coordinate for the folding process* [9, 13]. *Strikingly, it also provides a rigorous justification of the n = 5 coarse-wrapping motif* (Fig. 3.6) *through an implicit solvent model that reproduces the modulation of the dielectric environment in which intramolecular hydrogen bonds are formed. Finally, it establishes the fact that the progress of the folding process follows a reaction coordinate along the gradient of the dehydronic field, making the latter the dominant force driving chain compaction beyond the transition state.*

We start our theoretical treatment by defining a Cartesian coordinate system and placing the carbonyl oxygen atom effective charge q at the center of coordinates. We further define the x-axis as that along the carbonyl–amide hydrogen bond and place the amide hydrogen atom at position $\mathbf{r}$, 1.4–2.1 Å away along the positive x-axis. We assume the hydrogen bond to be surrounded by a discrete number of identical spherical hydrophobic units of radius $d/2$ (the parameter d is defined below) centered at fixed positions $\mathbf{r}_j, j = 1,2,\ldots, n$. This is an idealized picture but one that can be dealt with analytically.

Previously reported implicit solvent approaches [33] take into account the solvent structuring induced by the solvent–hydrophobe interface (Fig. 3.7), translate this

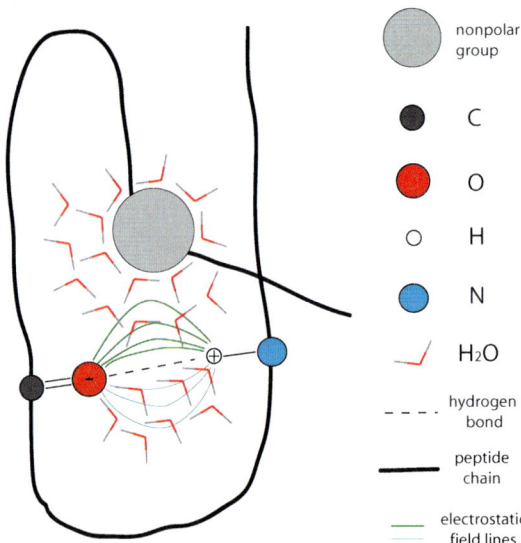

Fig. 3.7 Basic tenets of the nanoscale solvation theory. The solvent ordering promoted by the approaching hydrophobe enhances the electrostatics, an effect that must be captured by the model. The induced organization decreases the polarizability of the environment, preventing water dipoles from aligning with the electrostatic field lines. The *thicker lines* represent a stronger field. By contrast, the region exposed to bulk water facilitates dipole organization along the field lines, weakening the electrostatic field (*thin lines*). Reprinted from [35], with permission from Elsevier

effect into a distance-dependent permittivity, $\varepsilon(\mathbf{r})$, and quantify the effect on the Coulomb screening. A more heuristic, practical, and phenomenological approach is rooted in two pivotal components: (a) perturbation of the diffraction structure of bulk water as hydrophobes are incorporated at fixed positions and (b) recovery of their solvent-structuring effect by inverse Fourier transforming the previous result given in frequency space.

To propagate the solvent-structuring effect induced by the presence of the hydrophobic spheres, we replace the position-dependent dielectric by an integral kernel convoluted with the electric field at position $\mathbf{r}$ to represent the correlations with the field at neighboring positions $\mathbf{r}'$. This prompts us to replace the classical Poisson equation by the rigorous relation (see Chap. 14):

$$\text{div}\left[\int K\left(\mathbf{r},\mathbf{r}',\{\mathbf{r}_j\}\right) E\left(\mathbf{r}'\right) d\mathbf{r}'\right] = 4\pi q\delta(\mathbf{r}), \qquad (3.1)$$

where the kernel $K\left(\mathbf{r},\mathbf{r}'\{\mathbf{r}_j\}\right)$ is parametrically dependent on the fixed hydrophobe positions. In the absence of vicinal hydrophobic units, the correlations decay as $\exp\left(-\|\mathbf{r}-\mathbf{r}'\|/\xi\right)$ (ξ = characteristic correlation length). In the limit $\xi \to 0$, we get $K\left(\mathbf{r},\mathbf{r}'\right) \sim \delta\left(\mathbf{r}'-\mathbf{r}\right)$, yielding the standard Poisson equation.

The correlation kernel reflects the relationship between diffraction and dielectric. For bulk water, we get $K\left(\mathbf{r},\mathbf{r}'\right) = K\left(\mathbf{r}-\mathbf{r}'\right)$ by inverse transforming its frequency $\mathbf{k}$-vector representation:

$$K\left(\mathbf{r}-\mathbf{r}'\right) = \int \exp\left[\mathbf{ik}\cdot\left(\mathbf{r}-\mathbf{r}'\right)\right]L\left(\mathbf{k}\right) d\mathbf{k}. \qquad (3.2)$$

In (3.2), $L\left(\mathbf{k}\right) = \left[\left(\varepsilon_w - \varepsilon_0\right)/\left(1 + \varepsilon_w \|\mathbf{k}\|^2 \xi^2/\varepsilon_0\right) + \varepsilon_0\right], \xi \approx 5\,\text{Å}$ denotes the characteristic length, and ε_w, ε_0 are the permittivities of water and vacuum, respectively. To obtain the correlation kernel with n hydrophobic units, we need to incorporate their solvent-structuring effect:

$$K\left(\mathbf{r},\mathbf{r}'\{\mathbf{r}_j\}\right) = \left[\int \exp\left[\mathbf{ik}\cdot\left(\mathbf{r}-\mathbf{r}'\right)\right]L\left(\mathbf{k}\right) d\mathbf{k}\right]$$
$$\times\left[1 + \sum_{j=1,\dots,n} \Gamma_j\left(\mathbf{r},\mathbf{r}'\right)\right] \qquad (3.3)$$

On purely phenomenological grounds we get the following approximation [2, 33, 34]: $\Gamma_j\left(\mathbf{r},\mathbf{r}'\right) \sim \exp\left[-\left(\|\mathbf{r}-\mathbf{r}_j\| + \|\mathbf{r}'-\mathbf{r}_j\|\right)/\Lambda\right]$, for $\|\mathbf{r}-\mathbf{r}_j\|$ and $\|\mathbf{r}'-\mathbf{r}_j\| > d/2$ with characteristic length Λ subsuming the range of the water-structuring effect. This parameter is typically fixed at 1.8 Å, the effective thickness of a single water layer (cf. Fig. 3.7).

We now solve (3.1) and (3.3) by Fourier transformation, obtaining the electric field $E(\mathbf{r})$ by inverse Fourier transformation of the solution to (3.1) in $\mathbf{k}$-representation:

$$\int E(\mathbf{r})\,d\mathbf{r} = -(4\pi q)\int \exp(i\mathbf{k}\cdot\mathbf{r})\,\|\mathbf{k}\|^{-2}\left[K\left(\mathbf{k},\{\mathbf{k}_j\}\right)\right]^{-1} d\mathbf{k}. \qquad (3.4)$$

Direct residue evaluation at the first-order poles $k = \pm i\,(\varepsilon_0/\varepsilon_w)^{1/2}\xi^{-1}$ $(k = \|\mathbf{k}\|)$ and $\mathbf{k} = \mathbf{k}_j \pm i\Lambda^{-1}$ yields the electric field $E(\mathbf{r})$ by retaining only the real part in the residue calculation:

$$E(\mathbf{r}) = \left(q/r^2\right)\left[\left(\varepsilon_0^{-1} - \varepsilon_w^{-1}\right)\Omega\left(\{\mathbf{r}_j\}\right)(1 + r/\xi)\exp(-r/\xi) + \varepsilon_w^{-1}\right], \qquad (3.5)$$

where

$$\Omega\left(\{\mathbf{r}_j\}\right) = \prod_{j=1,\dots,n}\left[1 + \exp\left(-\|\mathbf{r}_j\|/\Lambda\right)\right] \\ \times \left[(1 + \exp\left(-\|\mathbf{r}-\mathbf{r}_j\|/\Lambda\right)\right] \qquad (3.6)$$

and $r = \|\mathbf{r}\|$. Equations (3.5) and (3.6) describe the net effect of the wrapping hydrophobic arrangement on the electric field. In accord with (3.5) and (3.6), the wrapping effect can be captured by replacing the permittivity constant ε_w for bulk water by an effective permittivity $\varepsilon = \varepsilon\left(\mathbf{r},\{\mathbf{r}_j\}\right)$ defined as

$$\varepsilon = \left[\left(\varepsilon_0^{-1} - \varepsilon_w^{-1}\right)\Omega\left(\{\mathbf{r}_j\}\right)(1 + r/\xi)\exp(-r/\xi) + \varepsilon_w^{-1}\right]^{-1}. \qquad (3.7)$$

This quantity tends to the bulk limit for long interaction distances $(r/\xi \to \infty)$.

We now turn to the problem of finding the optimal wrapping arrangement and contrasting our implicit solvent result with the phenomenological result emerging from the simulations (cf. Figs. 3.4 and 3.6). Since $\varepsilon_0^{-1} \gg \varepsilon_w^{-1}$, finding the wrapping cluster with the lowest dielectric in its interior is tantamount to finding the arrangement $\{\mathbf{r}_j\}$ that maximizes the function $\Omega\left(\{\mathbf{r}_j\}\right)$, in accord with (3.7). We thus find the optimal arrangement $\{\mathbf{r}_j\}$ of hydrophobes that yields the maximum value Ω^* for $\Omega\left(\{\mathbf{r}_j\}\right)$. First, we compute the maximum $\Omega^*(n)$ of $\Omega\left(\{\mathbf{r}_j\}\right)$ for each fixed n subject to the constraint of preserving a minimum distance d between any two hydrophobes. The d value is taken to be 5 Å in accord with typical minimal distances between α-carbon atoms in tertiary structure [22]. Our results are qualitatively invariant in the range 4.5 Å $\le d \le$ 6 Å. Using the Lagrange multipliers method to minimize the effective permittivity, we find that the optimal arrangement is invariably obtained by fixing $n-2$ hydrophobes at distance d from each other and equidistantly from the O and H atoms, and placing the remaining two along the x-axis at distance $(\Lambda + \eta)(1 + n^{-2})$ (to first approximation) away from the C and N atoms, with $\eta = $ C $-$ O distance in the carbonyl group. This gives for $n = 4$ (tetrahedron): $\Omega^*(4) = 3.419$; for $n = 5$ (trigonal bipyramid): $\Omega^*(5) = 4.144$; for

$n = 6$ (square bipyramid), $\Omega^*(6) = 3.952$; and for $n = 7$ (pentagonal bipyramid), $\Omega^*(7) = 3.421$. Similar calculations for all n allow us to establish the following order relations:

$$\Omega^*(3) < \Omega^*(4) < \Omega^*(5) > \Omega^*(6) > \Omega^*(7) > \ldots \tag{3.8}$$

Thus, $\Omega^*(n)$ has a single maximum at $n = 5$. This maximum is expected on the basis of the two conflictive tendencies in the stabilization of a hydrogen bond: (1) bringing close to the hydrogen bond as many hydrophobes as possible and (2) bringing them as close to the hydrogen bond as possible. However, both demands start becoming mutually incompatible due to the steric hindrances implicit in the Lennard–Jones repulsive terms.

In full agreement with the result described in Fig. 3.6 ($\rho_{coar} = 5$ is a constant of motion along the folding trajectory), a hydrogen bond is embedded in the lowest dielectric when surrounded by five wrapping residues, the optimal compromise between crowding and proximity to the hydrogen bond. Thus, we have benchmarked the solvation theory against a rigorous dynamic result.

3.6 Dehydronic Field Along the Folding Pathway and the Commitment to Fold

Equations (3.5), (3.6), and (3.7) enable us to compute the dehydronic field, that is, the mechanical equivalent of the dehydration propensity of hydrogen bonds made along the folding process. This computation requires the evaluation of the gradient $\mathbf{\Phi}(\mathbf{R}) = -\nabla_{\mathbf{R}}[4\pi\varepsilon(\mathbf{R})]^{-1} qq'/r$ of the electrostatic energy with respect to the position vector $\mathbf{R}$ of the test hydrophobe (in the simulations we adopted methane as test hydrophobe). This analysis is motivated by the need to support the two-state kinetic picture outlined in Fig. 3.1. To normalize for the number of hydrogen bonds formed at any given time, we computed the dehydronic field per hydrogen bond as $\vartheta(t) = \left\langle \left\| \mathbf{\Phi}(\mathbf{R})_{\|R\|=4\text{Å}} \right\| \right\rangle_t$, where $< >_t$ denotes the average over the hydrogen bonds formed by the chain at time t. The results along the folding trajectory described in Figs. 3.2, 3.3, 3.4, 3.5, 3.6, and 3.7 are presented in Fig. 3.8.

Contrasting Fig. 3.8 with Figs. 3.3 and 3.4 clearly validates the two-state folding scenario depicted in Fig. 3.1: The dehydronic field starts as a very minor contributor during the hydrophobic collapse of the chain. Yet, once hydrogen bonds are formed to compensate for the backbone burial and they become partially wrapped (transition state ensemble), the dehydronic force becomes the driving force behind chain compaction. Thus, the progress of the folding process follows a reaction coordinate dictated by the gradient of the dehydronic field, making the latter the dominant factor that commits the chain to fold.

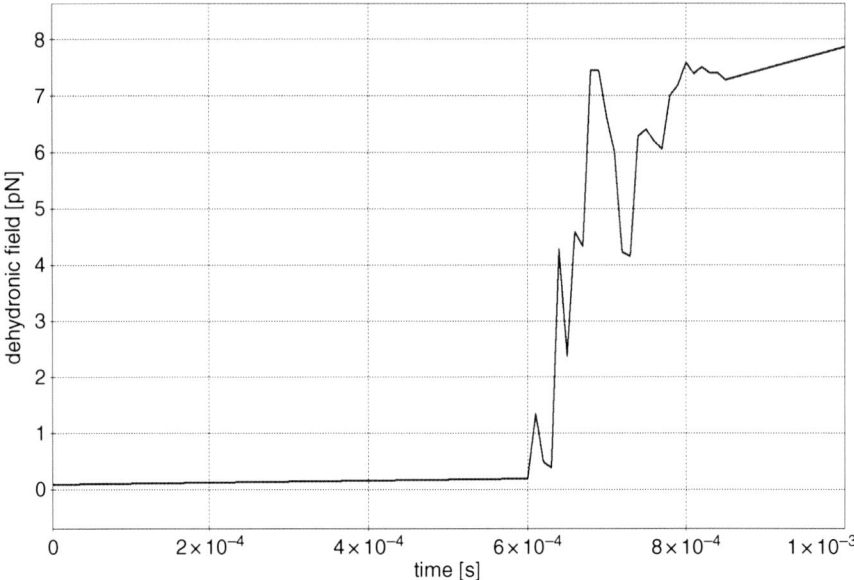

Fig. 3.8 Dehydronic field averaged over all backbone hydrogen bonds formed at time t for the protein G variant along the ab initio folding trajectory described in Figs. 3.3, 3.2, 3.3, 3.4, 3.5, 3.6, and 3.7

References

1. Anfinsen CB (1973) Principles that govern the folding of protein chains. Science 181:223–230
2. Fernández A, Sosnick TR, Colubri A (2002) Dynamics of hydrogen-bond desolvation in folding proteins. J Mol Biol 321:659–675
3. Chandler D (2005) Interfaces and the driving force of hydrophobic assembly. Nature 437:640–647
4. Jewett A, Pande VS, Plaxco KW (2003) Cooperativity, smooth energy landscapes and the origins of topology-dependent protein folding rates. J Mol Biol 326:247–253
5. Scalley-Kim M, Baker D (2004) Characterization of the folding energy landscapes of computer generated proteins suggests high folding free energy barriers and cooperativity may be consequences of natural selection. J Mol Biol 338:573–583
6. Fernández A, Colubri A, Berry RS (2002) Three-body correlations in protein folding: The origin of cooperativity. Physica A 307:235–259
7. Fernández A, Kostov K, Berry RS (1999) From residue matching patterns to protein folding topographies: General model and bovine pancreatic trypsin inhibitor. Proc Natl Acad Sci USA 96:12991–12996
8. Fernández A, Colubri A, Berry RS (2000) Topology to Geometry in protein folding: Beta-lactoglobulin. Proc Natl Acad Sci USA 97:14062–14066
9. Fernández A, Kardos J, Goto J (2003) Protein folding: Could hydrophobic collapse be coupled with hydrogen-bond formation? FEBS Lett 536:187–192
10. Fernández A (2001) Conformation-dependent environments in folding proteins. J Chem Phys 114:2489–2502
11. Avbelj F, Baldwin RL (2003) Role of backbone solvation and electrostatics in generating preferred peptide backbone conformations: Distributions of phi. Proc Natl Acad Sci USA 100:5742–5747

12. Fernández A (2004) Keeping dry and crossing membranes. Nat Biotechnol 22:1081–1084
13. Krantz BA, Moran LB, Kentsis A, Sosnick TR (2000) D/H amide kinetic isotope effects reveal when hydrogen bonds form during protein folding. Nat Struct Biol 7:62–71
14. Plaxco KW, Simmons KT, Baker D (1998) Contact order, transition state placement and the refolding rates of single domain proteins. J Mol Biol 277:985–994
15. Fersht A (2000) Transition-state structure as a unifying basis in protein-folding mechanisms: Contact order, chain topology, stability, and the extended nucleus mechanism. Proc Natl Acad Sci USA 97:1525–1929
16. Fernández A, Scott LR (2003) Adherence of packing defects in soluble proteins. Phys Rev Lett 91:018102
17. Fernández A (2003) What caliber pore is like a pipe? Nanotubes as modulators of ion gradients. J Chem Phys 119:5315–5319
18. Fernández A, Shen M, Colubri A, Sosnick TR, Freed KF (2003) Large-scale context in protein folding: Villin headpiece. Biochemistry 42:664–671
19. Duan Y, Kollman PA (1998) Pathways to a protein folding intermediate observed in a 1-microsecond simulation in aqueous solution. Science 282:740–744
20. Baldwin RL (2002) Making a network of hydrophobic clusters. Science 295:1657–1658
21. Nemethy G, Steinberg IZ, Scheraga HA (1963) The influence of water structure and hydrophobic contacts on the strength of side-chain hydrogen bonds in proteins. Biopolymers 1:43–69
22. Fernández A, Berry RS (2002) Extent of hydrogen-bond protection in folded proteins: A constraint on packing architectures. Biophys J 83:2475–2481
23. Novotny J, Bruccoleri R, Karplus M (1984) Analysis of incorrectly folded protein models. Implications for structure predictions. J Mol Biol 177:787–818
24. Daggett V, Levitt M (1992) A model of the molten globule state from molecular dynamics simulations. Proc Natl Acad Sci USA 89:5142–5146
25. Brooks CL, Case D (1993) Simulations of peptide conformational dynamics and thermodynamics. Chem Rev 93:2487–2502
26. Fernández A, Rogale K (2004) Sequence-space selection of cooperative model proteins. J Phys A: Math & Gen 37:197–202
27. Kuwata K, Shastry R, Cheng H, Hoshino M, Batt CA, Goto Y, Roder H (2001) Structural and kinetic characterization of early folding events in beta-lactoglobulin. Nat Struct Biol 8:151–155
28. Nymeyer H, Garcia AE, Onuchic JN (1998) Folding funnels and frustration in off-lattice minimalist protein landscapes. Proc Natl Acad Sci 95:5921–5928
29. Onuchic JN, Luthey-Schulten Z, Wolynes PG (1997) Theory of protein folding: The energy landscape perspective. Annu Rev Phys Chem 48:545–600
30. Chan HS, Dill KA (1997) From Levinthal to pathways to funnels. Nat Struct Biol 4:10–19
31. Fernández A, Colubri A, Berry RS (2001) Topologies to geometries in protein folding: Hierarchical and nonhierarchical scenarios. J Chem Phys 114:5871–5888
32. Shi Z, Krantz BA, Kallenbach N, Sosnick TR (2002) Contribution of hydrogen bonding to protein stability estimated from isotope effects. Biochemistry 41:2120–2129
33. Pietrosemoli N, Crespo A, Fernández A (2007) Dehydration propensity of order-disorder intermediate regions in soluble proteins. J Proteome Res 6:3519–3526
34. Despa F, Fernández A, Berry RS (2004) Dielectric modulation of biological water. Phys Rev Lett 93:228104
35. Fernandez A, Zhang X, Chen J (2008) Folding and wrapping soluble proteins: Exploring the molecular basis of cooperativity and aggregation. Prog Mol Biol Transl Sci 83:53–88

Chapter 4
Wrapping Deficiencies and De-wetting Patterns in Soluble Proteins: A Blueprint for Drug Design

This chapter unravels a link between wrapping deficiencies and defects in the hydration shell of soluble proteins. The hydration shell of a soluble protein is not uniform: its tightness, marked by the mobility of interfacial water, is site-dependent and modulates the propensity for protein associations. The most pronounced interfacial de-wetting propensity is promoted by dehydrons in the surface of the soluble proteins. Their dehydration is thermodynamically favored and hence may serve as a blueprint for drug design. The result has crucial implications for the designer as drugs may be engineered to expel the labile parts of the target hydration shell upon association. Hence, subtle differences in the location of hydration defects in proteins within the same family may be exploited to enhance drug specificity, as demonstrated in Chap. 8. The subject of interfacial water is again taken up in Chap. 14 that deals with the disruption of protein-protein interfaces with competing drugs.

4.1 Hydration Defects in Soluble Proteins

Solvation is not uniform on the surface of soluble proteins, rather it is marked by vastly different mobilities of interfacial water molecules [1–5]. This uneven distribution of mobilities of the hydrating molecules affects protein folding dynamics and associations. The hydration shell gets tighter or less mobile at interfacial regions where the hydrogen bond network of hydrating molecules is minimally disrupted and becomes looser in concave or flat nonpolar regions of the protein surface [3–5]. Thus, dehydration patterns have been associated with surface topography [5] and with fold topology [6], both determinants of protein associations. However, the pronounced de-wetting propensity of interfacial regions remains difficult to assess in general, as evidenced by the difficulty in predicting binding sites [7], in inferring propensities for aberrant aggregation [8], and in rationally designing ligands, notwithstanding some level of success in ligand docking [9]. This chapter explores de-wetting patterns determined by unburied or under-wrapped backbone hydrogen bonds, the *dehydrons* [10, 11] described in the preceding chapters. Since their levels of hydration may be modulated by protein associations, dehydrons are implicated in macromolecular recognition [8–11]. More precisely, dehydrons may become

A. Fernández, *Transformative Concepts for Drug Design: Target Wrapping*, DOI 10.1007/978-3-642-11792-3_4, © Springer-Verlag Berlin Heidelberg 2010

stabilized and electrostatically strengthened by the attachment of a ligand or binding partner that further contributes to their dehydration. As indicated in Chap. 1, the net gain in Coulomb energy associated with wrapping or protecting a dehydron has been experimentally determined to be $\sim$4 kJ/mol [12]. The adhesive force exerted by a dehydron on a hydrophobic group at 6 Å distance is $\sim$7.8 pN, a magnitude comparable to the hydrophobic attraction between two nonpolar moieties that frame unfavorable interfaces with water [12]. *Furthermore, the fact that dehydrons promote protein associations* (Chap. 1) *strongly suggests that they may signal sites with a loose hydration shell, that is, sites where the work needed to remove interfacial water is minimal and amply compensated by the enhancement of the Coulombic energy.*

As shown in Chap. 8 and again in Chap. 14, this result will prove to be of paramount relevance for the rational drug designer as it introduces a blueprint to guide ligand engineering: the de-wetting propensities in the target protein may be sculpted into the ligand so that water is expelled from the interface upon association wherever the work needed to do so is minimal and the dehydronic force is maximal. In other words, the drug/ligand may be engineered to tightly fit against the hot spots of labile interfacial water.

4.2 Wrapping as a Marker of Local De-wetting Propensity

In this section we introduce a descriptor of hydration tightness or de-wetting propensity in order to assess the functional role of dehydrons. Tightness is defined by the extent of mobility of hydrating molecules. Thus, the mean residence time of hydrating molecules within a domain around each residue on the protein surface serves as an adequate indicator. As an illustration, we describe in detail the interfacial water mobility for the autonomously folded SH3 domain (PDB.1SRL) [13], an all-beta protein involved in the regulation of cell signaling. We also analyze ubiquitin (PDB.1UBI) [14], a quintessential α/β fold, and finally explore the relationship between extreme dehydration propensity and amyloidogenic aggregation [9].

Our results lead us to identify dehydrons as the structural feature causative of the most dramatic decrease in residence times (highest mobility) of hydrating molecules. Dehydrons promote local de-wetting because the Coulomb energy of the intramolecular hydrogen bonds becomes magnified upon water removal, and the net gain in stability upon dehydration offsets the work required to remove surrounding water [10, 15, 16].

In order to assess the local mobility of water molecules in the hydration shell, the local mean residence time, $<\tau>_i$, of hydrating molecules around residue i is defined with respect to a microenvironment in the form of a spherical domain $D(i)$ of 6 Å radius ($\sim$width of three water layers [12]) centered at the α-carbon of residue i (Fig. 4.1a). The computations are performed for a range of radii (see below). The residence time is obtained as follows:

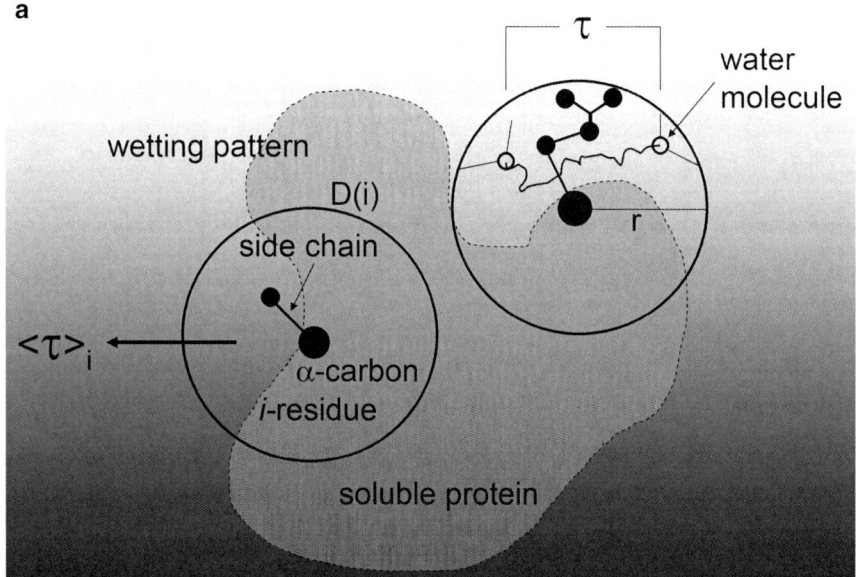

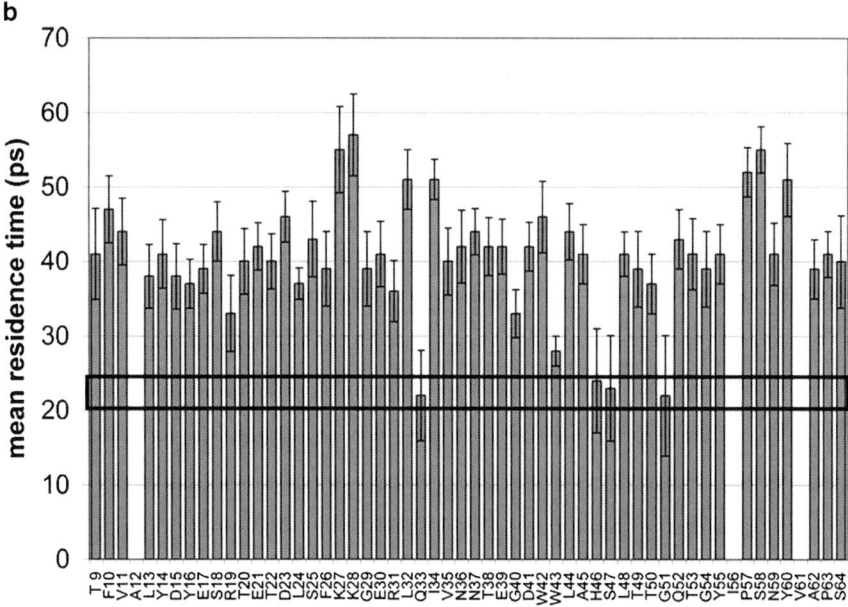

Fig. 4.1 (a) Basic scheme for the computation of de-wetting hot spots in soluble proteins. (b) Mean residence times of water molecules solvating the folded protein SH3 domain. Residue numbering follows PDB file 1SRL. The one-letter amino acid code was adopted for clarity. The mean residence times are computed using (4.1) and extracted from MD simulations of the hydration dynamics. The error bars represent variances and the *thick rectangle* represents the range of residence times for bulk water within a domain of the same dimensions of the one used to examine the protein surface. The range for bulk water serves as benchmark to single out hot spots of weak hydration. Three residues (Ala12, Ile56, and Val61) are fully buried in the folded structure and thus their hydration shells contain no water molecules. Figure 4.1b reprinted with permission from American Institute of Physics [20]. Copyright 2007

$$\langle \tau \rangle_i = \int \tau f_i(\tau) d\tau \Big/ \left(\int f_i(\tau) d\tau \right) ;$$

$$\int_0^\tau f_i(\tau') d\tau' = P_i(0) - P_i(\tau) ;$$

$$P_i(\tau) = \Theta^{-1} \int_0^\Theta \left[\sum_{\substack{v(t) \in U(i,t) \\ w(t+\tau) \in U(i,t+\tau)}} \delta\left(v(t), w(t+\tau)\right) \right] dt ;$$

(4.1)

where $f_i(\tau) d\tau / \int f_i(\tau) d\tau$ is the expected fraction of water molecules that exit $D(i)$ within time interval $[\tau, \tau + d\tau]$; $P_i(\tau)$ is the expected number of water molecules remaining in $D(i)$ at time $\tau(P_i(0) - P_i(\tau) =$ number of molecules that exited $D(i)$ in the time period $[0, \tau]$); $v(t), w(t + \tau)$ denote indexes labeling water molecules contained in $D(i)$ at times t and $t + \tau$, respectively; $U(i, t), U(i, t + \tau)$ denote the collection of indexes of water molecules contained in $D(i)$ at times t and $t + \tau$, respectively; δ is the Kronecker symbol ($\delta(v(t), w(t + \tau)) = 1$ if $v(t) = w(t + \tau)$ and 0, otherwise); and the integration over variable t is carried out over the interval of sampled times ($t = 0$ to $t = \Theta = 10$ ns) after 5 ns of prior equilibration (the sampling is considered exhaustive since $<\tau> << \Theta$ for all residues).

The mean residence times are obtained from classical trajectories generated by molecular dynamics (MD) simulations starting from the PDB structure embedded in a pre-equilibrated cell of explicitly represented water molecules and counterions [17, 18]. Computations are performed by integration of Newton's equations of motion with time step 2 fs using the GROMACS program [19] in an NPT ensemble with box size $8 \times 8 \times 8$ nm^3 and periodic boundary conditions maintained at 300 K and 1 atm. The box size is calibrated so that the solvation shell extended at least 12 Å from the protein surface at all times. Specific details on the generation of these classical trajectories may be found in [20].

4.3 Dehydrons Are Loosely Hydrated

We start by providing a detailed analysis of the de-wetting propensities in the SH3 domain ($N = 55$, PDB.1SRL) and their correlation with dehydrons in the structure. The mean residence times of water molecules at the protein–water interface has an overall average value of 44 ps, nearly twice the residence time for bulk water ($\sim$21 ps, Fig. 4.1b) [21]. The hydration shell reveals some standard features. Thus, short residence times correspond to de-wetting propensity associated with hydrophobic hydration at cavities with relatively large curvature radius (~ 6 Å) [22], like the hydrophobic pocket containing Trp43 (Figs. 4.1b and 4.2).

By contrast, exposed hydrophobic residues like Val35 may be effectively "clathrated" or accommodated within a water cavity that introduces a minimal perturbation of the tetrahedral hydrogen bonding network of water. In fact, clathration actually tightens the hydration shell (Figs. 4.1b and 4.2).

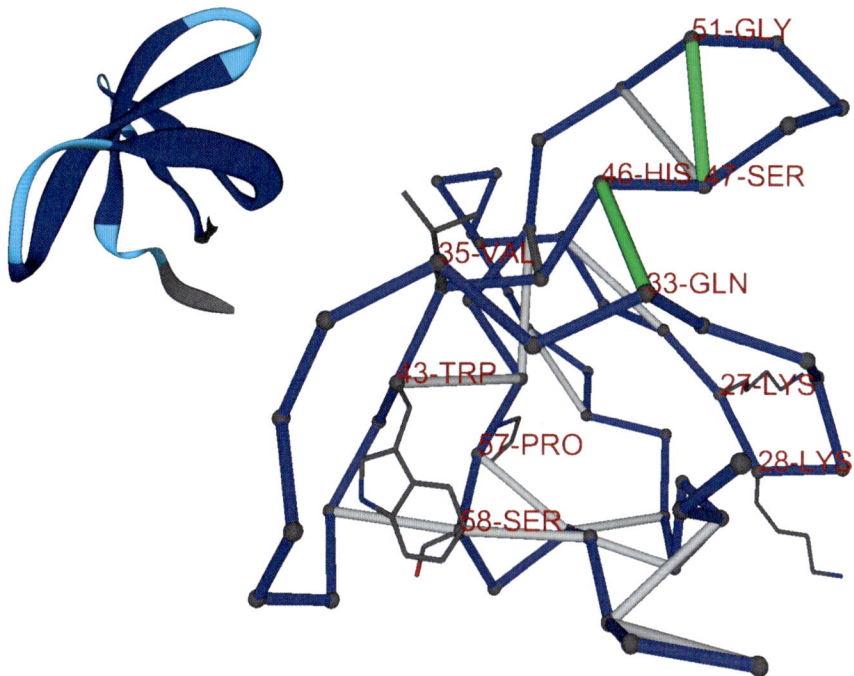

Fig. 4.2 PDB structure of SH3 domain indicating the packing defects in the protein. The backbone is displayed by *blue segments* representing virtual bonds joining α-carbons, well-wrapped hydrogen bonds are shown as *white segments* joining α-carbons of paired residues, and dehydrons are shown in *green*. Some residue labels and side chains are displayed to illustrate hydration patterns. Reprinted with permission from American Institute of Physics [20]. Copyright 2007

Side chains with small exposed cations (Lys27 and Lys28) have the longest residence times since they are tightly hydrated as are polar residues in cavities (like Ser58), since their hydrogen bonding to water is protected. By contrast, exposed residues with delocalized positive charges like Arg19 have looser hydration [23].

The residues with the shortest residence times ($<\tau> < 24$ ps) and also with the largest variances ($<(\tau - <\tau>)^2 >^{1/2} > 6$ ps) are Gln33, His46, Ser47, and Gly51, indicating a highly fluctuating environment and a high de-wetting propensity. As shown below, these residues are paired by the only backbone amide–carbonyl hydrogen bonds which are incompletely wrapped or shielded from water by nonpolar groups of the protein chain. In accord with their shortest water residence times (Fig. 4.1b), dehydrons become favorably dehydrated, a process that decreases the overall polarizability of their microenvironments. This dehydration strengthens the electrostatic contribution and stabilizes the hydrogen bond by destabilizing the nonbonding or unbound state to the point of overcoming the work needed to remove surrounding water [8]. These packing defects and the decrease in medium polarizability or dielectric modulation they promote differ from packing-density variations in nonpolar interiors [24]: They relate to the exposure of the polar backbone due to incomplete wrapping by nonpolar side chains.

To explain the local weaknesses in the hydration shell of the protein (Fig. 4.1b), the de-wetting propensity of dehydrons is quantified and compared with that of well-wrapped hydrogen bonds. As shown in Chaps. 1 and 3, the de-wetting propensity has a mechanical equivalent [12, 25]: The attractive force exerted on a nonpolar test group (i.e., methane, CH_4) by a pre-formed hydrogen bond that may be strengthened upon removal of surrounding water. As the nonpolar group approaches a dehydron, it displaces water thus decreasing the environmental polarity, thereby enhancing and stabilizing the bond. The wrapping microenvironment of a hydrogen bond may be described by the set of position vectors $\{\mathbf{R}_j\}_{j=1,\dots,K} = \{\mathbf{R}_j(t)\}_{j=1,\dots,K}$, relative to the hydrogen atom (origin of coordinates), of K nonpolar groups from the protein chain within the two spheres of radius 6 Å centered at the α-carbons of the paired residues. All time-dependent coordinates are taken from MD trajectories.

To compute the de-wetting of intramolecular hydrogen bonds, we take into account the modulation of local permittivity determined by a change in the "environmental coordinates": $\{\mathbf{R}_j\}_{j=1,\dots,K} \rightarrow \{\mathbf{R}_j, \mathbf{R}\}_{j=1,\dots,K}$, where $\mathbf{R}$ is the position vector of the test methane that approaches the bond orthogonally to the Coulombic field. The energy change ΔE associated with the change in wrapping is then given by [12, 25]

$$\Delta E(t) = -(qq'/4\pi r)\left[\varepsilon^{-1}\left(\{\mathbf{R}_j, \mathbf{R}\}_{j=1,\dots,K}\right) - \varepsilon^{-1}\left(\{\mathbf{R}_j\}_{j=1,\dots,K}\right)\right], \qquad (4.2)$$

where q, q' are the effective charges at the hydrogen and oxygen atoms of the amide–carbonyl hydrogen bond, r is the hydrogen bond length ($r = r(t) = \|\mathbf{r}(t)\|$, with $\mathbf{r} =$ position vector of oxygen), and the reciprocal permittivity ε^{-1}, quantifying the Coulomb screening due to wrapping by nonpolar groups, is generically given by [25]

$$\varepsilon^{-1}(\{\mathbf{R}_n\}) = (\varepsilon_0^{-1} - \varepsilon_w^{-1})\Omega(\{\mathbf{R}_n\})[(1 + r/\xi)\exp(-r/\xi)] + \varepsilon_w^{-1}, \qquad (4.3)$$

where, as in Chap. 3, ε_0, ε_w denote, respectively, the permittivity of vacuum and bulk water, $\Omega(\{\mathbf{R}_n\}) = \Pi_{n=1,\dots}\left[1 + \exp(-\|\mathbf{R}_n\|/\Lambda)\right]\left[1 + \exp(\|\mathbf{r} - \mathbf{R}_n\|/\Lambda)\right]$, where ξ, fixed at 5 Å, is the characteristic length for water dipole reorientation influence, and $\Lambda = 1.8$Å is the characteristic length for water structuring around a nonpolar group [25].

Applying (4.2) and (4.3) to the MD trajectories, we quantify the enhancement in the Coulombic contribution to the hydrogen bond resulting from the decrease in permittivity as the test hydrophobe approaches the protein surface (Fig. 4.3). Thus, the de-wetting field $\mathbf{\Phi}(\mathbf{R})$ at position $\mathbf{R}$ of the test hydrophobe generated by a hydrogen bond wrapped by K nonpolar groups is given by

$$\mathbf{\Phi}(\mathbf{R}) = (qq'/4\pi r)\nabla_{\mathbf{R}}(\varepsilon^{-1}(\{\mathbf{R}_j, \mathbf{R}\}_{j=1,\dots,K})). \qquad (4.4)$$

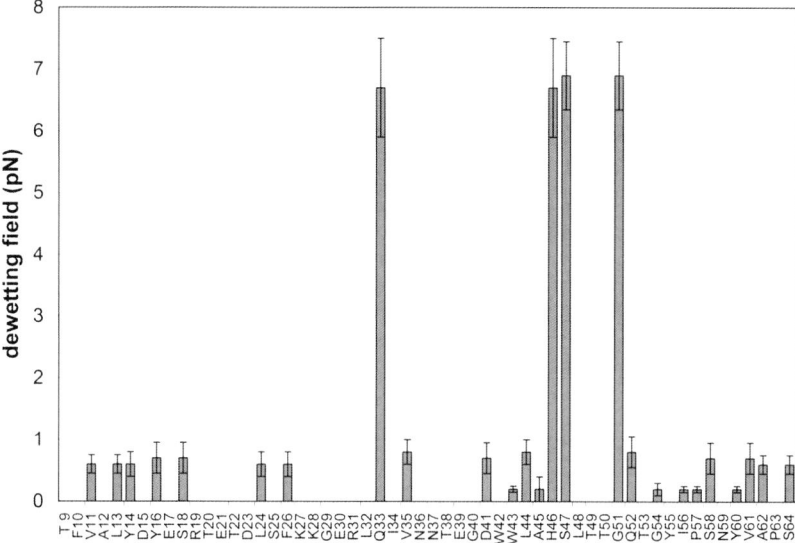

Fig. 4.3 De-wetting field exerted by residues paired by hydrogen bonds in the SH3 domain. The dehydration propensity of a hydrogen bond, $\|\Phi(\mathbf{R})\|$, is determined at $\|\mathbf{R}\| = 6$ Å and is assigned to the two residues paired by the hydrogen bonds. The data displayed show a time average and variance over the interval of sampled times ($t = 0$ to $t = \Theta = 10$ ns) after 5 ns of prior equilibration. Reprinted with permission from American Institute of Physics [20]. Copyright 2007

The time averages of the magnitude and variance of the de-wetting field were computed for each residue in the protein (Fig. 4.3).

Comparison of Figs. 4.1b and 4.3 prompt the following conclusions: (a) the tightness of the hydration shell of a soluble protein is not uniform, yielding an uneven distribution of mobilities for hydrating molecules; (b) dehydrons generate the most pronounced loosening of the hydration shell; and (c) dehydrons are endowed with dehydration propensities.

A similar analysis was conducted for other single-domain soluble proteins, unrelated evolutionarily or topologically to the study case presented and the same conclusions regarding the de-wetting propensity of dehydrons were shown to hold [20]. For example, we focused on the highly conserved ubiquitin ($N = 76$, 11 dehydrons, PDB.1UBI). In this case dehydrons also generate the most intense de-wetting field in the protein (Fig. 4.4), in accord with their role as structural markers for protein associations. The de-wetting hot spots account for 9 out of the 11 dehydrons identified for this structure, and only partially account for the remaining two. The two dehydrons which do not fully promote de-wetting are Glu24–Asp52 and Pro19–Ser57. The larger residence times in these cases may be attributed to the pronounced hydration demands of the two charged side chains in the former case and the special steric hindrance promoted by Pro19, which precludes water approach to the backbone in the latter.

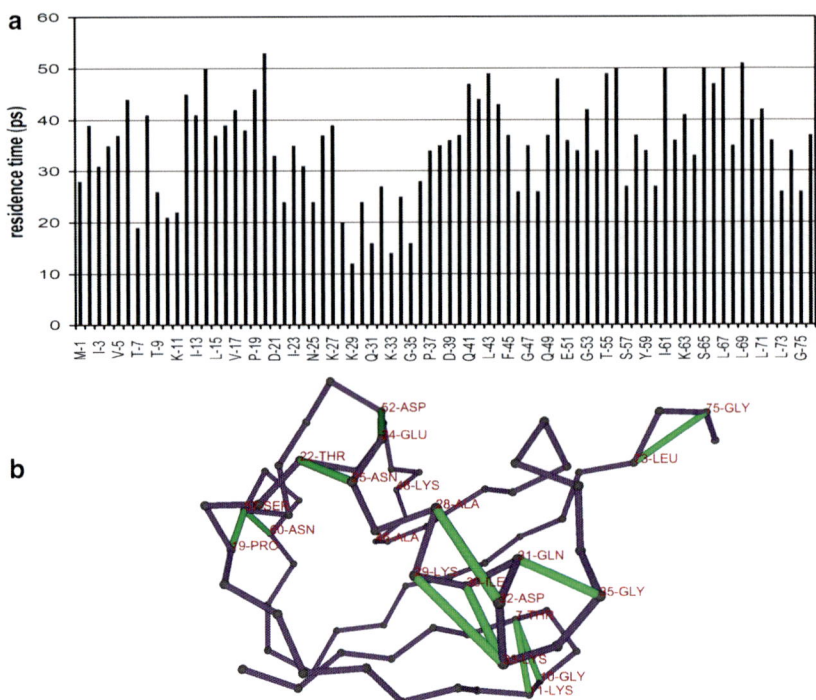

Fig. 4.4 (a) Mean residence times of water molecules solvating the folded protein *ubiquitin*. Residue numbering follows the file PDB.1UBI. **b** PDB structure of ubiquitin highlighting the dehydrons (displayed in *green*). Reprinted with permission from American Institute of Physics [20]. Copyright 2007

4.4 Displacing Loose Hydrating Molecules: A Blueprint for the Drug Designer

In conclusion, the identification of structural determinants of protein/water interfacial behavior is vital to understand protein associations and to design drugs/ligands with better affinity. The findings reported in this chapter represent a constructive step in this direction, since we have singled out structural singularities in soluble proteins – the dehydrons – that behave as de-wetting hot spots. The identification of these sites with defective and loose hydration opens up an engineering avenue to understand and manipulate protein/ligand associations. Thus, by sculpting on the drug the de-wetting hot spots of the target, we take advantage of the minimal work associated with removal of dehydron-solvating water molecules, amply compensated by the resulting enhancement of the electrostatic interaction (see Chap. 3). In this way, we may enhance the affinity for the protein target by adopting the pattern of hydration defects in the target as a blueprint for molecular engineering (Fig. 4.5). This concept is brought to fruition in the reengineering of the powerful anticancer

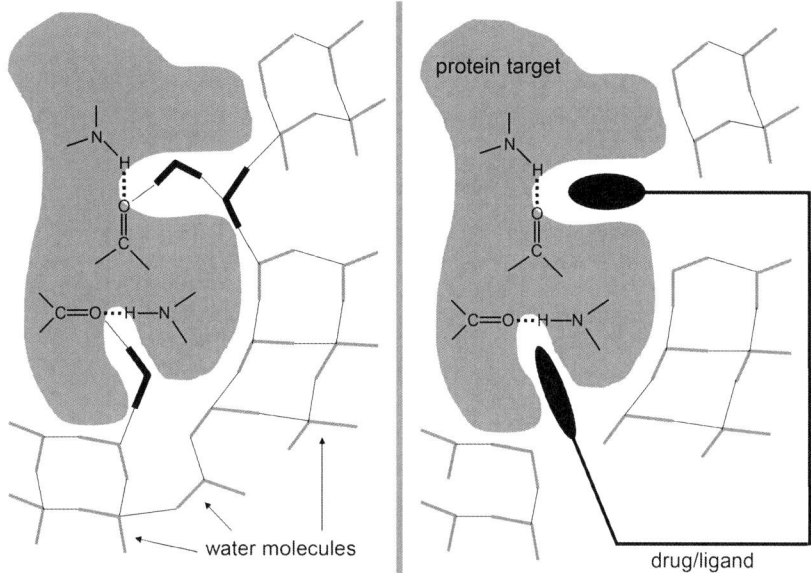

Fig. 4.5 The pattern of defects in the hydration shell of a target protein is a blueprint for drug design. Interfacial water molecules solvating dehydrons have short residence time and hence are easily displaced upon association with a purposely designed ligand. Their loss of hydrogen bonding partnerships results from their partial confinement. Fast interfacial water molecules are represented as thick *black angular lines*, while bulk-like water molecules are indicated as *gray angular lines*. Hydrogen bonds are shown as *thin black lines*, while dehydrons are indicated as *dotted lines*

drug *imatinib* (Gleevec), as described in Chap. 8. Furthermore, it heralds the advent of a new generation of drugs based on an assessment of the architecture of the protein fold in relation to the solvating environment, as shown in Chap. 14. The implications of this concept for drug specificity will be explored in Chap. 8, where differences in de-wetting patterns across purported targets enable us to engineer a selective therapeutic impact.

References

1. Otting G, Liepinsh E, Wüthrich K (1991) Protein hydration in aqueous solution. Science 254:974–980
2. Qiu W, Kao Y, Zhang L et al (2006) Protein surface hydration mapped by site-specific mutations. Proc Natl Acad Sci USA 103:13979–13984
3. Makarov V, Andrews BV, Smith PE, Pettitt BM (2000) Residence times of water molecules in the hydration sites of myoglobin. Biophys J 79:2966–2974
4. Chandler D (2005) Interfaces and the driving force of hydrophobic assembly. Nature 437: 640–647
5. Cheng Y, Rossky PJ (1998) Surface topography dependence of biomolecular hydrophobic hydration. Nature 392: 696–699

6. Liu P, Huang X, Zhou R, Berne BJ (2005) Observation of a dewetting transition in the collapse of the melitin tetramer. Nature 437:159–162
7. Fernández A, Scheraga HA (2003) Insufficiently dehydrated hydrogen bonds as determinants of protein interactions. Proc Natl Acad Sci USA 100:113–118
8. Fernández A, Kardos J, Scott R, Goto Y, Berry RS (2003) Structural defects and the diagnosis of amyloidogenic propensity. Proc Natl Acad Sci USA 100:6446–6451
9. Hetenyi C, van der Spoel D (2006) Blind docking of drug-sized compounds to proteins with up to a thousand residues. FEBS Lett 580:1447–1450
10. Fernández A, Berry RS (2004) Molecular dimension explored in evolution to promote proteomic complexity. Proc Natl Acad Sci USA 101:13460–13465
11. Fernández A, Scott RL (2003) Dehydron: A structurally encoded signal for protein interaction. Biophys J 85:1914–1928
12. Fernández A, Scott LR (2003) Adherence of packing defects in soluble proteins. Phys Rev Lett 91:018102
13. Yu H, Rosen MK, Schreiber SL (1993) 1H and 15 N assignments and secondary structure of the Src SH3 domain. FEBS Lett 324:87–92
14. Vijay-Kumar S, Bugg CE, Cook WJ (1987) Structure of ubiquitin refined at 1.8 angstrom resolution. J Mol Biol 194:531–544
15. van der Spoel D, van Maaren P, Larsson P, Timneanu N (2006) Thermodynamics of hydrogen bonding in hydrophilic and hydrophobic media. J Phys Chem B 110:4393–4398
16. Sheu S, Yang D, Selzle H, Schlag EW (2003) Energetics of hydrogen bonds in peptides. Proc Natl Acad Sci USA 100:12683–12687
17. Rizzo RC, Jorgensen WL (1999) OPLS All-atom model for amines: Resolution of the amine hydration problem. J Am Chem Soc 121:4827–4836
18. Jorgensen WL, Chandrasekhar J, Madura J, Impey RW, Klein ML (1983) Comparison of simple potential functions for simulating liquid water. J Chem Phys 79:926–935
19. Lindahl E, Hess B, Van der Spoel D (2001) GROMACS 3.0: A package for molecular simulation and trajectory analysis. J Mol Model 7:302–317
20. Fernández A, Chen J, Crespo A (2007) Solvent-exposed backbone loosens the hydration shell of soluble folded proteins. J Chem Phys 126:245103
21. Denisov V, Halle B (1995) Protein hydration dynamics in aqueous solution. A comparison of bovine pancreatic trypsin inhibitor and ubiquitin by oxygen-17 spin relaxation dispersion. J Mol Biol 245:682–697
22. Lum K, Chandler D, Weeks JD (1999) Hydrophobicity at small and large length scales. J Phys Chem B 103:4570–4577
23. Mason PE, Neilson GW, Dempsey CE, Barnes AC, Cruickshank JM (2003) The hydration structure of guanidinium and thiocyanate ions: Implications for protein stability. Proc Natl Acad Sci USA 100:4557–4561
24. Kocher JP, Prevost M, Wodak S, Lee B (1996) Properties of the protein matrix revealed by the free energy of cavity formation. Structure 4:1517–1529
25. Fernández A, Sosnick TR, Colubri A (2002) Dynamics of hydrogen-bond desolvation in folding proteins. J Mol Biol 321:659–675

Chapter 5
Under-Wrapped Proteins in the Order–Disorder Twilight: Unraveling the Molecular Etiology of Aberrant Aggregation

Soluble folded proteins maintain their structural integrity by properly shielding their backbone amides and carbonyls from full hydration. Thus, a poorly wrapped backbone hydrogen bond or *dehydron* constitutes an identifiable structural deficiency. In this chapter we describe the physical properties of proteins that possess large clusters of dehydrons in their soluble structure. We show that these clusters represent unique structural singularities belonging to an order–disorder twilight and generate a sharp local quenching of the dielectric permittivity of the surrounding medium. The functional roles of these singularities are explored across natural proteins and related to their physical properties, with special emphasis on the molecular etiology of amyloidogenic aggregation. Our analysis required a description of interfacial water that cannot be properly captured by conventional continuous models, where solvent degrees of freedom are typically averaged out.

5.1 Dehydron Clusters and Disordered Regions

As noted in the preceding chapters, the structural integrity of a soluble protein is contingent on its capacity to exclude water from its amide–carbonyl hydrogen bonds [1, 2]. Thus, under-wrapped intramolecular hydrogen bonds or dehydrons constitute structural singularities representing packing defects that have been extensively characterized as implicated in protein associations and macromolecular recognition [2–7, 8–13]. The strength and stability of dehydrons may be modulated by an external agent. More precisely, intramolecular hydrogen bonds which are not "wrapped" by a sufficient number of nonpolar groups may become stabilized and strengthened by the attachment of a ligand or binding partner that further contributes to their dehydration.

In this chapter we identify functional indicators of PDB-reported soluble proteins with clusters of dehydrons. Such regions, rich in structural vulnerabilities, may be characterized as belonging to a "twilight zone" between order and native disorder [4]. This characterization is already suggested by a strong correlation between wrapping of intramolecular hydrogen bonds (ρ) and propensity for structural disorder (f_d), as shown in Fig. 5.1. The correlation reveals that the inability to exclude

A. Fernández, *Transformative Concepts for Drug Design: Target Wrapping*,
DOI 10.1007/978-3-642-11792-3_5, © Springer-Verlag Berlin Heidelberg 2010

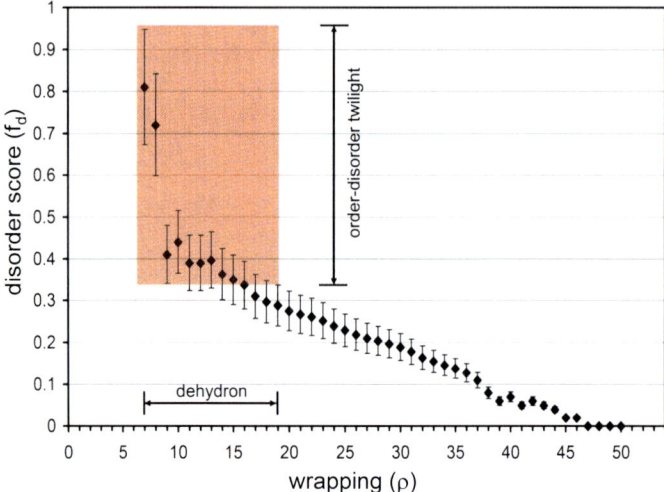

Fig. 5.1 Correlation between the disorder score of a residue and the extent of wrapping (ρ) of the backbone hydrogen bond engaging that particular residue (if any). The disorder score on each individual residue was obtained for 2,806 nonredundant nonhomologous PDB domains. Residues have been independently grouped in 45 bins of 400 residues each, according to the extent of wrapping ($7 \leq \rho \leq 52$). The average score has been determined for each bin (*square*) and the *error bars* represent the dispersion of disorder scores within each bin. The strong correlation between the disorder score and the extent of wrapping and the dispersions obtained imply that dehydrons can be safely inferred in regions where the disorder score is $f_d > 0.35$. The *red rectangle* represents the region of order–disorder twilight where the existence of dehydrons may be inferred from disorder score with absolute certainty. No hydrogen bond in monomeric domains reported in PDB was found to possess less than seven wrappers, implying a threshold for structural sustainability in soluble proteins. Reprinted with permission from [19] copyright 2007 American Chemical Society

water intramolecularly from pre-formed hydrogen bonds is causative of the loss of structural integrity.

The disorder propensity may be accurately quantified by a sequence-based score generated by the program PONDR-VLXT [14–16], a predictor of native disorder that takes into account residue attributes such as hydrophilicity, aromaticity, and their distribution within the window interrogated. The disorder score ($0 \leq f_d \leq 1$) is assigned to each residue within a sliding window, representing the predicted propensity of the residue to be in a disordered region ($f_d = 1$, certainty of disorder; $f_d = 0$, certainty of order). Only 6% of 1,100 nonhomologous PDB proteins gave false-positive predictions of disorder in sequence windows of 40 amino acids. The strong correlation (over 2,806 nonredundant nonhomologous PDB domains, Fig. 5.1) between disorder score of a residue and extent of wrapping of the hydrogen bond engaging the residue (if any) implies that clusters of dehydrons correspond to extended regions of structural vulnerability within a structured domain. Hence, the characterization of dehydron clusters as belonging to an order–disorder twilight zone appears to be warranted.

Both order and native disorder are well-characterized structural attributes of protein chains [16]. However, the highly vulnerable regions in a soluble fold described in this chapter belong to the novel category of "tamed disorder" because they can acquire and maintain a structured state only upon association. Neither order nor disorder is an adequate category to describe such dehydron-rich protein regions.

In this chapter we explore the functional role of regions with the largest dehydron clusters. A cluster is defined as the maximal set of dehydrons with intersecting desolvation domains or overlapping microenvironments. These regions actually belong to an order–disorder twilight and are shown to be strong dielectric modulators, i.e., quenchers of the local dielectric permittivity, thus providing a local enhancement of Coulomb fields nearby. This property arises since clusters of packing defects promote local dehydration of the protein interface promoted by a significant reduction in hydrogen bonding partnerships of solvating water molecules.

These discrete effects relating to local water structuring around packing defects cannot be captured properly by existing continuous models of the interfacial electrostatics [17]. This is mainly because such models are based on mean-force potential approximations to solvent interactions, where solvent degrees of freedom are averaged out, and this is clearly an inappropriate ansatz to deal with cavities of the size of solvent molecules themselves.

5.2 Discrete Solvent Effects Around Dehydrons

The dehydrons in a soluble protein ($\rho \leq 19$, for desolvation radius 6 Å) are partially exposed to solvent. As shown in the previous chapters, these bonds promote the removal of hydrating molecules as a means to enhance the underlying polar pair electrostatics. Furthermore, the resulting bond stabilization overcomes the amount of work needed to remove solvating water molecules [3, 5]. To describe this dehydration propensity, we first compute the extent of constraint of interfacial water molecules. This parameter is identified with the thermal average of Γ, the average number of hydrogen bond partnerships involving water molecules within the desolvation domain of each residue along the chain ($0 \leq \Gamma \leq 4$, Fig. 5.2). As an illustration we focus on the DNA-binding domain of antitumor gene p53 [18]. This domain was selected because it contains three of the largest dehydron clusters to be found in PDB (Fig. 5.3). The functional significance of these clusters relates to their interplay with the electrostatics of DNA recognition by this transcription factor [19].

Figure 5.2 shows the thermal average, $<\Gamma>$, of the average number of hydrogen bond partnerships involving water molecules within the desolvation domain for each residue. Three dehydration hot spots are apparent, comprising residues 171–181, 236–246, and 270–289, respectively. The location of these hot spots corresponds to the three major dehydron clusters shown in Fig. 5.3.

The thermal average, $<\Gamma>$, of the number of hydrogen bond partnerships involving water molecules hydrating the p53 DNA-binding domain was obtained from

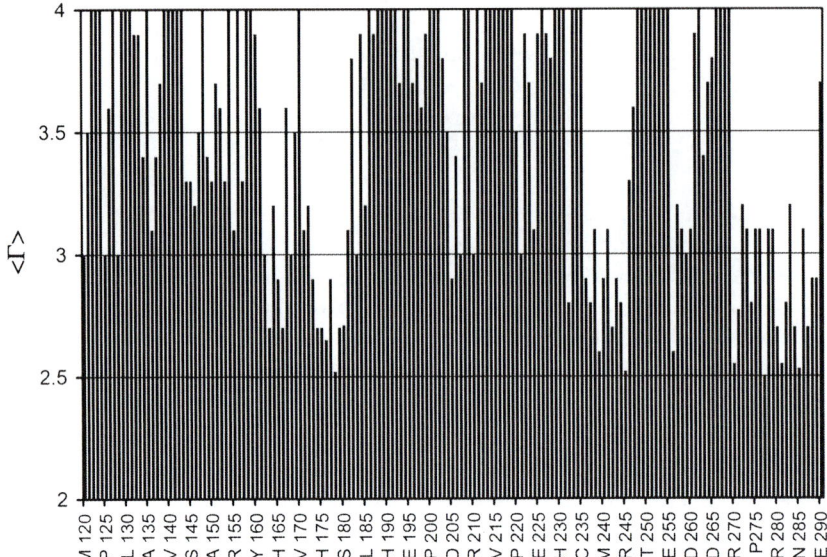

Fig. 5.2 Thermal average of the average number of hydrogen bond partnerships, Γ, for water molecules located within the desolvation domain of each residue in the DNA-binding domain of p53. If no water is found in the desolvation domain (buried residue), the bulk water value Γ=4 is adopted. Reprinted with permission from [19] copyright 2007 American Chemical Society

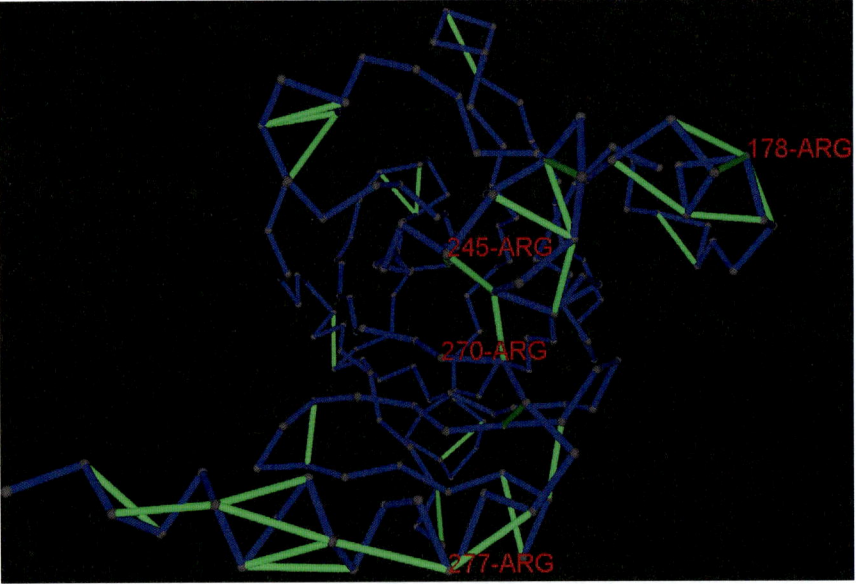

Fig. 5.3 Dehydrons for p53 DNA-binding domain. The backbone is indicated by *blue virtual bonds* joining α-carbons and dehydrons are shown as *green segments* joining the α-carbons of residues paired by backbone hydrogen bonds. Reprinted with permission from [19] copyright 2007 American Chemical Society

classical trajectories generated by 5 ns molecular dynamics (MD) simulations (see Chap. 4). The Γ-values were determined for each water molecule within a 6 Å radius spherical domain centered at the α-carbon of the residues. The adopted starting geometry is the monomeric structure from PDB entry 2GEQ. The starting conformation was embedded in a pre-equilibrated cell of explicitly represented water molecules and counterions [19]. The entire system was initially equilibrated for 5 ns. Computations were performed by integration of Newton's equations of motion with time step 2 fs using the GROMACS program [20] in the NPT ensemble with box size $8 \times 8 \times$ nm^3 and periodic boundary conditions, as described in the preceding chapter.

This computation was repeated to include representatives of the three major fold topologies: all-β (SH3 domain, $N = 55$, 2 dehydrons, PDB.1SRL), all-α (λ-repressor, $N = 86$, 26 dehydrons, PDB.1LMB), and α/β (ubiquitin, $N = 76$, 16 dehydrons, PDB.1UBI). Consistency with the p53 calculation was obtained, as dehydrons proved to become the dehydration hot spots on the protein interface [19].

The confinement of solvating water determined by a packing defect is illustrated in Fig. 5.4. The figure displays a snapshot of a water molecule within the desolvation domain of Arg277, a residue paired by a dehydron to Arg280. This water molecule has three hydrogen bond partners, two with vicinal water and one with the Arg277 backbone carbonyl. A 3.6 Å threshold is adopted for hydrogen bond distance between the heavy atoms. The snapshot was extracted after 1 ns of MD simulations equilibrating the protein chain with surrounding water. Due to the incomplete wrapping of dehydron (Arg277 and Arg280) in the p53 domain, the closest water molecule is found at a distance of 2.8 Å between carbonyl and water oxygen atoms. While electrostatically engaged with the Arg277 backbone carbonyl, this water molecule is deprived of one hydrogen bond partnership when compared with bulk water (Figs. 5.3 and 5.4).

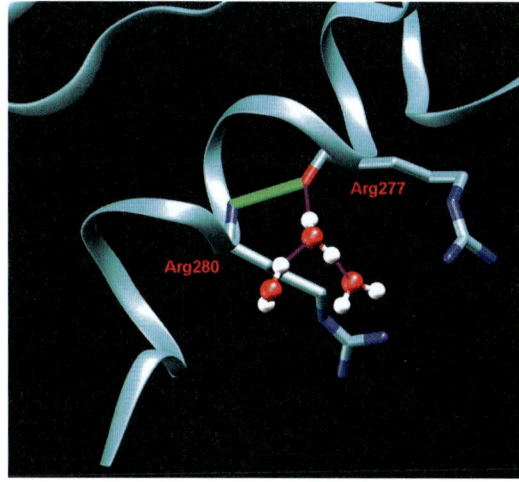

Fig. 5.4 Selected snapshot (after 1 ns of MD) of a solvating water molecule and its hydrogen bond partnerships (*purple bonds*) within the desolvation domain of Arg277 in the DNA-binding domain of tumor antigen p53 (ribbon representation, fragment). The backbone amide–carbonyl dehydron Arg277–Arg280 is shown in *green*. Reprinted with permission from [19] copyright 2007 American Chemical Society

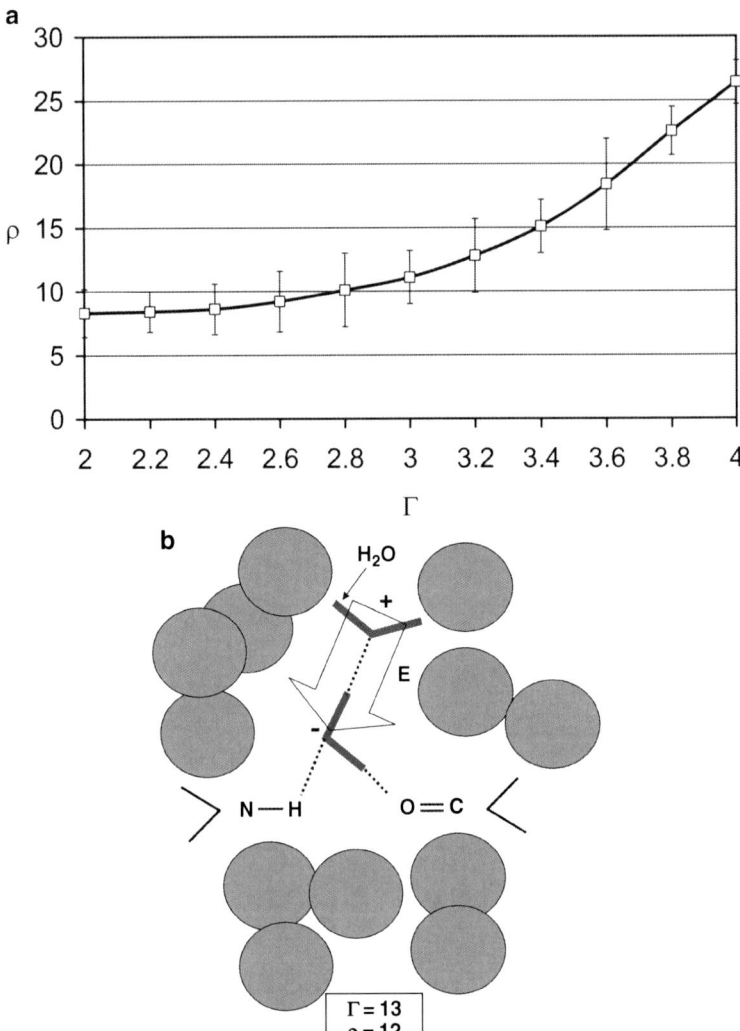

Fig. 5.5 (**a**) Correlation between hydrogen bond wrapping ρ and wetting parameter Γ. Each residue is assigned a ρ-value averaged over all backbone hydrogen bonds in which it is engaged. The data were extracted from the wetting computation on the p53 DNA-binding domain and three additional folds: the SH3 domain ($N = 55$, 2 dehydrons, PDB.1SRL), ubiquitin ($N = 76$, 16 dehydrons, PDB.1UBI), and λ-repressor ($N = 86$, 26 dehydrons, PDB.1LMB). Reprinted with permission from [19], copyright 2007 American Chemical Society. (**b**) Schematic representation of the relation between protein wrapping, interfacial water confinement, restrictions on hydrogen bonding partnerships for interfacial water molecules, and net internal electrostatic field arising from such restrictions

In order to determine a generic relation between ρ and Γ, three additional single-domain folds representative of protein topologies were also analyzed: SH3-domain, ubiquitin, and λ-repressor, as indicated above.

The correlation between wrapping and dehydration propensity (Fig. 5.5a) has the following characteristics: (a) dehydrons ($\rho \leq 19$) generate Γ-values in the range $2 \leq \Gamma \leq 3.6$; (b) the upper wrapping bound, $\rho = 28$, corresponds to bulk-like water ($\Gamma = 4$) in the desolvation domain; and (c) all solvating water is excluded from the desolvation domain for $\rho > 28$.

5.3 Dielectric Modulation of Interfacial Water Around Dehydrons

In this section we show how dielectric modulation is promoted by discrete solvent effects arising from local imperfections in the wrapping of the protein backbone. These discrete effects cannot be captured properly by conventional continuous models, which should in principle be adapted to deal with local dielectric modulations. As demonstrated in this section, the dielectric modulation, i.e., the quenching in the local dielectric permittivity is caused by the local reduction in hydrogen bonding partnerships of solvating water molecules [21]. To quantify this effect, we compute the polarizability associated with restricted interfacial water under the influence of a nonzero net internal electrostatic field $\mathbf{E}$ ($\mathbf{E} = \mathbf{0}$ in bulk water as there is no preferred dipole orientation and therefore no collective net effect). The relation between wrapping-based confinement of interfacial water, its hydrogen bonding partnerships, and the net internal field $\mathbf{E}$ resulting from water confinement is illustrated in Fig. 5.5b.

For convenience, we define the entropy loss of a water molecule associated with the reduction of hydrogen bond exchange possibilities from 4 in bulk water (the parameter for the tetrahedral lattice of hydrogen bonds) to Γ at the interface [22] as follows: $\Delta s(\Gamma) = -k \ln (\Gamma/4)$ (k = Boltzmann constant). In addition, we introduce the dimensionless parameter $\lambda(\Gamma) = T\Delta s(\Gamma)/(Ld)$, where d = dipole moment of a single water molecule and L = Lorentz field = $\eta d/3\varepsilon_0$ (η = bulk water density and ε_0 = vacuum permittivity). The thermal average of the dipole moment vector $\mathbf{d}$ is given by $<\mathbf{d}> = d<\boldsymbol{\mu}>$, where $\boldsymbol{\mu} = \boldsymbol{\mu}(\mathbf{E})$ is the dipole moment unit vector. The thermal average of the dipole projection μ_E onto the net internal field may be readily determined from the Boltzmann average [23]:

$$< \mu_E >= [\,\coth (\beta Ed) - (\beta Ed)^{-1}\,], \tag{5.1}$$

$\beta = (kT)^{-1}$ where $E = \|\mathbf{E}\|$. The scattering $\sigma(\mathbf{E})$ of the net field produced by dipole interactions within volume V is the real part (Re) of

$$S = V - \int \exp[\,-\,i v(\mathbf{r})\rho\cdot <\boldsymbol{\mu}>\,]\,d\mathbf{r}, \tag{5.2}$$

where $v(\mathbf{r}) = d/(4\pi\varepsilon_0 r^3)$ is the dipole interaction field and ρ = net field unit vector. Integration in (5.2) is carried out in the region of correlation: $r_w \leq \|\mathbf{r}\| \leq r_\Gamma$, with $r_\Gamma = r_w[1-\lambda(\Gamma)]^{1/3}$, where r_w is the bulk dipole–dipole correlation distance. Thus, the field scattering is given as follows:

$$\sigma\,(E) = \mathrm{ReS} = \lambda(\Gamma)L^2\left(|<\mu_E>|^2\right)/18$$

$$= \eta kT[27\varepsilon_0]^{-1}\ln\,(4/\,\Gamma)[\,\coth\,(\beta Ed) - (\beta Ed)^{-1}]^2 \tag{5.3}$$

The probability distribution of net internal fields is then

$$P_\Gamma(\mathbf{E}) = [\pi\sigma^2(E)]^{-3/2}\exp[\,-E^2/\sigma^2(E)], \tag{5.4}$$

satisfying

$$\lim_{\Gamma\to 4} P_\Gamma(\mathbf{E}) = \delta(\mathbf{E}), \tag{5.5}$$

in accord with the fact that there is no net internal field in bulk water. The dielectric susceptibility χ along a preferential direction z is directly computed as a function of Γ by introducing a perturbation $\mathbf{\Delta}_z$ of the net internal field $\tilde{\mathbf{E}} = \mathbf{E} + \mathbf{\Delta}_z$.

By definition

$$\chi_z = \lim_{\|\mathbf{\Delta}\|\to 0} \partial\Xi_z/\partial\Delta_z, \tag{5.6}$$

where $\Xi_z = \eta d\int[\,\coth\,(\beta\tilde{E}_z d) - (\beta\tilde{E}_z d)^{-1}]P_\Gamma(\mathbf{E})d(\mathbf{E})$ is the polarizability along the z-direction. This gives

$$\chi_z = \chi_z(\Gamma) = \eta d^2\beta/3\varepsilon_0 - \eta d^2\beta/3\varepsilon_0\int[\,\coth^2\,(\beta E_z d) - (\beta E_z d)^{-2}]P_\Gamma(\mathbf{E})d(\mathbf{E}), \tag{5.7}$$

where $\eta d^2\beta/3\varepsilon_0 = \chi_{zw}$ is the bulk water susceptibility. The bulk limit is obtained substituting (5.5) into (5.7) in the limit $\Gamma\to 4$:

$$\lim_{\Gamma\to 4} \chi_z = \eta d^2\beta/3\varepsilon_0 - \eta d^2\beta/3\varepsilon_0\int[\,\coth^2\,(\beta E_z d) - (\beta E_z)^{-2}]\delta(\mathbf{E})d(\mathbf{E}) \tag{5.8}$$

$$= \eta d^2\beta/3\varepsilon_0 = \chi_{zw},$$

since $\lim_{\xi\to 0}[\coth^2(\xi) - (\xi)^{-2}] = 0$.

Figure 5.6 displays the rigorously derived Γ-dependence of the dielectric permittivity $\varepsilon = 1 + \chi(\Gamma)$ (the subindex z denoting a generic direction has been dropped). The dielectric quenching is extreme upon moderately small losses in hydrogen bond partnerships. Thus, the most dramatic decrease is marked by a drop in ε-values from 50 to 7 as Γ is reduced from 3.5 to 2.5.

The combination of Figs. 5.5 and 5.6 leads us to the conclusion that *clusters of packing defects act as dramatic enhancers of the electric fields generated at*

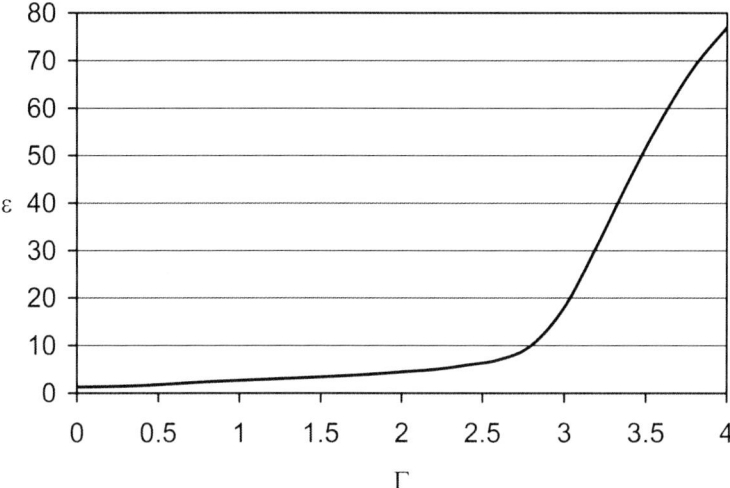

Fig. 5.6 Analytical dependence of the dielectric permittivity ε on the parameter Γ. Reprinted with permission from [19] copyright 2007 American Chemical Society

the protein interface. The typical loss in hydrogen bonding partnerships associated with dehydron solvation places Γ in the range $2 \leq \Gamma \leq 3.6$. This range contains the region of most dramatic dielectric quenching, decreasing the permittivity by an order of magnitude with respect to bulk water. In turn, this effect translates in an order of magnitude increase in electrostatic interactions, hence rationalizing the functional role of these dehydron-rich regions.

5.4 A Study Case: Dielectric Quenching in the p53 DNA-Binding Domain

The functional significance of the three dielectric modulators in the DNA-binding domain of p53 may be understood by examining its dimeric state and its role as transcription factor. Thus, a major cluster involving the five dehydrons (173, 176), (174, 178), (175, 178), (176, 179), and (178, 180) is found at the dimer interface (Figs. 5.3 and 5.7). This cluster fosters dimerization in accord with the dehydration propensity of dehydrons and their role as promoters of protein associations [2, 5, 10–13]. The dimerization involves a resonant pairing of the Arg178 from each monomer (Fig. 5.7) likely, to promote supramolecular charge delocalization with distal charge separation at all times. Significantly, $<\Gamma>$ reaches a minimum precisely at Arg178 (Fig. 5.2), in accordance with the low dehydration penalty for the guanidinium ion [24] and with the fact that the Arg–Arg resonant association requires guanidinium dehydration (cf. Fig. 5.7).

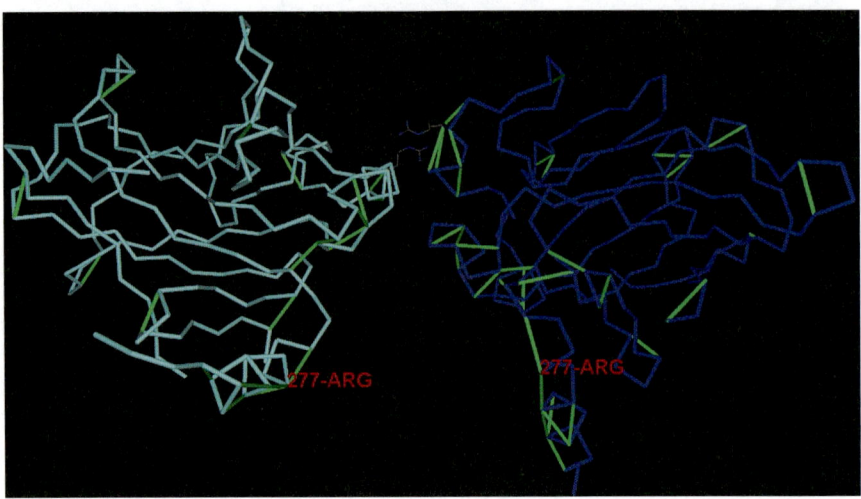

Fig. 5.7 Backbone/dehydron representation of the dimer interface for the DNA-binding domain of p53 (PDB.2GEQ). The side chains of the Arg178 of each monomer involved in a resonance pair are shown. Reprinted with permission from [19] copyright 2007 American Chemical Society

Besides Arg 178, there are three additional minima in $<\Gamma>$, corresponding to residues Arg245, Arg270, and Arg277 (Fig. 5.2). The latter three arginines play a pivotal role in DNA recognition [18], as discussed below. Residue Arg245 is engaged in the dielectric quenching region 236–246 and is part of the dehydron cluster involving pairs (236, 239), (237, 245), (237, 271), (239, 242), (239, 244), (240, 242) (Fig. 5.3). Residues 270 and 277 lie within the dielectric quenching region 270–289 and are part of the dehydron cluster involving pairs (237, 271), (274, 277), (277, 281), (280, 284), (281, 285), (282, 285), (285, 288) (Figs. 5.3 and 5.8). The latter cluster is one of the largest to be found in a PDB-reported soluble protein, as discussed below.

Direct examination of the protein–DNA complex (PDB.2GEQ) reveals that the three residues directly implicated in DNA recognition are precisely Arg245, Arg270, and Arg277 (Fig. 5.8). Residue Arg277 acts as intrabase intercalator, while the other two interact with the negatively charged backbone phosphates (Fig. 5.8). The electrostatics of protein–DNA recognition is not merely the result of matching charges along the geometrically compatible interfaces, otherwise ion pairs would prevail in water. Rather, electrostatic recognition of the DNA polyelectrolyte requires a device to promote dehydration at the protein–nucleic acid interface. The large dehydron clusters surrounding the three arginines directly implicated in the protein–DNA association (Figs. 5.3 and 5.8) provide such an expedient, as they quench the local dielectric (Fig. 5.2), thus enhancing the electrostatic recognition. Thus, the fact that the three arginines involved in DNA recognition are also dehydration hot spots satisfies a functional imperative for the transcription factor.

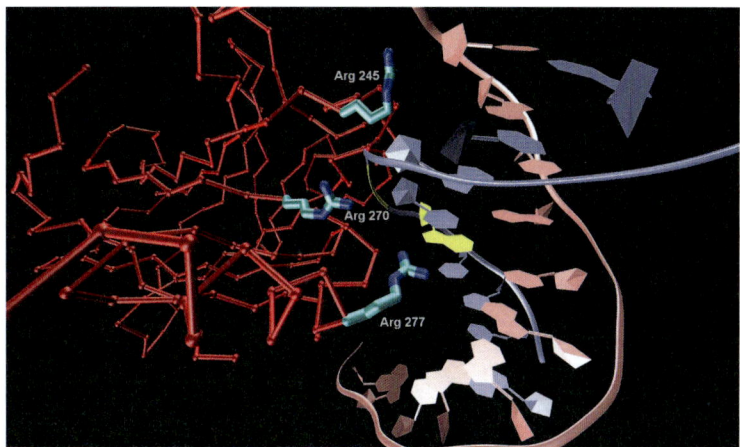

Fig. 5.8 Protein–DNA complex of the DNA-binding domain of p53 (PDB.2GEQ). Side chains of the key residues directly implicated in DNA recognition, Arg245, Arg270, and Arg277 are shown. The pyridine base recognized by Arg277 is shown in *yellow*, while the individual DNA strands are shown in *lilac* and *light magenta*. Reprinted with permission from [19] copyright 2007 American Chemical Society

5.5 Proteins with Dehydron Clusters

A curated PDB-derived database was constructed free of redundancy for single-domain proteins with contour length $N > 50$. The database is comprised of 2,982 entries with <50% pairwise identity in aligned sequences. Proteins were grouped according to n, the size of the dehydron cluster in the structure. The groups intersect to a considerable extent [19].

Each n-group has been lumped into five nondisjoint functional categories: biosynthesis, enzymology, cell signaling, cytoskeleton, and cancer. The contribution of each category to each n-group is quantified through suitable normalization, taking into account the relative abundance of each category in the curated database. Thus, the number of PDB domains within an n-cluster in a functional category is divided by the total number of PDB domains in the category. After normalization, percentages indicating representativity from each functional category are computed for each n-group (Fig. 5.9).

The distribution of dehydron clusters becomes a marker of the functional category. Thus, biosynthesis peaks at 2, enzymology at 3, cell signaling at 6, while cytoskeleton and cancer are monotonically increasing with dehydron cluster size. *Intriguingly, the dominance of the cancer category becomes apparent for the poorest packed proteins.* The dearth of domains for $n \geq 7$ facilitates individual identification and assessment. Thus, there are only five domains with $n = 7$: calmodulin (cell signaling, Fig. 5.10) [25], actin (cytoskeleton) [26], p53 DNA-binding domain (cancer, Figs. 5.3 and 5.7) [18], BRCT, the terminal repeat domain of breast cancer gene BRAC1 [27], and the cellular prion protein (Fig. 5.11) [28]. The group with

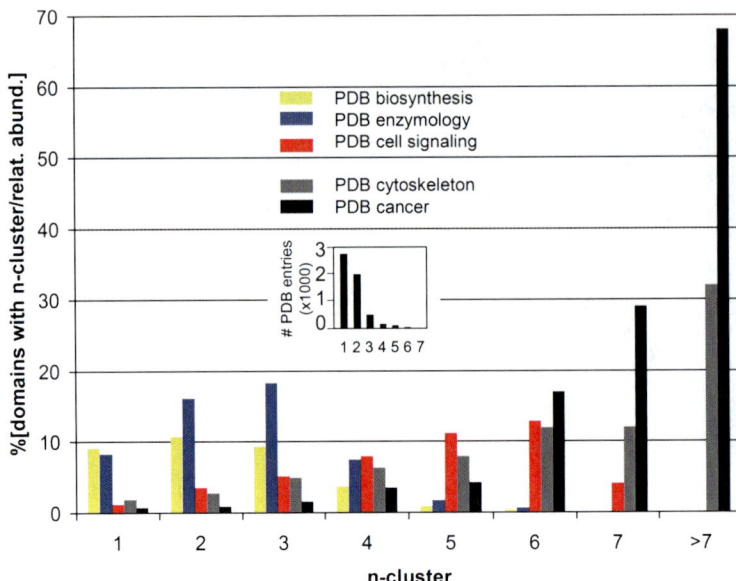

Fig. 5.9 Percentages of PDB domains in functional categories binned into groups determined by dehydron cluster size n. Each cluster size group is divided into five nondisjoint functional categories: biosynthesis, enzymology, cell signaling, cytoskeleton, and cancer. The number of PDB domains in each group is normalized to the relative abundance of the functional category. Thus, the number of PDB domains in a cluster size group and functional category is divided by the total number of PDB domains in the category. The *inset* indicates the number of domains in each cluster size group. Reprinted with permission from [19] copyright 2007 American Chemical Society

$n > 7$ only contains three members: severin (cytoskeleton) [29] and two oncogenic transcription factors with DNA-stabilizing induced fit, jun/fos (Fig. 5.12, [30]) and myc/max [31].

The eight protein domains with unusually large dehydron clusters are highly interactive proteins. In spite of their functional diversity, a common functional motif is discerned: as soluble proteins, they all possess a major weakness in the hydration shell. Thus, a dehydron cluster plays at least three interrelated roles: (a) promoter of protein associations (calmodulin, actin, severin), (b) dielectric modulator enhancing intermolecular electrostatic interactions (cancer-related transcription factors), and (c) a structural weakness promoting water attack on backbone hydrogen bonds with concurrent refolding leading to aggregation (cellular prion protein).

To summarize, proteins with severe weaknesses in their hydration shell resulting from extended regions with poor wrapping are capable of creating a significant dielectric quenching of interfacial water. This property becomes most apparent for proteins possessing the largest dehydron clusters in the PDB (seven or more

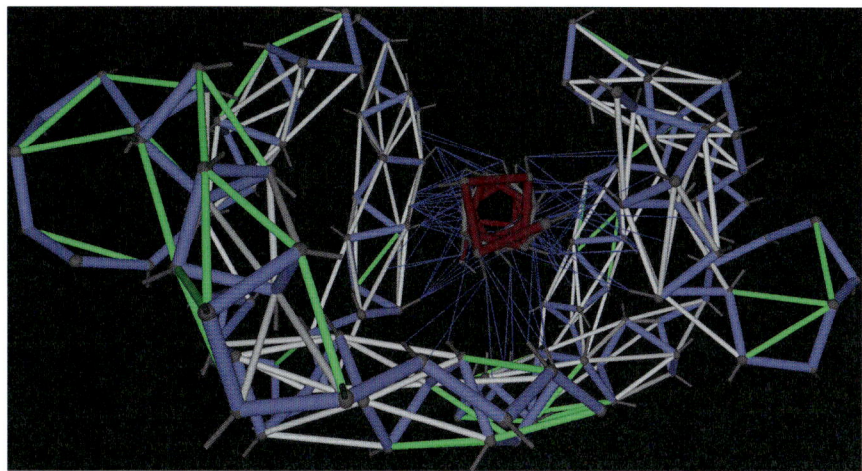

Fig. 5.10 Dehydron pattern of calmodulin (PDB. 1CDM) complexed with the calmodulin-binding domain of calmodulin-dependent protein kinase II. The backbone is indicated by *blue* virtual bonds joining α-carbons and dehydrons are shown in *green*. The intermolecular wrapping of calmodulin hydrogen bonds by the ligand side chains is indicated by *thin blue lines*. The extremities of these lines are the β-carbon of the residue contributing to the desolvation shell of the hydrogen bond and the midpoint of the hydrogen bond that is being wrapped intermolecularly. Reprinted with permission from [19] copyright 2007 American Chemical Society

Fig. 5.11 Dehydron pattern
of the cellular prion protein
(PDB.1QM0). Reprinted with
permission from [19]
copyright 2007 American
Chemical Society

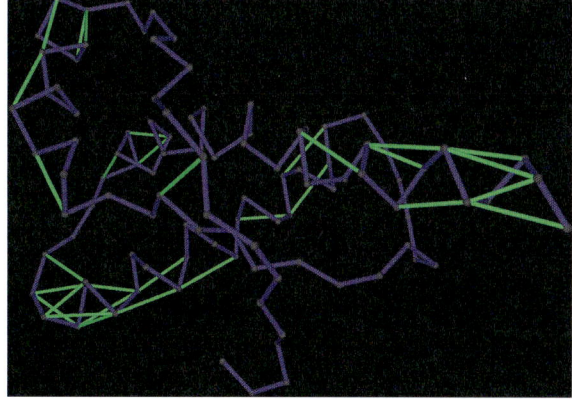

dehydrons). We found that there are eight such proteins reported in PDB, possessing structural regions in the order–disorder twilight. These regions belong to cancer-related (oncogenic or antitumor) proteins and also to highly interactive proteins and to a cellular prion that promotes misfolding and aberrant aggregation.

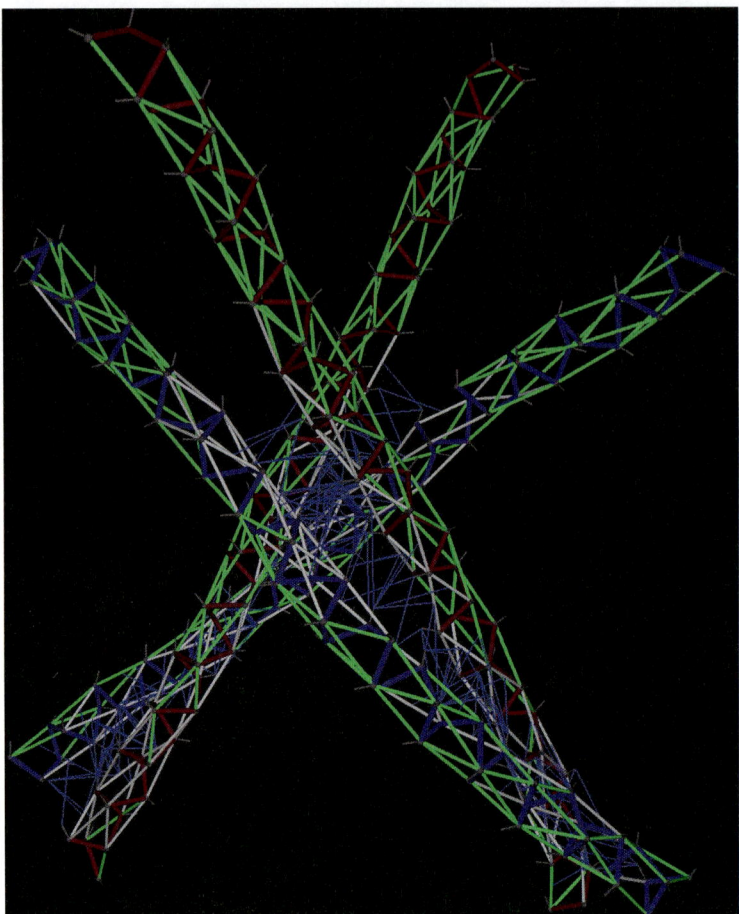

Fig. 5.12 Dehydron pattern of oncogenic transcription factor jun/fos (tetramer, PDB.1FOS). Intermolecular wrapping is displayed as in Fig. 5.10. Monomeric *jun* chains are in *blue* and *fos* chains in *dark red*. Reprinted with permission from [19] copyright 2007 American Chemical Society

5.6 Misfolding and Aggregation: Consequences of a Massive Violation of Architectural Constraints

In Chap. 2 we investigated the architectural constraints imposed by the extent of under-wrapping of a soluble protein. In view of the characterization of proteins with large dehydron clusters given in the previous section, time is ripe to pose the question *What is physically the fate of a soluble protein with a large dehydron cluster and with $Y >> 5X + 20$, that is, with poor disulfide bridge buttressing for its high level of structural deficiency?* The answer obviously depends on the extent to which the protein is capable of recruiting binding partners to maintain its structural integrity.

As noted in Chap. 2, the worse wrapped proteins are potassium channel neurotoxins but their buttressing is actually adequate for an extracellular protein with such a level of structural deficiency.

On the other hand, the cellular form of the human prion (PDB.1QM0) is very poorly buttressed for its extensive under-wrapping, belongs to the reducing cytosolic space, contains spots of large dehydron density, and makes no obligatory complex. We know that prions misfold and aggregate aberrantly into highly organized amyloidogenic fibrils. Is this then a generic behavior of proteins sharing the same molecular attributes? This question prompts us to investigate the amyloid phase more closely.

There is significant evidence supporting the view that amyloidogenic aggregation is a generic phase of peptide chains [32–34]. The term generic phase refers to a 3D organization dominated by main chain interactions which is essentially insensitive to the information encoded in the primary sequence. Such intermolecular associations appear to be dominated by a basic structural motif: the cross-β structure [35], an intermolecular sheet-pleated pattern ubiquitous in fibrillogenic aggregation. This assertion remains conjectural, as no crystal of the fiber for natural prions has been obtained. By contrast, the folded state of the peptide chain is relatively well understood: natural soluble proteins tend to adopt single-molecule conformations of marginal stability.

While the folding process and its final stable outcome are crucially dependent on the amino acid composition of the chain, the amyloid state appears to be fairly insensitive to this information: At first sight, amyloidogenic aggregation does not seem to require an "aggregation code," as some recent claims would have it [32–34]. But further analysis reveals that it must place severe constraints on the primary sequence, as some proteins tend to be prone to aggregate under physiological conditions [8], while others require extreme conditions to do so or simply do not aggregate reproducibly [32, 36]. In addition, negative design features of the folded state purposely encoded in the protein primary sequence may avert aggregation [37]. *Thus, it is not entirely correct to characterize the aberrant aggregation as a "polymer physics phase," shared by polypeptides with arbitrary, suboptimal, or random sequence, in contrast with the folded state, determined unambiguously by the primary sequence.*

Proteins are subject to selection pressure and adapt to become good folders, i.e., expeditious structure seekers with a reproducible and stable soluble structure. This optimization is needed to prevent the functionally competent fold from reverting to a primeval amyloid phase. On the other hand, certain sequences are better optimized to escape aggregation than others even under conditions known to sustain the native fold [8].

While amyloidogenic aggregation has been shown to be always plausible provided sufficiently stringent denaturation conditions are applied [32–36], a marked amyloidogenic propensity has been detected on a number of proteins even under physiological or near-physiological conditions, particularly if the monomeric folding domain is deprived of its natural interacting partners [8]. *Such findings imply that not all soluble structures have been optimized to the same degree in order to avert aggregation, and that the more reliant the structure is on binding part-nerships or complexations, the more vulnerable it becomes in regard to reverting to the primeval phase.* Thus, an over-expression of a folding domain with high complexation requirements in vivo, or the modification of its binding partners as a result of genetic accident, or any factor that distorts its natural interactive context is likely to bolster a transition to an amyloidogenic state.

These observations lead us to the following question: *What type of deficiency in the native fold constitutes a signal for aberrant aggregation?* A recent assessment of the *wrapping* of soluble structure might prove critical to address this problem. For highly under-wrapped proteins ($\sim$50% dehydrons or more), densities higher than four dehydrons per 1,000 Å^2 on the protein surface become inducers of protein aggregation [8].

Thus, the condition of "keeping the structure dry in water" becomes a require-ment to preserve the structural integrity of soluble proteins and imposes a severe building constraint (and thereby an evolutionary pressure) on such proteins. It is expected that the optimization of the structures resulting from this type of evolu-tionary constraint would be uneven over a range of soluble proteins, resulting in marked differences in aggregation propensity.

This wrapping analysis clarifies the physical picture put forth by Dobson [32–34], in which amyloidogenic propensity depends crucially on the fact that main chain interactions become dominant in detriment of the amino acid sequence that encodes the folded state. Precisely, main chain interactions may dominate as the main chain of the folded state is not properly protected from water attack.

It is instructive to compare this statement with the local analysis of Avbelj and Baldwin [38] in the sense that backbone solvation is a determinant of β-sheet propensity. Thus, an over-exposed backbone hydrogen bond in the native fold is an indicator of a failure in folding cooperativity, as it reveals an inability to remove water from an interactive polar pair by means of a many-body correlation, and at the same time, it is a signal enabling the diagnosis of amyloidogenic propensity. Thus, the wrapping concept enables us to discern why some soluble proteins may have been better optimized to avoid amyloidogenic aggregation than others.

Direct inspection of the pattern of desolvation of the main chain clearly reveals that the cellular fold of the human prion [39] (Fig. 5.11) is too vulnerable to water attack and at the same time too sticky to avert aggregation. Clearly, its sequence has not been optimized to "keep the backbone hydrogen bonds dry" in the folded state. In fact, *their extent of exposure of backbone hydrogen bonds is the highest among soluble proteins in the entire PDB*, with the sole exception of some toxins whose stable fold is held together by a profusion of disulfide bridges, as shown in Chap. 2.

It is suggestive that an inability to protect the main chain is precisely the type of deficiency that best correlates with a propensity to reverse to a primeval aggrega-tion phase determined by main chain interactions. The actual mechanism by which such defects induce or nucleate the transition is still opaque, although the inherent adhesiveness of packing defects obviously plays a role.

Recently, an atomic-detail structure of a fibrillogenic aggregate, with its β-sheets parallel to the main axis and the strands perpendicular to it, was reported and revealed a tight packing of β-sheets [40]. The cross-β spine of the fibrillogenic peptide GNNQQNY reveals a double-parallel β-sheet with tight packing of side chains leading to the full dehydration of intrasheet backbone–backbone and side chain–side chain hydrogen bonds. However, *there is not a single pairwise interaction between the β-sheets, no hydrogen bond, and no hydrophobic interaction.* Instead, a direct examination of the crystal structure reveals that the association is driven by the dehydration propensity of pre-formed intrasheet dehydrons, as depicted in Fig. 5.13.

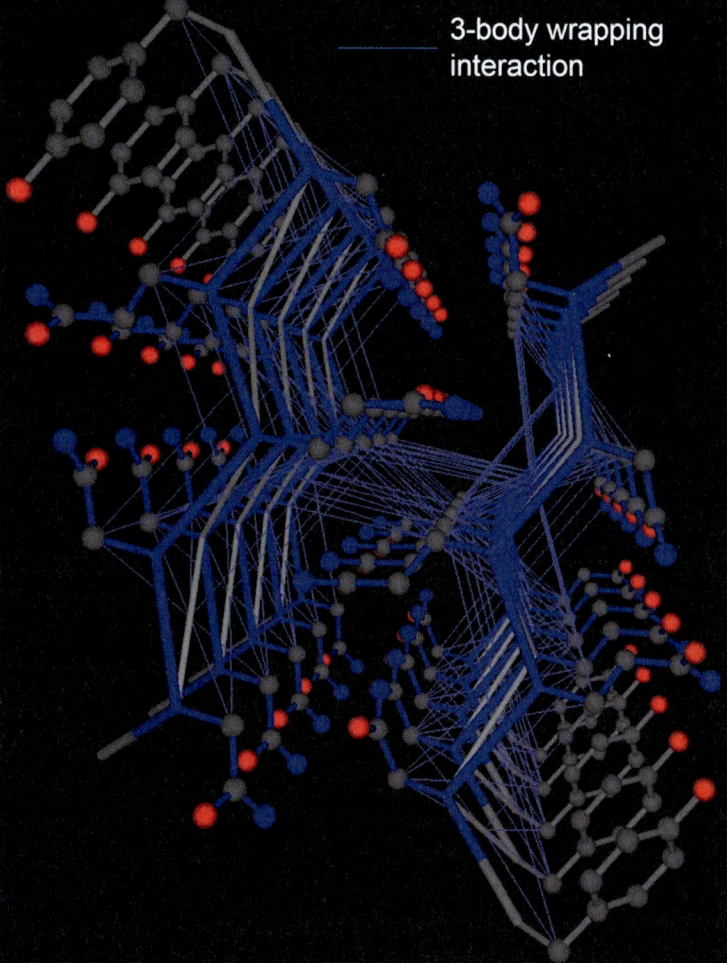

Fig. 5.13 Pattern of intersheet wrapping of backbone–backbone hydrogen bonds in the fibrillogenic state of peptide GNNQQNY. The strand backbone is represented as virtual bonds (*blue*) joining α-carbons and backbone hydrogen bonds are displayed as *light gray* lines joining the α-carbons of the paired residues. A *thin blue line* from the β-carbon of a residue to the baricenter of a hydrogen bond indicates wrapping of the bond by the residue: at least one non-polar group from the residue is contained in the desolvation sphere of the bond

This discussion naturally leads us to some evolutionary considerations. A paradigmatic discovery in biology revealed that folds are conserved across species to perform specific functions. However, the wrapping of such folds is clearly not conserved [4]. This fact suggests how complex physiologies may be achieved without dramatically expanding genome size, a standing problem in biology. Considerable network complexity may be achieved by actually fostering a higher level of complexation or binding partnership, as promoted by a more precarious wrapping of the isolated protein domains. According to our previous analysis, such complex design also entails an inherent danger: the reversal of highly under-wrapped folding domains to an amyloidogenic phase even under physiological conditions. Amyloidosis is thus likely to be a consequence of high complexity in proteomic connectivity, as dictated by the structural fragility of highly interactive proteins.

Prion-like aggregation has been widely recognized as a causative of pathogenic dysfunction [39], but recent work [41] has revealed that there may be also epigenetic consequences to prion-like aggregation. These novel insights lead to a significant extension of the prion hypothesis. Furthermore, while a connection between protein under-wrapping and propensity for aberrant amyloidogenic aggregation has been unraveled, no misfolding inference can be made for cases when the extent of under-wrapping is so severe that no soluble structure can be maintained by the monomeric chain. This case appears to be of biological interest in light of recent research [41] that focused on nonpathogenic yeast prions. These prions are capable of promoting phenotypic polymorphism through a transmittable conformational change that they regard as having epigenetic consequences, thus defining a protein-based element of inheritance. The yeast prions studied, especially *PSI* in yeast gene Sup35, are unlikely to sustain a monomeric structure.

From the wrapping perspective, this is so because the prion sequences contain large windows (>30 residues) containing mostly under-wrapping residues (especially G, N, Q, S, A, P). In turn, these residues are under-wrappers because they contain few nonpolar side chain groups, thereby exposing the backbone to the solvent, while some (i.e., S, N, Q) even prevent other nonpolar groups from clustering around the backbone hydrogen bonds. The other under-wrapping residue, D, is unlikely to be found in such prions as its charge promotes hydration, thus introducing a negative design element for aggregation.

These observations are in accordance with the structural characterization by Krishnan and Lindquist [41], which places yeast prions in the twilight between order and disorder, in consonance with the wrapping-based characterization of the prion described in this chapter.

Thus, the epigenetic consequences associated with misfolding of severely under-wrapped proteins incapable of sustaining monomeric structure should be mandatory subjects of any research agenda built on the premises of this chapter. The focus of such endeavors are proteins endowed with large windows (>30) of under-wrapping residues. When combined with tyrosine (probably needed for stacking), such motifs are likely promoters of self-aggregation leading to pathology [39] or epigenetic prion-based phenotype polymorphism [41]. Thus, future research efforts should be directed at identifying highly under-wrapped human genes containing the sequence motifs indicated and assessing the epigenetic consequences of their transformation into sequestered aggregates. A preliminary scanning of ca. 16,000 human proteins in Swiss-Prot reveals 13 genes with the severely under-wrapped motif. Among them, RNF12, AF9, MAML2 are implicated in cancer promotion, while seven are involved in transcription regulation, hence with likely epigenetic consequences associated with a conformational switching into sequestered aggregate states.

References

1. Fernández A, Sosnick TR, Colubri A (2002) Dynamics of hydrogen-bond desolvation in folding proteins. J Mol Biol 321: 659–675
2. Fernández A, Scheraga HA (2003) Insufficiently dehydrated hydrogen bonds as determinants of protein interactions. Proc Natl Acad Sci USA 100:113–118
3. Fernández A, Scott LR (2003) Adherence of packing defects in soluble proteins. Phys Rev Lett 91:018102
4. Fernández A, Berry RS (2004) Molecular dimension explored in evolution to promote proteomic complexity. Proc Natl Acad Sci USA 101:13460–13465
5. Fernández A, Scott RL (2003) Dehydron: A structurally encoded signal for protein interaction. Biophys J 85:1914–1928
6. Fernández A (2004) Keeping dry and crossing membranes. Nat Biotechnol 22:1081–1084
7. Fernández A, Scott RL, Berry RS (2006) Packing defects as selectivity switches for drug-based protein inhibitors. Proc Natl Acad Sci USA 103:323–328
8. Fernández A, Kardos J, Scott R, Goto Y, Berry RS (2003) Structural defects and the diagnosis of amyloidogenic propensity. Proc Natl Acad Sci USA 100:6446–6451
9. Fernández A, Berry RS (2003) Proteins with hydrogen-bond packing defects are highly interactive with lipid bilayers: Implications for amyloidogenesis. Proc Natl Acad Sci USA 100:2391–2396
10. Deremble C, Lavery R (2005) Macromolecular recognition. Curr Opin Struct Biol 15:171–175
11. Ma B, Elkayam T, Wolfson H, Nussinov R (2003) Protein-protein interactions: Structurally conserved residues distinguish between binding sites and exposed protein surfaces. Proc Natl Acad Sci USA 100:5772–5777
12. Ma B, Pan Y, Gunasekaran K, Venkataraghavan RB, Levine AJ, Nussinov R (2005) Comparison of the protein-protein interfaces in the p53-DNA crystal structures: Towards elucidation of the biological interface. Proc Natl Acad Sci USA 102:3988–3993
13. Rajamani D, Thiel S, Vajda S, Camacho CJ (2004) Anchor residues in protein-protein interactions. Proc Natl Acad Sci USA 101:11287–11292
14. Dunker AK, Cortese MS, Romero P, Iakoucheva LM, Uversky VN (2005) Flexible nets: The roles of intrinsic disorder in protein interaction networks. FEBS J 272:5129–5148
15. Iakoucheva LM, Dunker AK (2003) Order, disorder, and flexibility: Prediction from protein sequence. Structure (London) 11:1316–1317
16. Dunker AK, Obradovic Z (2001) The protein trinity-linking function and disorder. Nat Biotechnol 19: 805–806
17. Wade RC, Gabdoulline RR, Lüdemann SK, Lounnas V (1998) Electrostatic steering and ionic tethering in enzyme-ligand binding: Insights from simulations. Proc Natl Acad Sci USA 95:5942–5949
18. Ho WC, Fitzgerald MX, Marmorstein R (2006) Structure of the p53 core domain dimer bound to DNA. J Biol Chem 281:20494–20502
19. Pietrosemoli N, Crespo A, Fernández A (2007) Dehydration propensity of order-disorder intermediate regions in soluble proteins. J Prot Res 6:3519–3526
20. Lindahl E, Hess B, Van der Spoel D: GROMACS 3.0: A package for molecular simulations and trajectory analysis. J Mol Model 2001, 7: 302–317

21. Fernández A (2003) What caliber pore is like a pipe? Nanotubes as modulators of ionic gradients. J Chem Phys 119:5315–5319
22. Ben-Naim A (1980) Hydrophobic Interactions. Plenum Press, New York
23. Debye P (1929) Polar Molecules. Dover, New York
24. Mason PE, Neilson GW, Dempsey CE, Barnes AC, Cruickshank, JM (2003) The hydration structure of guanidinium and thiocyanate ions: Implications for protein stability in aqueous solution. Proc Natl Acad Sci USA 100:4557–4561
25. Kabsch W, Mannherz HG, Suck D, Pai EF, Holmes KC (1990) Atomic structure of the actin: DNase I complex. Nature 347:37–44
26. Meador WE, Means AR, Quiocho FA (1993) Modulation of calmodulin plasticity in molecular recognition on the basis of x-ray structures. Science 262:1718–1721
27. Williams RS, Green R, Glover JN (2001) Crystal structure of the BRCT repeat region from the breast cancer-associated protein BRCA1. Nat Struct Biol 8:838–842
28. Zahn R, Liu A, Luhrs T et al (2000) NMR solution structure of the human prion protein. Proc Natl Acad Sci USA 97:145–150
29. Schnuchel A, Wiltscheck R, Eichinger L, Schleicher M, Holak, TA (1995) Structure of severin domain 2 in solution. J Mol Biol 247:21–27
30. Glover JN, Harrison SC (1995) Crystal structure of the heterodimeric bZIP transcription factor c-Fos-c-Jun bound to DNA. Nature 373:257–261
31. Lavigne P, Crump MP, Gagne SM et al (1998) Insights into the mechanism of heterodimerization from the 1H-NMR solution structure of the c-Myc-Max heterodimeric leucine zipper. J Mol Biol 281:165–181
32. Dobson CM (1999) Protein misfolding, evolution and disease. Trends Biochem Sci 24:329–332
33. Dobson CM (2001) The structural basis of protein folding and its links with human disease. Philos Trans R Soc Lond B 356:133–145
34. Fändrich M, Dobson CM (2002) The behavior of polyamino acids reveals an inverse side chain effect in amyloid structure formation. EMBO J 21:5682–5690
35. Sunde M, Blake CCF (1998) From the globular to the fibrous state: protein structure and structural conversion in amyloid formation. Q Rev Biophys 31:1–39
36. Dobson CM (2002) Protein misfolding diseases: Getting out of shape. Nature 418:729–730
37. Richardson JS, Richardson DC (2002) Natural β-sheet proteins use negative design to avoid edge-to-edge aggregation. Proc Natl Acad Sci USA 99:2754–2759
38. Avbelj F, Baldwin RL (2003) Role of backbone solvation and electrostatics in generating preferred peptide backbone conformations: Distributions of phi. Proc Natl Acad Sci USA 100:5742–5747
39. Prusiner SB (1998) Prions. Proc Natl Acad Sci USA 95:13363–13383
40. Nelson R, Sawaya M, Balbirnie M et al (2005) Structure of the cross-beta spine of amyloid-like fibrils. Nature 435:773–778
41. Krishnan R, Lindquist SL (2005) Structural insights into a yeast prion illuminate nucleation and strain diversity. Nature 435:765–772

Chapter 6
Evolution of Protein Wrapping and Implications for the Drug Designer

Proteins with common ancestry (homologs) typically share a common fold. This structural similarity introduces major problems for rational drug design since a major goal in the field is the control of specificity. Yet, as shown in this chapter, while the topology of the native structure is extremely similar across homologs, their wrapping patterns and expression regulation tend to be different, offering a novel opportunity to funnel the impact of a drug on clinically relevant targets. The evolutionary root of these dissimilarities across homologous proteins is carefully dissected in this chapter both across species and within the human species. Furthermore, as hinted in this and demonstrated in subsequent chapters, the wrapping variations across homologs have profound consequences for drug design as we aim at engineering target-specific and species-specific therapeutic agents and build insightful animal models for disease or malignancy.

6.1 An Evolutionary Context for the Drug Designer

From a structural biology perspective, one of the most striking observations regarding protein evolution is the uncanny similarity of the 3D structure of proteins with common ancestry. This similarity across homologs is often quantified by an RMSD $< 1.5\text{Å}$ for the aligned backbone atoms of the proteins and holds even at low levels of sequence identity ($\sim 30\%$) that are nevertheless sufficient to establish homology.

From the perspective of a drug designer, this structural similarity across homologs of the protein target is particularly troublesome. Thus, major goals in drug design are related or entangled with this evolutionary aspect. They are as follows: (a) the control of drug specificity; (b) the engineering of species-selective therapeutic impact in the treatment of infectious diseases; and (c) the building of insightful animal models to properly assess drug efficacy. In all three contexts, the existence of homologs of the protein target within the species (paralogs) or across species (orthologs) can introduce undesired cross-reactivities due to their high level of structural similarity with the target. These cross-reactivities, in turn, can result

A. Fernández, *Transformative Concepts for Drug Design: Target Wrapping*,
DOI 10.1007/978-3-642-11792-3_6, © Springer-Verlag Berlin Heidelberg 2010

in adverse and even health-threatening side effects or can render inconsequential or irrelevant the animal models for a particular disease or malignancy.

The fold may be highly conserved across homologs, but as shown in this chapter, the wrapping is not conserved and neither is the regulation of expression of the homologs. As described below, we should distinguish orthologs from paralogs when assessing the evolutionary origins of these dissimilarities. Thus, in the case of orthologs, the variations arise mostly from differences in the level of efficiency of natural selection across different species. On the other hand, in paralogs the dissimilarity is rooted in the imperative to curb dosage imbalances that would arise if two copies of the same gene would be retained undifferentiated. As shown in the subsequent chapters, these subtle variations of proteins that essentially share the same 3D-fold have paramount consequences to engineer drugs that can fulfill the three goals described above.

The picture that emerges from this chapter is essentially that wrapping constitutes a biological dimension utilized by evolution to promote complexity and at the same time that wrapping differences within a fold offer escape routes to the fitness consequences introduced by paralog retention. Both evolutionary aspects will be harnessed in the subsequent chapters as we herald the new generation of wrapping-based drugs.

6.2 Wrapping Across Species: Hallmarks of Nonadaptive Traits in the Comparison of Orthologous Proteins

Proteins with common ancestry tend to share the same fold or at least the same structural topology [1, 2]. Thus, essentially the same molecular machinery is operative in organisms that diverged from each other billions of years ago. For example, the metabolic enzymatic function "dehydrofolate reductase" (DHFR) is exerted by a molecular machinery that has hardly changed in structure as we compare species in different kingdoms of life, like *archea* (i.e., *Haloferax volcanii*), bacteria (i.e., *Escherichia coli*), and higher eukaryotes (i.e., *Homo sapiens*). When closely examined, the wrapping of the DHFR has gotten worse (richer in dehydrons) in species with small population size, that is, those where natural selection became less efficient [3] (Fig. 6.1a). This trend is apparent even when examining highly conserved protein domains, such as ubiquitin: *There is a progressive enrichment in dehydrons as the species population size decreases* (Fig. 6.1b).

For operational reasons, the under-wrapping or packing deficiency of a protein may be defined as $\nu = \%$ dehydrons in the set of backbone hydrogen bonds. As we focus on any functionally competent fold, the anecdotal examples we harvested suggest a clear trend: $\nu(\text{archea}) < \nu(\text{bacteria}) < \nu(\text{unicellular eukaryotes}) < \cdots < \nu(\text{higher eukaryotes})$. This trend probably follows from the fact that mildly deleterious mutations have a better chance to prevail and get fixed in species with smaller

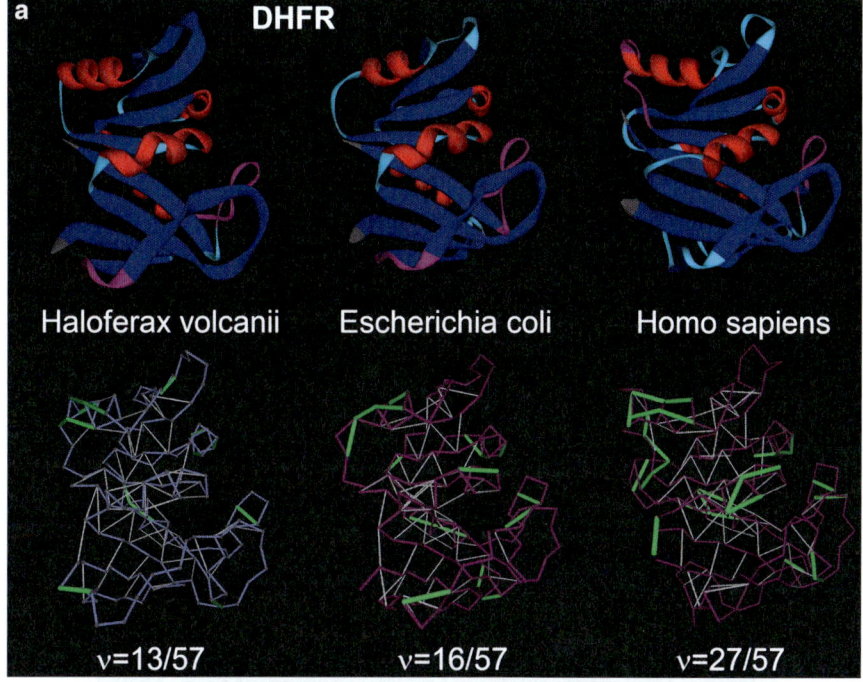

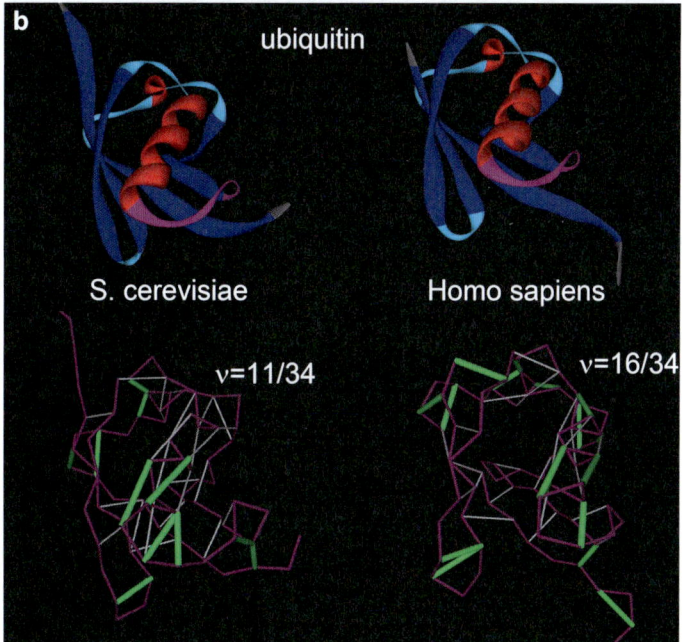

Fig. 6.1 (continued)

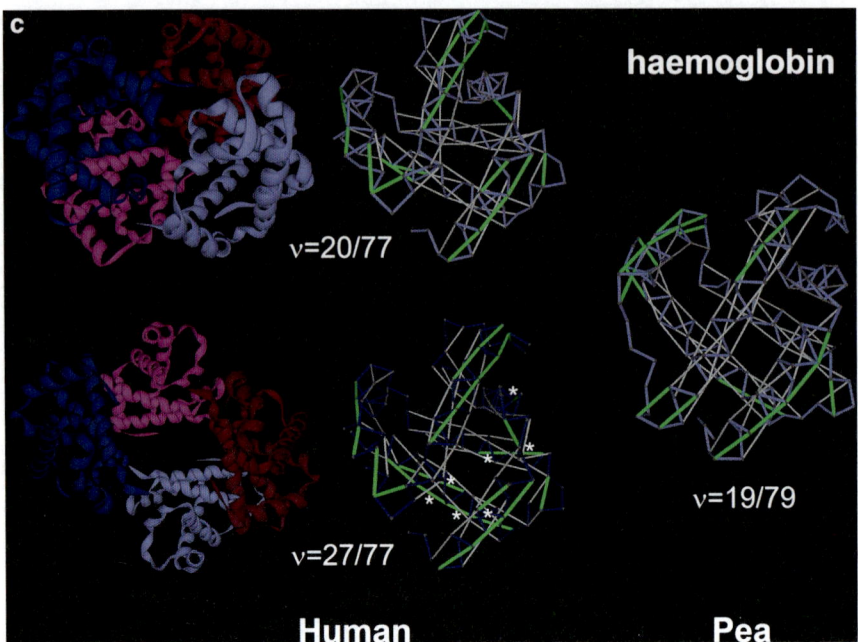

Fig. 6.1 (**a**) Under-wrapping of three orthologous versions of dehydrofolate reductase (DHFR) from three kingdoms of life archea, bacteria, and eukarya, represented by species *Haloferax vol-canii*, *Escherichia coli*, and *Homo sapiens*, respectively. (**b**) Wrapping patterns of ubiquitin from *Saccharomyces cerevisiae* (*yeast*) and *Homo sapiens*. (**c**) Wrapping patterns of hemoglobin in humans and of its ortholog leghemoglobin in pea (*Pisum sativum*). In humans, the monomer within the tetramer (*upper left panels*) contains less dehydrons that the monomer in isolation (*lower left panels*) due to intermolecular wrapping. As the monomer is taken in isolation, seven additional dehydrons (*marked by asterisks*) occur. These dehydrons promote the quaternary structure and become well wrapped hydrogen bonds in the homomeric complex. By contrast the pea leghe-moglobin ortholog is better wrapped and lacks the dehydrons that would trigger oligomerization and promote allostery

populations [3]. In this regard, the reader should note that most mutations are detri-mental and hardly ever beneficial. Thus it is more likely that a mutation will loose up the fold (create a dehydron), rather than tighten it. These mildly detrimental mutations that prevail manage to degrade the wrapping of the protein, enriching its dehydron content, thereby making the protein more reliant on binding part-nerships to maintain its structural integrity [1, 2]. *Thus, as protein interactivity, regulation, and allostery increases with decreasing species population, one may say that complexity is in this sense promoted by nonadaptive forces.*

An illustration of the probable nonadaptive origin of allostery is furnished by the wrapping of hemoglobin across orthologs in species with vastly different pop-ulation size. Thus, this protein becomes richer in dehydrons and more prone to oligomerization in species with smaller population, with the majority of the new dehydrons located at the interface that promotes the quaternary structure of the homomer (Fig. 6.1c).

6.3 Wrapping and Natural Selection

Dosage imbalances occur when protein concentration levels at specific locations in tissues or metabolic/developmental phases do not fit the stoichiometry of the complexes in which the proteins are involved [4–6]. The complexes may be transient, adventitious, or obligatory with regard to maintaining the structural integrity of the protein [7] and hence the effects of the imbalances may vary widely. Therefore, dosage sensitivity, that is, the impact of dosage imbalances on fitness, must be influenced not only by whether the protein is part of a complex but also by the extent of reliance of the protein on its binding partners to maintain structural integrity and functional competence.

While overexpression, gene duplication, misfolding and self-aggregation may all cause dosage imbalance, the structural or molecular properties determining the magnitude of the resulting effects remain largely unknown. For example, as we focus on gene duplication, we notice that paralog proteins, identical when they initially diverge, are subject to higher or lower selection pressure depending on their dosage sensitivity [5].

Cross examination of genetic and structural information revealed that the under-wrapping or packing deficiency of a protein, v, is a molecular quantifier of dosage sensitivity [5] precisely because it constitutes a measure of the reliance of the protein on binding partnerships to maintain the integrity of the native fold. As expected, a deficiently packed protein is more likely to be engaged in an obligatory complex [7] and its concentration imbalances relative to the complex stoichiometry are more likely to impact fitness than those of a well packed protein.

In unicellular organisms, the packing quality $(100–v)$ of soluble gene products correlates with the number of paralogs or family size [5]. That is, the better the packing quality, the less likely that gene duplication would introduce a fitness disadvantage, hence the higher the chance that the duplicate gene would be retained. However, this correlation becomes less significant in higher eukaryotes, as shown in Fig. 6.2. Thus, paralog survival is dependent on the packing quality of protein

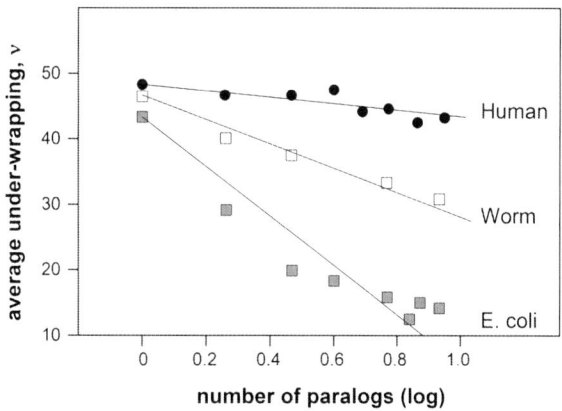

Fig. 6.2 Negative correlation between protein under-wrapping, v, and gene duplicability or family size represented by the number of paralogs in the gene family. Structural and genetic information was obtained for different species and v-values were averaged over all families with a fixed number of paralogs, as described in [5]

structure with $p < 10^{-16}$ in *Escherichia coli* and $p > 6.7 \times 10^{-3}$ in humans [5]. This contrast between simple and complex organisms is hard to interpret due to wide differences at the proteome level. However, alternative measures point to a similar trend. For example, the average difference in packing deficiency between singletons and duplicate genes is 18% in *E. coli*, 6% in worm (*Caenorhabditis elegans*), and ~2% in humans (*Homo sapiens*).

In humans, this insensitivity to dosage imbalance may be attributed in part to selection inefficiency arising from smaller population size [3], implying that the selection pressure exerted on paralogs of deficiently packed proteins has simply not become operative. Alternatively, the higher complexity of expression regulation in higher eukaryotes may introduce a tolerance to dosage imbalance not found in unicellular organisms. This chapter explores this latter possibility, focusing on evolution-related dosage imbalances and the mechanisms that humans possess to cope with the troubling aspects of selection inefficiency.

6.4 How Do Humans Cope with Inefficient Selection?

If selection is indeed inefficient in humans and this inefficiency enabled duplicate genes to stay around awaiting further specialization, significant dosage imbalances must have occurred during human evolution. How did we cope with such imbalances, which are likely to be significant when the gene products are very under-wrapped? The key to this problem lies in the ability of humans to regulate the expression of both gene copies dissimilarly through a plausible process that involves random deleterious mutations.

To address this problem, we first assess the selection pressure on gene duplicates exerted as paralogs are co-expressed at the mRNA (messenger RNA) level and hence are likely to compete for their interactive partners. Then we relate packing deficiency (v) with differences in post-transcriptional regulation patterns within families. Thus, we investigate how differences in miRNA (microRNA) target patterns [8], telling apart paralogs through different patterns of translational repression, impinge on the selection pressure on duplicate genes by mitigating dosage imbalances. In humans these patterns are significantly dissimilar across paralogs of poorly packed proteins while nearly coincident across paralogs of well packed proteins [17], thus under-scoring a means to buffer dosage imbalance effects arising from gene duplication.

This miRNA-based capacitance is not expected to be nearly as significant in species with larger effective population size due to the higher efficiency of evolutionary forces in such organisms when compared with humans [3]. Thus, the selection pressure affecting the retention of gene duplicates is likely to be more efficient in these organisms promoting adaptation through functional innovation or purifying selection.

The next two sections are devoted to provide further evidence in support of the claim that protein wrapping quality is a determinant of dosage sensitivity while upholding the view that resilience to dosage imbalance is achieved in humans by diversifying miRNA regulatory patterns across paralogs.

6.4.1 Regulatory Patterns for Paralog Proteins

As indicated in Chap. 1, soluble proteins may further protect their backbone hydrogen bonds and improve packing quality through binding partnerships by increasing the number of nonpolar groups in their hydrogen bond microenvironments [7, 9]. Hence, the extent of intermolecular protection determines whether the complex is obligatory, ephemeral or adventitious, and thus packing quality may be recognized as an important factor in determining dosage sensitivity [5].

Gene duplication introduces dosage imbalance and the resulting selection pressure on paralogs [6, 10, 11] appears to depend on the packing deficiency of the parental gene [5]. This trend is clear in *E. coli* and *S. cerevisiae*, but not so apparent in higher eukaryotes. This observation suggests that expression dissimilarities at the mRNA level and at post-transcriptional levels may be exploited to separate paralogs and avoid competition for the binding partners of the parental gene. Thus, to study human capacitance to dosage imbalance arising from gene duplication, we examined families with paralog co-expression at the mRNA level [12] and assessed post-transcriptional miRNA regulation patterns in relation to the packing quality of the proteins in the family.

To assess the role of miRNA regulation in the human capacitance to dosage imbalance, we selected human genes from an exhaustive set of 583 non-singleton families for which genetic [13], evolutionary [14], structural [5], expression [12], and post-transcriptional [15, 16] data are available for at least two paralogs (Table 6.1). We obtained human gene information from the Ensembl Genome Database (NCBI36). Using the Ensembl gene family annotation [13], 22,357 human genes were grouped into 12,394 families [17]. Gene expression data were obtained from Novartis Gene Expression Atlas [12] in the form of expression levels across a panel of 73 normal human tissues. We then mapped the putative miRNA target patterns for each classified gene. Putative conserved target sites in the 3'-UTR (untranslated region) of each gene for 156 conserved microRNA families were identified using TargetScanS (Version 5.1). Thus, to determine co-expression and co-regulation patterns across paralogs, each gene i is represented by two vectors:

(1) A normalized mRNA expression vector $\Phi_i / \|\Phi_i\|$, where the vector Φ_i has 73 entries indicating mRNA expression levels in 73 normal tissues [12] and $\|\Phi_i\|$ is the norm of the vector.

(2) A normalized miRNA vector $\Psi_i / \|\Psi_i\|$ of 156 entries representing the pattern of miRNA-related repression efficacy on gene i, with $\|\Psi_i\|$=vector norm. This representation is in accord with the identified target sites for the 156 conserved miRNA families (broadly conserved, intermediately conserved, and mammalian specific) in 17,444 human genes [8, 15]. The nth entry in Ψ_i is $\Psi_i(n) = 1 - 2^{s(i,n)}$, where $s(i, n) \leq 0$ is the context score of conserved miRNA-binding site n in the 3'UTR of gene i [18]. Thus, $\Psi_i(n) = 1$ indicates full repressive efficacy of the nth miRNA conserved site on gene i ($s(i, n) = -\infty$), while $\Psi_i(n) = 0$ (or $s(i, n) = 0$) indicates absolute lack of repressive regulatory power. The context score is known to correlate tightly

Table 6.1 PDB representation of selected human gene families of different sizes. Human gene information is from Ensembl Genome Database (NCBI36)

Ensembl ID	Family size	Swiss prot ID	Gene name	PDB ID
>ENSF00000000393	7	Q05066	SRY	1HRY_A
>ENSF00000000393	7	P48431	SOX2	1O4X_B
>ENSF00000000784	4	P08048	ZFY	1KLR_A
>ENSF00000002256	4	Q8IUE1	TGIF2LX	2DMN_A
>ENSF00000000640	6	Q9HC56	PCDH9	2EE0_A
>ENSF00000000504	6	P38919	EIF4A3	2HXY_A
>ENSF00000000661	16	Q01105	SET	2E50_A
>ENSF00000001155	23	Q15414	RBMY1B	2FY1_A
>ENSF00000001155	23	Q14011	CIRBP	1X5S_A
>ENSF00000000702	5	P17612	PRKACA	2GU8_A
>ENSF00000000530	19	P13501	CCL5	1B3A_A
>ENSF00000000530	19	O00175	CCL24	1EIG_A
>ENSF00000001027	4	Q92565	RAPGEF5	1WGY_A
>ENSF00000000301	14	Q05923	DUSP2	1M3G_A
>ENSF00000000301	14	Q16690	DUSP5	2G6Z_A
>ENSF00000001036	2	O00763	ACACB	2DN8_A
>ENSF00000002731	2	P61923	COPZ1	2HF6_A
>ENSF00000000164	18	Q9Y3D6	CLDN15	1NZN_A
>ENSF00000000212	11	Q9UHX3	EMR2	2BO2_A
>ENSF00000002211	2	Q8WUA7	TBC1D22A	2QFZ_A
>ENSF00000002844	2	Q8IWR0	ZC3H7A	2D9M_A
>ENSF00000001386	3	Q9UQB8	BAIAP2	1WDZ_A
>ENSF00000001211	4	P01100	FOS	1A02_F
>ENSF00000000285	19	Q8NHL6	LILRB1	1G0X_A
>ENSF00000001565	4	P50616	TOB1	2D5R_B
>ENSF00000002661	2	Q6XZF7	DNMBP	1UG1_A
>ENSF00000000647	6	P55201	BRPF1	2D9E_A
>ENSF00000001800	2	O96006	DHRSX	2CT5_A
>ENSF00000000849	6	Q9H8S9	MOBK1B	1PI1_A
>ENSF00000000030	2	P03999	OPN1SW	1KPN_A
>ENSF00000000192	21	P15157	TPSB2	1LTO_A
>ENSF00000000192	21	P20231	TPSAB1	1A0L_A
>ENSF00000000254	9	Q14832	GRM3	1S8M_A
>ENSF00000000521	7	Q05586	GRIN1	2NR1_A
>ENSF00000000521	7	Q13224	GRIN2B	1S11_A
>ENSF00000001623	2	P30291	WEE1	1X8B_A
>ENSF00000000001	458	Q8WTR7	ZNF473	2EMB_A
>ENSF00000000001	458	Q96SE7	ZNF347	2EMA_A
>ENSF00000000001	458	Q8NHY6	ZFP28	2EM2_A
>ENSF00000000001	458	Q9H4T2	ZSCAN16	2COT_A
>ENSF00000000822	3	Q86UL8	MAGI2	1UEP_A
>ENSF00000001433	3	Q9H2H8	PPIL3	1XYH_A
>ENSF00000000610	5	P84022	SMAD3	1MHD_A
>ENSF00000000610	5	Q15797	SMAD1	1KHU_A
>ENSF00000000153	12	Q96JP2	MYO15B	2DLP_A

with the post-transcriptional downregulation efficacy, $2^{s(i,\, n)}$, of the predicted binding site for the nth miRNA family within the 3'-UTR of gene i [18]. Thus, $2^{s(i,n)} \approx g(i)/g(i, n)$, where $g(i)$ is the translation level for gene i and $g(i, n)$ is the i-translation level with knockout of the nth miRNA family. If the n-site in not predicted in the 3'-UTR of gene i, $g(i) = g(i, n)$ and therefore $s(i, n) = 0$.

Only paralogs that are significantly co-expressed are likely to produce dosage imbalances if the genes have not diverged significantly. Thus, similarities between mRNA expression profiles of two genes i, j will be assessed by the Pearson correlation coefficient $\eta(i, j)$ of their expression vectors Φ_i and Φ_j. In general, for two expression vectors X and Y, the Pearson coefficient is given by

$$\eta(\mathbf{X}, \mathbf{Y}) = \frac{<(X- <X>)(Y- <Y>)>}{\sqrt{<X^2> - <X>^2}\sqrt{<Y^2> - <Y>^2}}, \tag{6.1}$$

where X, Y are generic coordinates in the vectors X and Y, respectively, and $<>$ indicates mean over cell types.

For paralogs with significant co-expression, a tolerance to dosage imbalance may still arise through differences in translational repression patterns. Thus, orthogonal miRNA repression patterns for paralogs with high dosage sensitivity may introduce an escape route to the selection pressure introduced by the dosage imbalance. To test this hypothesis, we introduce the extent of miRNA target coincidence $\tau(i, j)$, defined as the scalar (dot) product of the two miRNA target vectors: $\tau(i, j) = \Psi_i / \|\Psi_i\| \cdot \Psi_j / \|\Psi_j\|$.

To determine the dosage sensitivity we calculated the packing deficiency of each gene-encoded protein based on its PDB coordinates, if available. Otherwise, packing deficiency was determined based on homology-threaded structure coordinates adopting as templates PDB-reported paralogs [19]. The input for the computation consists of the set of constraints applied to the spatial structure of the amino acid sequence to be modeled and the output is the 3D structure that best satisfies these constraints.

The resulting homology model was validated by comparing its inferred dehydron pattern with the dehydron pattern predicted from a sequence-based computation of disorder score, as described in the previous chapter [17]. The inability of an isolated protein fold to protect specific intramolecular hydrogen bonds from water attack may lead to a structure-competing backbone hydration with concurrent local or global dismantling of the structure. This view of under-wrapping implies a strong correlation between the degree of solvent exposure of intramolecular hydrogen bonds and the local propensity for structural disorder. Hence, the latter parameter was used to validate the former.

6.4.2 Wrapping Deficiency Causes Dosage Imbalance and Regulation Dissimilarity

To assess the selection pressure imposed by dosage imbalance we first consider an exhaustive set of 457 non-singleton human gene families with paralog co-expression at the mRNA level: $<\eta> \gg 0, <> = $ family average. This condition is essential since paralogs expressed in different cell types cannot introduce dosage imbalance, regardless of their extent of identity. The families with paralog co-expression are selected to discern the factors that buffer dosage imbalance caused by gene duplication. Thus, the selection pressure may be assessed at the post-transcriptional level in terms of dissimilarities in miRNA-targeting patterns across paralogs.

The families with significant mRNA co-expression were deemed likely to generate dosage imbalance. To assess how these imbalances impinge on the degree of divergence in post-transcriptional repression patterns across paralogs, we must compare families with similar divergence time of gene duplicates. This is so since significant regulatory dissimilarities across paralogs may simply result from long divergence times. Thus, we adopt K_s, the synonymous nucleotide divergence [14], as a proxy for divergence time [20] and bin human families with $< \eta >> 0$ according to their respective maximum K_s over paralog pairs. Each class contains families whose duplicate divergence is located in time vis-á-vis particular speciation events. Thus, we construct four classes of human families with significantly co-expressed paralogs [17]: class I: $K_s > 2.98$ (378 families); class II: $2.98 > K_s > 1.48$ (68 families); class III: $1.48 > K_s > 1.11$ (6 families) and class IV: $K_s < 1.11$ (5 families), in accord with the K_s values between humans and orangutan (*Pongo pygmaeus*) ($K_s = 2.98$), humans and gorilla (*Gorilla gorilla*) ($K_s = 1.48$), and humans and chimpanzee (*Pan troglodytes*) ($K_s = 1.11$) [21]. All K_s values are given as percentages and the data for each orthology class are compiled in Table 6.2.

Table 6.2 Data set of evolutionary, expression, miRNA-based regulation, and structural information on human families binned in different K_s classes

Ensembl ID	Size	Max. K_s	Expression correlation	miRNA target coincidence	Packing deficiency
Class I: Representative families from 378 with $K_s > 2.98$					
>ENSF00000000529	7	4.7368	0.092	0	58.611
>ENSF00000000100	20	4.3361	0.105	0	57.905
>ENSF00000000139	4	3.5898	0.009	0	57.905
>ENSF00000001722	3	4.5429	0.112	0	57.235
>ENSF00000001167	4	4.257	0.031	0.100156612	49.166
>ENSF00000000095	26	4.8626	0.091	0.100505579	41.711
>ENSF00000001308	3	4.5669	0.137	0.100723554	53.574
>ENSF00000000711	4	4.0037	0.058	0.100919215	43.801
>ENSF00000000137	10	5.618	0.054	0.146230992	40.942
>ENSF00000000284	10	4.2759	0.099	0.146538636	41.175
>ENSF00000001426	4	3.7137	0.026	0.183940909	31.682
>ENSF00000000564	5	4.4951	0.052	0.184436777	38.162
>ENSF00000002030	3	4.5383	0.032	0.203391736	44.664
>ENSF00000001001	2	3.1208	0.201	0.204847727	41.439
>ENSF00000001947	3	4.5458	0.355	0.206692562	44.336
>ENSF00000000469	6	4.2126	0.369	0.207894326	47.726
>ENSF00000001542	3	4.9462	0.088	0.20806405	33.63
>ENSF00000001394	3	4.4438	0.185	0.215967149	40.265
>ENSF00000000676	4	5.1652	0.052	0.220155165	32.679
>ENSF00000000750	4	4.555	0.032	0.249786777	31.144
>ENSF00000000662	6	4.1415	0.148	0.252284504	33.377
>ENSF00000000678	3	4.398	0.077	0.259919421	29.562
>ENSF00000001387	4	4.6489	0.132	0.265183678	33.705

Table 6.2   (continued)

Ensembl ID	Size	Max. K_s	Expression correlation	miRNA target coincidence	Packing deficiency
>ENSF00000001289	2	3.2296	0.234	0.289453099	37.129
>ENSF00000002412	2	3.5859	0.504	0.291411364	34.575
>ENSF00000000546	7	4.4044	0.314	0.29621157	37.76
>ENSF00000001974	3	5.605	0.116	0.301903926	31.719
>ENSF00000001090	4	3.8111	0.092	0.30237686	38.952
>ENSF00000002334	3	4.0352	0.117	0.305980165	26.024
>ENSF00000001936	2	5.4117	0.076	0.309117355	26.752
>ENSF00000000747	3	4.4318	0.239	0.326588636	39.481
>ENSF00000001530	3	4.1787	0.1	0.326981198	36.141
>ENSF00000001820	2	3.007	0.048	0.337169628	34.925
>ENSF00000001225	3	4.5539	0.042	0.436954752	35.533
>ENSF00000004001	2	3.5857	0.042	0.467541529	21.877
>ENSF00000001160	3	3.8174	0.393	0.746814463	19.7721

Class II: The 68 human families with 2.98 > K_s > 1.48

Ensembl ID	Size	Max. K_s	Expression correlation	miRNA target coincidence	Packing deficiency
>ENSF00000001825	2	2.9144	0.041	0	58.601
>ENSF00000001266	2	2.6079	0.155	0	57.999
>ENSF00000000708	2	1.7241	0.157	0	58.999
>ENSF00000003280	2	2.9426	0.201	0	57.981
>ENSF00000001173	3	2.6772	0.23	0	56.613
>ENSF00000001409	3	2.0711	0.473	0	56.132
>ENSF00000004229	2	2.0857	0.062	0	55.729
>ENSF00000000656	5	2.736	0.471	0	55.62
>ENSF00000001587	2	2.6153	0.452	0	55.192
>ENSF00000000881	2	2.4164	0.12	0	54.574
>ENSF00000000675	5	2.4562	0.009	0	54.434
>ENSF00000001679	2	2.3438	0.48	0	54.33
>ENSF00000001375	4	2.2318	0.217	0	54.074
>ENSF00000001081	2	2.864	0.392	0	54.036
>ENSF00000000178	7	1.9737	0.146	0	53.456
>ENSF00000000572	2	2.1986	0.045	0	53.121
>ENSF00000003334	2	1.7844	0.035	0	52.671
>ENSF00000002313	2	1.8049	0.018	0	52.216
>ENSF00000000436	5	2.8531	0.29	0	51.945
>ENSF00000000074	21	2.7213	0.061	0	51.689
>ENSF00000003192	3	2.3561	0.4	0	51.646
>ENSF00000001401	3	2.8501	0.041	0	51.446
>ENSF00000000271	16	2.2647	0.292	0	51.438
>ENSF00000001153	4	2.6993	0.155	0	51.215
>ENSF00000001694	3	2.0918	0.258	0	51.198
>ENSF00000003861	2	1.6126	0.243	0	50.846
>ENSF00000001802	2	1.6984	0.537	0	50.768
>ENSF00000000847	4	2.8478	0.376	0	50.687
>ENSF00000001880	2	1.9662	0.219	0	50.687
>ENSF00000003481	2	1.8106	0.446	0	50.687
>ENSF00000003653	2	2.2285	0.183	0	50.332
>ENSF00000001645	3	2.2403	0.206	0	50.135
>ENSF00000000637	8	1.9319	0.041	0	50.062
>ENSF00000002038	2	1.8469	0.109	0	52.797

Table 6.2 (continued)

Ensembl ID	Size	Max. K_s	Expression correlation	miRNA target coincidence	Packing deficiency
>ENSF00000002767	2	2.2562	0.07	0	52.795
>ENSF00000001162	5	2.0596	0.097	0	53.704
>ENSF00000001747	2	1.7805	0.019	0	52.532
>ENSF00000000570	6	2.1163	0.013	0	52.377
>ENSF00000000671	3	2.2928	0.178	0	52.279
>ENSF00000000389	5	2.4549	0.22	0	53.107
>ENSF00000002053	2	1.8579	0.411	0	58.107
>ENSF00000001405	2	2.2754	0.042	0	58.055
>ENSF00000002395	4	2.5819	0.365	0	57.152
>ENSF00000000402	5	2.2446	0.561	0.056	57.012
>ENSF00000001931	5	2.3206	0.035	0.091	45.743
>ENSF00000003072	2	2.0413	0.093	0	54.973
>ENSF00000000492	6	1.9765	0.195	0.141	44.787
>ENSF00000002120	2	2.6622	0.02	0	54.427
>ENSF00000004883	4	2.7168	0.14	0	53.174
>ENSF00000002772	4	1.7219	0.132	0.094	51.132
>ENSF00000001313	3	2.5443	0.099	0.064	50.351
>ENSF00000001227	3	2.1405	0.285	0.17	33.99
>ENSF00000001199	2	2.9308	0.411	0.098	38.351
>ENSF00000001760	2	1.8274	0.123	0.088	37.322
>ENSF00000001365	3	1.5327	0.112	0.162	37.295
>ENSF00000000850	3	1.6986	0.193	0.335	22.644
>ENSF00000001242	3	1.8311	0.58	0.204	36.633
>ENSF00000002009	3	2.1064	0.339	0.267	31.05
>ENSF00000003693	2	2.8605	0.132	0.256	34.582
>ENSF00000000931	3	1.9106	0.07	0.35	24.055
>ENSF00000001181	3	2.5267	0.093	0.279	28.222
>ENSF00000000463	4	1.9925	0.134	0.152	32.358
>ENSF00000004286	2	1.869	0.564	0.385	29.439
>ENSF00000001127	2	2.7223	0.688	0.367	29.419
>ENSF00000000265	14	2.0125	0.699	0.213	28.91
>ENSF00000001157	4	2.3618	0.064	0.169	27.24
>ENSF00000002286	2	2.4737	0.046	0.41	17.284
>ENSF00000000715	9	2.4495	0.237	0.707	24.718
Class III: The 6 human families with 1.48 > K_s > 1.11					
>ENSF00000001699	2	1.4313	0.223	0.408	21.703
>ENSF00000001755	4	1.4269	5.00E–001	0.136	31.305
>ENSF00000001091	4	1.4047	0.112	0.333	21.279
>ENSF00000000239	5	1.3547	0.442	0.298	34.019
>ENSF00000001198	3	1.3243	0.335	0	55.438
>ENSF00000000360	6	1.3127	0.18	0.343	27.938
Class IV: The 5 human families with 1.11 > K_s					
>ENSF00000000399	13	1.102	0.138	0	43.174
>ENSF00000002309	2	1.0081	0.551	0	49.846
>ENSF00000000530	19	0.615	0.23	0	49.235
>ENSF00000002885	2	0.4883	0.012	0	47.305
>ENSF00000000786	6	0.28	0.107	0	43.32

The conservation-based reliability of miRNA site prediction [8] is the highest in class I and decreases with lower divergence times for duplicate genes. This is so since the condition K_s (duplicate genes) > K_s (speciation) implies that orthologs of the paralog human genes are likely to be found in the diverging species [22]. Thus, paralogs for families in class I are likely to have orthologs in orangutan, gorilla, and chimpanzee; those in class II, only in gorilla and chimpanzee; etc.

Human families with paralog co-expression and the most reliable miRNA site inference (class I) exhibit a tight anticorrelation ($R^2 = 0.697$) between packing deficiency and miRNA target coincidence (Fig. 6.3): *paralogs with deficient packing are more likely to be localized separated from each other as dictated by their dissimilar miRNA target patterns of post-transcriptional regulation:* $< \tau > \rightarrow 0$ as $< v > \rightarrow$ *maximum* $\approx 58\%$. These disjoint localization patterns reduce paralog competition for binding partners, thereby buffering the evolution-related dosage imbalance. This result highlights the role of miRNA regulation as a capacitor for dosage imbalance.

An even tighter anticorrelation between packing deficiency and miRNA target coincidence is found for family class II ($R^2 = 0.792$, Fig. 6.3). The slope of the linear fit obtained by the least-squares linear regression is now significantly larger in magnitude (–69.34 versus –57.54 for class I). This implies that for a fixed level of packing deficiency, a more effective buffer (lower miRNA target coincidence) is needed for the newer families (K_s class II) than for the older ones (K_s class I). This result is expected since a longer exposure of surviving paralogs to the selection pressure promoted by dosage imbalance is likely to promote higher level of adaptation through functional divergence, and hence, as older paralogs become more differentiated, a capacitance to dosage imbalance becomes less necessary. The same trend is apparent as we examine class III (slope –72.53, $R^2 = 0.786$), although the scarcity of the data precludes a reliable statistical analysis. Class IV consists of only five families and hence no trend can be established, except that all families have zero miRNA target coincidence irrespective of their packing deficiency. This fact is clearly indicative of a pressing need to buffer dosage imbalances arising from duplicates that have not yet undergone sufficient functional differentiation.

The trends in terms of tighter v–τ-anticorrelation and steeper slope as classes with lower K_s are considered (Fig. 6.3) imply that an miRNA-based capacitance to dosage imbalance is more operative for younger families (classes II–IV versus class I). This result is compatible with the fact that selection pressure on more recent paralogs has had comparably less time to promote adaptation through functional divergence and hence duplication-related dosage imbalances are more significant than those in older families.

These results reveal that the human capacitance to dosage imbalance is in part required due to the inefficiency of the selection pressure on duplicate genes, precluding sufficient differentiation over the evolutionary times of the latest speciations, thereby maintaining an evolutionarily related dosage imbalance.

The results of Fig. 6.3 imply that miRNA target dissimilarity across paralogs may be assimilated to a capacitance to dosage imbalance effects arising from gene

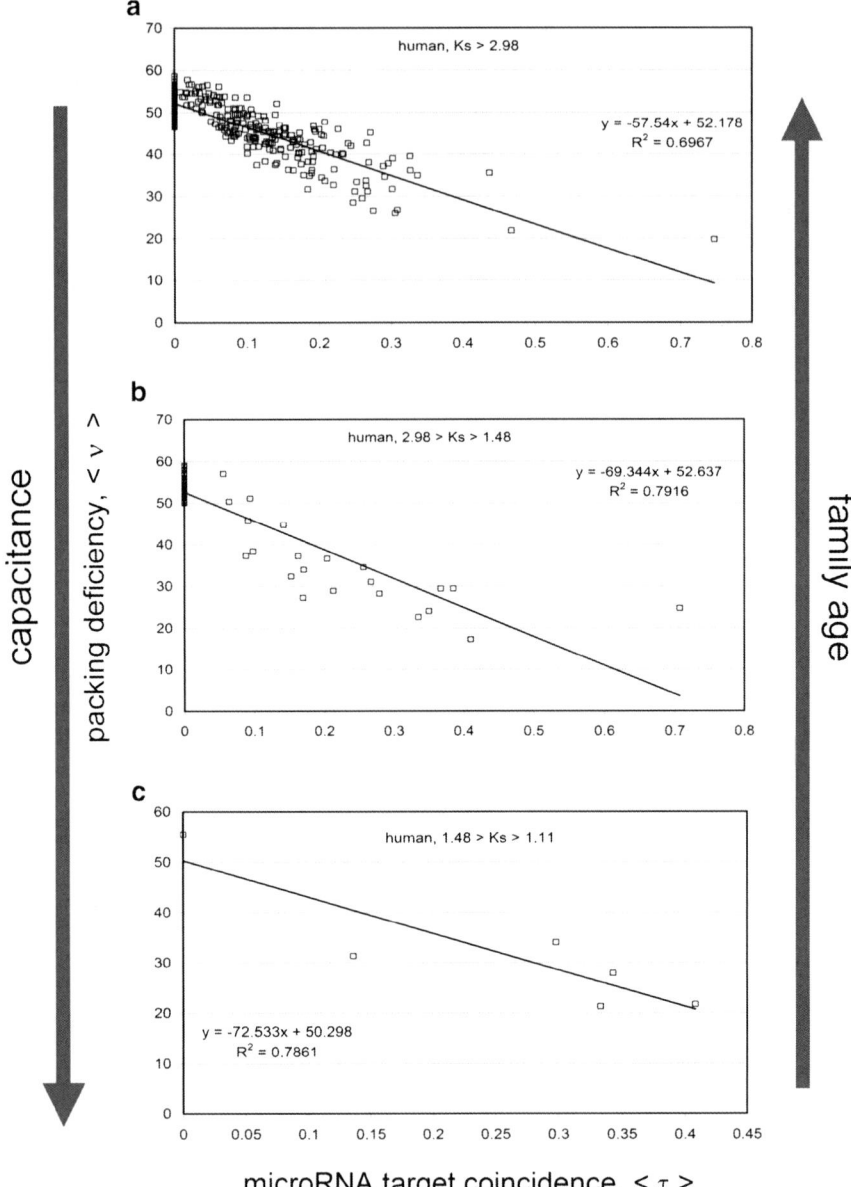

Fig. 6.3 Negative correlation between under-wrapping or packing deficiency (ν) and miRNA target coincidence (τ) for human families in K_s classes I (**A**), II (**B**), and III (**C**). The linear fits were obtained by least-squares linear regression

duplication. The severity of such effects is in turn quantified by packing deficiency: Dosage imbalances are less tolerated for deficiently packed proteins forcing paralogs to be localized separately from each other.

6.5 Human Capacitance to Dosage Imbalances in the Concentrations of Under-Wrapped Proteins

In this Section we examined dosage imbalances that have an evolutionary origin. Thus, gene duplication events generate dosage imbalances that impose selection pressure on paralogs and the magnitude of the effects of this pressure depend on the extent of under-wrapping or packing deficiency of the gene product. However, this dependence varies widely from unicellular to higher eukaryotes, with humans being particularly insensitive to dosage imbalances. In humans, there is a significant amount of genes with packing deficiency which are nevertheless extensively duplicated. This suggests that humans are resilient to evolution-related dosage imbalances, a capacitance that may be rationalized in terms of escape routes available to humans but not to unicellular organisms, where dosage imbalances have clear deleterious effects.

In the absence of expression dissimilarity, the initially identical paralogs of deficiently packed proteins are subject to high selection pressure because they compete for binding partners needed to maintain structural integrity. Conversely, tight protein packing reduces dosage sensitivity, thereby curbing selection pressure. Cross examination of genetic and structural data reveals that humans have a built-in resilience or capacitance to dosage imbalances. The determinant of this human capacitance is traced in this chapter to the paralog-discriminatory power of miRNA regulatory patterns. In this way, dissimilar paralog localization governed by post-transcriptional regulation of protein levels mitigates the competition of paralogs for common binding partners that become obligatory for proteins of low packing quality. In other words, dissimilarity in paralog localization operative through miRNA control offers an escape route to dosage imbalances created by gene duplication, and this escape route becomes more necessary as protein packing deficiency makes these dosage imbalances less tolerable.

If selection is indeed inefficient in humans as Lynch and co-workers proposed [3], one may wonder how miRNA-based capacitance could be achieved through random genetic drift. The removal of an miRNA-binding site is readily achievable through a single deleterious mutation in one paralog and is unlikely to occur at the same binding site in another paralog. For instance, if M non-overlapping miRNA-binding sites are present in the $3'$-UTR of two paralogs ($1 << M < 156$), the probability that a pair of mutations (one in each paralog) will discriminate paralogs is $(M-1)/M$, while the probability that they both occur at the same site in the two paralogs and hence do not contribute to increase the capacitance is $1/M$. Thus, dissimilarity and even orthogonality in post-transcriptional repressive patterns is readily achievable through widespread random mutation and hence may be the result of the nonadaptive forces prevalent in humans.

6.6 Why Should the Drug Designer Be Mindful of Molecular Evolution?

As surveyed in this chapter, in spite of an uncanny structural similarity arising from common ancestry, both ortholog and paralog proteins exhibit vast dissimilarities in their wrapping patterns. However, the root of these variations is very different depending on whether orthologs or paralogs are considered. As indicated in the previous sections, differences in selection efficiency across species appear to be responsible for wrapping differences across orthologs. On the other hand, in the case of paralogs, significant dosage imbalances would arise if paralogs were identical, simply because paralogs coexist in the same individual. To avert such imbalances and their fitness consequences, sufficient divergence of the original duplicate genes must take place through dissimilar expression regulation. Thus, in the previous sections we showed how paralog segregation may be achieved.

An alternative evolutionary route to differentiate paralogs as needed to buffer the dosage imbalances they create is through dissimilar wrapping of their otherwise highly similar 3D structures. *As demonstrated in the subsequent chapters, this phenomenon has profound consequences for drug design because it enables designers to discriminate paralogs guided by their different dehydron patterns, exploiting the idea that a drug may function as an external wrapper of the packing defects in the target protein.*

Differences in wrapping patterns may be significant even for close paralogs with high levels ($\sim$88%) of sequence identity, such as the kinase domains of the insulin receptor (IR) and that of the insulin-like growth factor receptor (IGF1R), as shown in Fig. 6.4. From the perspective of molecular targeted cancer therapy, there is a vested interest in engineering drugs able to discriminate between these two paralogs: IGF1R is probably a major target in the treatment of breast cancer [23], while inhibition of the IR kinase is likely to have devastating consequences, even the risk of inducing a diabetic coma in the patient. Yet, harnessing on the wrapping design concept, it should be possible to target differences in the wrapping patterns of both kinases (Fig. 6.4) to enhance therapeutic efficacy against breast cancer.

On the other hand, the observation that there are wrapping differences across orthologous proteins should also have a significant impact for the drug designer. *For example, the therapeutic efficacy of drugs is routinely tested in animal models through xenografts and other assays, always under the unwarranted assumption that the affinity and specificity of the drug for its target in humans would be the same or very close to its affinity and specificity for the animal ortholog of the targeted protein. In light of the wrapping differences across orthologs, there is simply no assurance that this scenario is indeed valid. This observation could potentially undermine the very foundation on which animal testing is predicated.*

Yet, there is a positive implication of the observed wrapping differences across orthologs, and it relates to the drug-based treatment of infectious diseases such as AIDS. It is indeed highly advantageous for the drug designer to detect wrapping differences between the target protein in the infecting species and its human

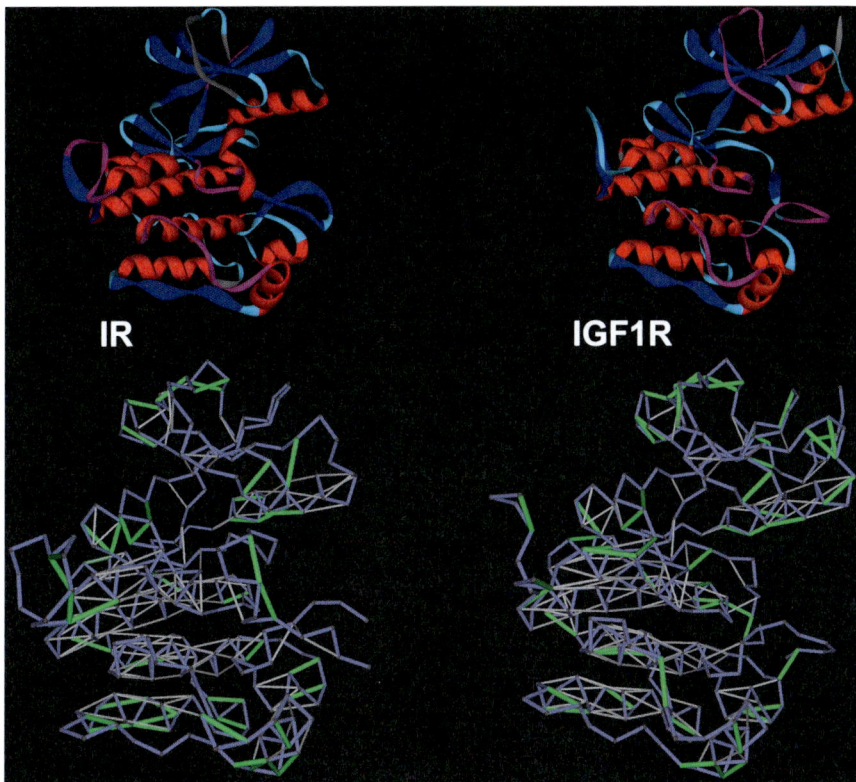

Fig. 6.4 Comparison of structures (*ribbon representation*) and wrapping patterns of the human paralog kinases for the IR (insulin receptor) and the IGF1R (insulin-like growth factor 1 receptor)

ortholog, if any: *A drug purposely designed to wrap specific dehydrons in the protein of the infecting species that are not present in the human ortholog is likely to be far less toxic than a drug that cannot tell apart the two orthologs.* Species selectivity is hardly a concern in an AIDS treatment geared at targeting the reverse transcriptase of the HIV-1 retrovirus, simply because this protein is not encoded in the human genome. On the other hand, a molecular design rooted in wrapping differences would be very helpful if the goal is to target a retroviral protein that is also represented in the human proteome. Species selectivity matters in this last context.

References

1. Fernández A, Scott R, Berry RS (2004) The nonconserved wrapping of conserved folds reveals a trend towards increasing connectivity in proteomic networks. Proc Natl Acad Sci USA 101:2823–2827
2. Fernández A, Berry RS (2004) Molecular dimension explored in evolution to promote proteomic complexity. Proc Natl Acad Sci USA 101:13460–13465

3. Lynch M, Conery JS (2003) The origins of genome complexity. Science 302:1401–1404
4. Kondrashov FA, Koonin EV (2004) A common framework for understanding the origin of genetic dominance and evolutionary fates of gene duplications. Trends Genet 20:287–290
5. Liang H, Rogale-Plazonic K, Chen J, Li WH, Fernández A (2008) Protein under-wrapping causes dosage sensitivity and decreases gene duplicability. PLoS Genet 4:e11
6. Papp B, Pal C, Hurst LD (2003) Dosage sensitivity and the evolution of gene families in yeast. Nature 424:194–197
7. Fernández A, Scheraga H (2003) Insufficiently dehydrated hydrogen bonds as determinants for protein interactions. Proc Natl Acad Sci USA 100:113–118
8. Bartel D (2009) MicroRNAs: Target recognition and regulatory functions. Cell 136:215–233
9. Fernández A (2004) Keeping dry and crossing membranes. Nat Biotechnol 22:1081–1084
10. Veitia RA (2002) Exploring the etiology of haploinsufficiency. Bioessays 24:175–184
11. Veitia RA (2004) Gene dosage balance: Deletions, duplications and dominance. Trends Genet 21: 33–35
12. Su AI, Wiltshire T, Batalov S et al (2004) A gene atlas of the mouse and human protein-encoding transcriptomes. Proc Natl Acad Sci USA 101:6062–6067
13. Birney E, Andrews D, Caccamo M et al (2006) Ensembl 2006. Nucleic Acids Res 34: D556–D561
14. Yang Z, Nielsen R (2000) Estimating synonymous and nonsynonymous substitution rates under realistic evolutionary models. Mol Biol Evol 17:32–43
15. Friedman RC, Farth KK, Burge CB, Bartel DP (2009) Most mammalian mRNAs are conserved targets of microRNAs. Genome Res 19:92–105
16. Lewis B, Burge C, Bartel D (2005) Conserved seed pairing, often flanked by adenosines, indicates that thousands of human genes are microRNA targets. Cell 120:15–20
17. Fernández A, Chen J (2009) Human capacitance to dosage imbalance: coping with inefficient selection. Genome Res 19:2185–2192
18. Grimson A, Farth KK, Johnston WK et al (2007) MicroRNA target specificity in mammals: Determinants beyond seed pairing. Mol Cell 27:91–105
19. Aloy P, Ceulemans H, Stark A, Russell RB (2003) The relationship between sequence and interaction divergence in proteins. J Mol Biol 332:989–998
20. Gu Z, Nicolae D, Lu HH, Li W-H (2002) Rapid divergence in expression between duplicate genes inferred from microarray data. Trends Genet 18:609–613
21. Chen F, Li W-H (2001) Genomic divergences between humans and other hominoids and the effective population size of the common ancestor of humans and chimpanzees. Am J Hum Genet 68:444–456
22. Gao L, Innan H (2004) Very low gene duplication rate in the yeast genome. Science 306: 1367–1370
23. Surmacz E, Bartucci M (2005) Role of estrogen receptor alpha in modulating IGF-I receptor signaling and function in breast cancer. J Exp Clin Cancer Res 23:385–394

Chapter 7
Wrapping as a Selectivity Filter for Molecular Targeted Therapy: Preliminary Evidence

The conservation of structure across homolog proteins often diffuses the impact of drug-based inhibition by promoting alternative protein–ligand associations that may ultimately lead to toxic side effects. This problem becomes particularly acute when attempting to interfere with signaling pathways involved in cell fate and cell proliferation, the type of molecular intervention often exploited in molecular anticancer therapy. This is because the clinically relevant targets, the kinases, evolved from each other and hence share an uncanny structural resemblance. However, as shown in the previous chapter, the sticky wrapping defects or dehydrons are typically *not* conserved across proteins of common ancestry, making them valuable a priori targets to enhance specificity. Thus, nonconserved packing defects may be utilized as selectivity switches across homolog targets. This chapter explores this paradigmatic concept and its ramifications for the rational design of drugs with controlled specificity.

7.1 The Specificity Problem in Drug Design

Creating inhibitors with controlled specificity and without undesirable side effects is a major goal of drug design. Drug-induced enzymatic inhibition may have at least two kinds of side effects: the drug may interfere with pathways other than the one of therapeutic relevance because of the manifold roles of the target protein or paralog proteins, being structurally similar, may alternatively associate with the inhibitor. Building on the evolutionary knowledge acquired from the previous chapter, this chapter introduces new and paradigmatic concepts to minimize drug interactions with toxicity-related targets. In principle, the control of specificity is enabled as we exploit characteristics of protein–ligand interactions that are not preserved across paralogs, despite the similarity of their 3D structures [1–3]. Thus, the insights from the previous chapter prompt us to focus on a singular feature, the dehydron, a sticky packing defect shown to be of paramount importance in promoting protein associations [4–6]. Quantifiable features of these interactions allow us to introduce a target phylogeny that we can exploit to infer drug cross-reactivities and modulate the inhibitory impact of a drug, funneling it onto the realm of clinical relevance.

A. Fernández, *Transformative Concepts for Drug Design: Target Wrapping*, DOI 10.1007/978-3-642-11792-3_7, © Springer-Verlag Berlin Heidelberg 2010

Considerable research has helped clarify many aspects of protein–ligand inter-action [7]. Various methods enable one to predict such interactions, such as the assessment of packing defects [8], analysis of protein interface geometry [9], dock-ing dynamics [10], modeling free energy [11], computational molecular probing [12], sequence threading [13], and learning theory approaches [14]. The first four of these are based on properties of protein interfaces; the latter two are based only on sequences and attempt to "learn" features encoded in those sequences. Combined methodologies are also being developed [14]. Some of these techniques seek fea-tures of protein surfaces that promote protein–protein interactions. In this chapter we show how features with a quantifiable metric can be used to compare paralogs in a way that provides a guidance to develop more selective drugs.

As described in the previous chapters, to function, soluble proteins must maintain relatively stable structures, a condition often requiring associations [15–17]. Isolated structures with packing defects arising as poorly protected hydrogen bonds typically do not prevail in water [3, 18]. On the other hand, such defects are inherently sticky, promoting removal of surrounding water through protein associations [8], as indicated in Chap. 1. Accordingly, in this chapter we show that *dehydrons may be targeted in a new generation of highly selective drug-based inhibitors.* Known structures of protein–inhibitor complexes [19–24] provide the foundation to design inhibitors that in effect become wrappers or protectors of dehydrons.

As shown in Chap. 6, while folds are typically conserved across homolog pro-teins, wrapping defects are typically not conserved [3], and this property provides the main motivation for advocating a paradigmatic shift in drug design. While every targetable protein family may in principle be subject to the same analysis, we shall narrow down the scope of the discussion for the sake of clarity. Thus, we shall restrict ourselves to the goal of designing molecular therapeutic agents that selectively interfere with cell signaling. In this context, structure-based design becomes particularly daunting because the basic signal transducers in the cell, the kinases, share a common molecular ancestry and hence possess an uncanny structural similarity. This fact turns drug specificity into a major issue.

Kinase inhibitors are designed to impair the ATP-dependent signal transduction (Fig. 7.1a, b). Hence they should be capable of binding to the target competitively or noncompetitively vis-à-vis the natural kinase ligand ATP. The main structural fea-tures of the binding of ATP to a kinase are illustrated in Fig. 7.1a, b, where an active (phosphorylated) tyrosine kinase, that of the insulin receptor (IR), is shown bound to ATP and to a substrate peptide. The kinase substrate becomes phosphorylated at a tyrosine residue through a *trans*-esterification reaction involving transference of the γ-phosphate of ATP (Fig. 7.1b). Thus, a major challenge to design ATP competitive inhibitors is the fact that the ATP-binding regions of most kinases are structurally very similar (with an RMSD <1.2 Å in backbone coordinates) and have a very high level of amino acid conservation.

Contrasting with this troublesome aspect, the lack of conservation of the wrap-ping (dehydron) pattern across kinases has paramount potential to promote drug specificity, as illustrated in Fig. 7.1c, d. Thus, the structural alignment of the focal adhesion kinase (FAK, PDB.2J0L) and the insulin receptor kinase (PDB.1GAG)

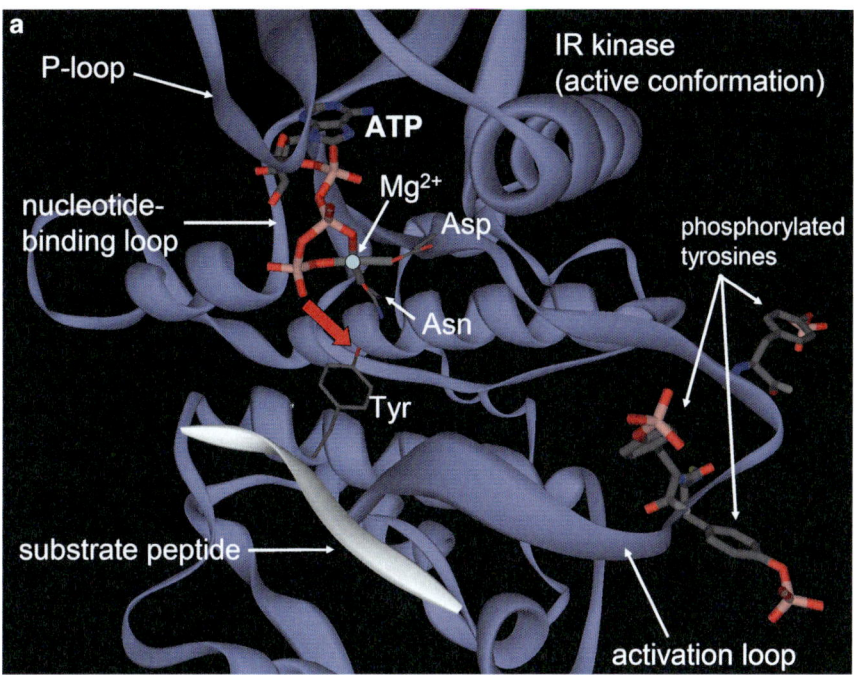

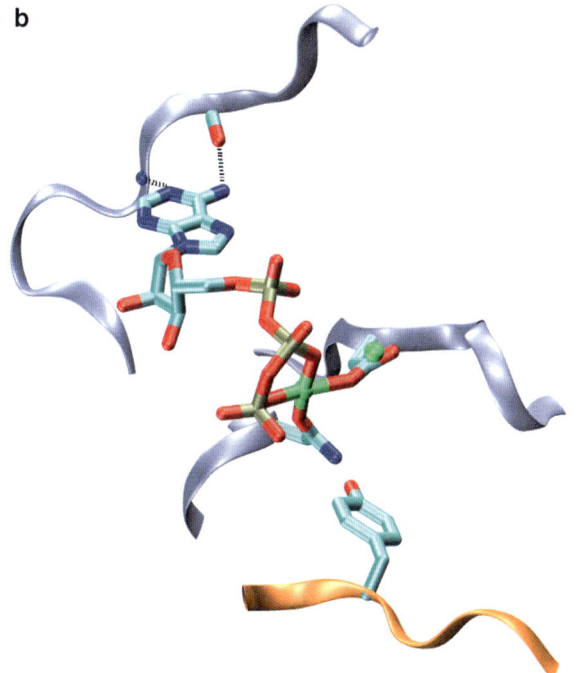

Fig. 7.1 (continued)

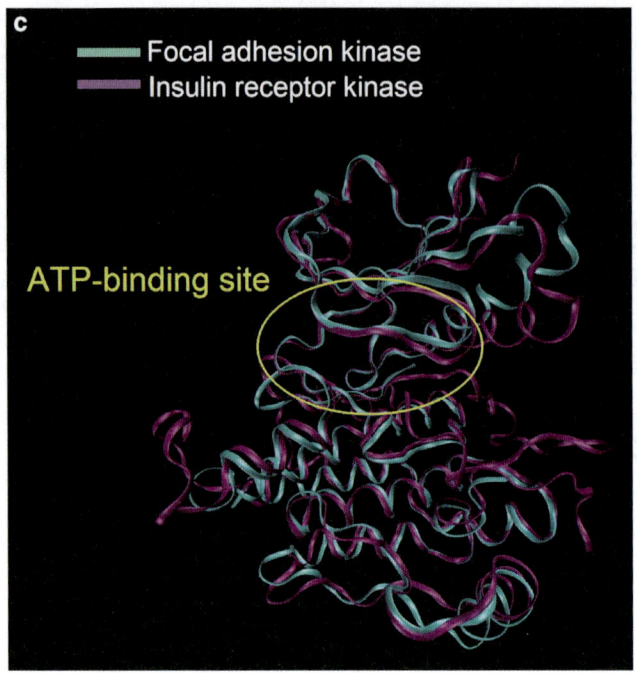

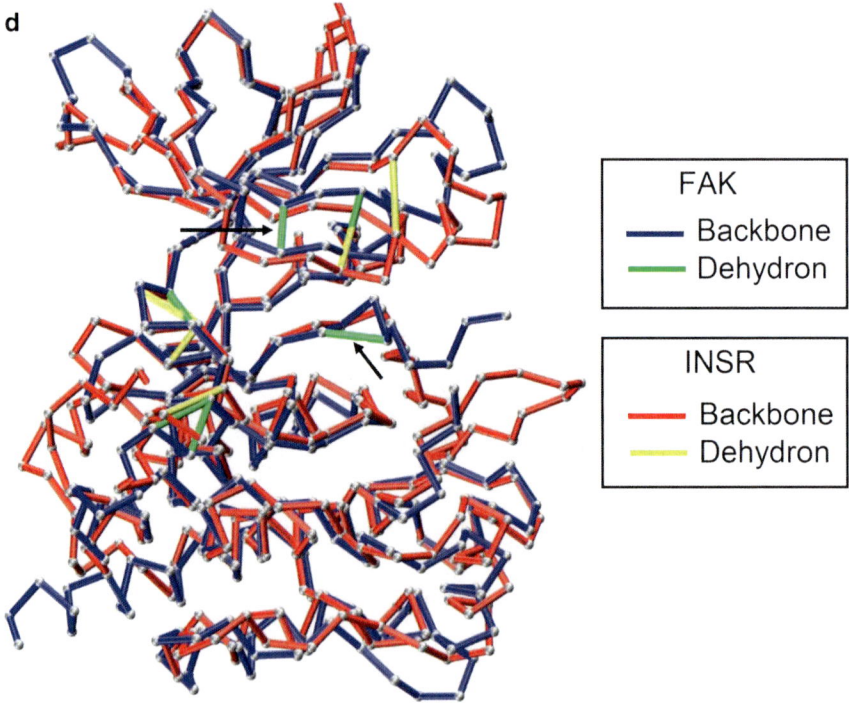

Fig. 7.1 (continued)

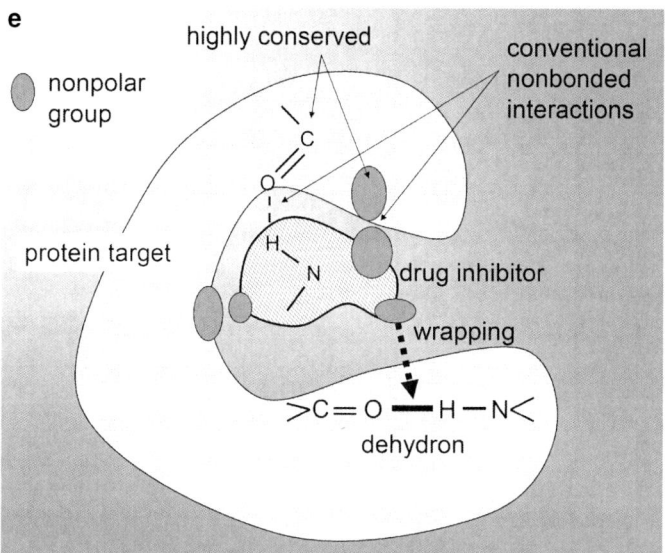

Fig. 7.1 (**a**) Generic structural features of a ternary complex involving a phosphorylated (active) kinase, its natural ligand ATP, and its phosphorylation substrate peptide (*white*). The phosphorylation event (*red arrow*) involves *trans*-phosphoesterification of the substrate tyrosine with transference of the γ-phosphate from ATP to the substrate, concurrent dismantling of the anchoring Mg(II)-coordination complex, and finally, release of ADP. (**b**) Detailed rendering of the ATP position with respect to the phosphorylation substrate. Atoms are conventionally colored with Mg(II) in *green*, substrate peptide in *gold*, and IR kinase backbone in *light blue*. Structural (**c**) and wrapping (**d**) comparison between the focal adhesion kinase (FAK, PDB.2J0L) and the insulin receptor kinase (INRK, PDB.1GAG). The dehydrons in the ATP-binding region that are unique to FAK are marked by *arrows*. (**e**) Drug paradigmatically represented as a wrapper of pre-formed dehydrons in the target protein. Conventional nonbonded interactions between drug and target such as hydrophobic contacts and intermolecular hydrogen bonds typically foster drug promiscuity as they involve conserved features in the target protein

reveal a significant similarity, with RMSD (root mean square displacement) <1.3 Å (Fig. 7.1c). Yet, there are crucial differences in the dehydron patterns of both proteins (Fig. 7.1d), particularly in their ATP-binding pockets, where the ATP competitive inhibitors bind [18]. These differences can be exploited with considerable therapeutic advantage: FAK is a target of major clinical relevance to treat cancer metastasis and yet the probable activity of a purported FAK inhibitor against the structurally related insulin receptor kinase is likely to have potentially devastating effects, arising mainly from the induction of a diabetic coma in the patient [18]. However, *there are dehydrons unique to FAK and not found in its paralog (arrows in Fig.* 7.1d) *which may be targeted to satisfy the therapeutic imperative of specificity.* This peculiar way of targeting a molecular feature materializes as intermolecular wrapping, as illustrated in Fig. 7.1e. The example shown in Fig. 7.1c, d dramatically reveals the difficulty in telling apart targets of clinical relevance from toxicity-related targets in molecular targeted therapy and unravels a possible avenue to solve this problem.

Thus, in principle, side effects resulting from off-target ligand binding may be minimized by selectively targeting nonconserved dehydrons unique to the desired target protein with the guidance of a measure of wrapping proximity between structures. In this regard, this chapter reveals that packing defects may act not only as selectivity filters operative across homolog proteins but also as selectivity switches, to re-engineer a protein–inhibitor to redirect its impact toward another target. This latter role requires taking advantage of the spatial displacement of packing defects across aligned homologous structures.

Functional differences across homolog proteins sharing a common fold are very frequent in biology. Previous studies revealed that the wrapping constitutes a molecular dimension explored when evolution is constrained to preserve the fold for functional reasons. Thus, it is likely that wrapping differences across paralogs are indicative of the functional fine-tuning needed to avoid dosage imbalances (see Chap. 6). It is precisely this fine-tuning that needs to be discriminated in drug-based molecular therapy. *As this work reveals, such wrapping differences offer a new control in constructing the selectivity filters and selectivity switches that may endow the wrapping technology with a high therapeutic value.*

Structural descriptors of protein-binding sites, such as hydrophobicity [25], curvature [26], and accessibility [27], are routinely used to guide inhibitor design. However, examination of the 814 nonredundant protein–inhibitor PDB complexes reveals that 488 of them have binding cavities with only average hydrophobicity [25], not significantly higher than the rest of the surface. In such cases, ligand affinity is attributed to the intermolecular hydrogen-bonding propensies of the inhibitor, inferred from protein–substrate transition state mimetics [21–24]. However charge screening in water renders putative intermolecular hydrogen bonds unlikely promoters of protein–ligand association, unless water removal is favored at the interface by some pre-existing feature on the surface of the protein [8].

One obvious such feature is the solvent-exposed nonpolar moiety, except that, as stated above, such features occur infrequently on binding sites and if they do, they are highly conserved across homologs. A more common and less troublesome water-excluding feature has been recently identified. We have reported [3, 8, 18, 28, 29] that dehydrons constitute sticky sites with a propensity to become dehydrated as a result of protein associations.

To explore the potential of dehydrons as promoters of drug selectivity, we start by noting that *in most PDB protein–inhibitor complexes, the ligand is in effect a wrapper of protein dehydrons, although it was not purposely designed to fulfill this role.* In this way, the available bioinformatics data support the novel design concept. Ultimately, we shall develop a wrapping technology and provide evidence supporting the idea that targeting dehydrons that are not conserved across paralogs constitute a useful strategy to enhance the selectivity of the inhibitor. Thus, the notion of wrapping (packing) similarity will be rigorously introduced and used selectively to modify a multiple-target inhibitor to achieve a higher specificity toward a particular target and subsequently to redirect its impact toward an alternative target.

7.2 Ligands as Wrappers of Proteins in PDB Complexes: Bioinformatics Evidence

Is intermolecular wrapping a hitherto undetected factor contributing to ligand affinity and selectivity? To answer this question, the interfaces of the 814 protein–inhibitor PDB complexes were re-examined to determine whether inhibitors were "dehydron wrappers," that is, whether nonpolar groups of inhibitors penetrated the desolvation domain of dehydrons. This feature was found in 631 complexes and *invariably* found in the 488 complexes whose binding cavities presented average or no surface hydrophobicity [25]. Figure 7.2 illustrates such wrapping for the HIV-1 protease [21, 30]. The inhibitor contribution to improve the protein packing is not fortuitous since the substrate must be anchored and water must be expelled from the enzymatic site. *Strikingly, the wrapping of dehydrons is not purposely targeted in current drug design.*

The Merck inhibitor Indinavir (Crixivan) bound to the functionally dimeric HIV-1 protease (PDB.2BPX) is shown in Fig. 7.2 [21, 30]. The dehydrons in the protease are marked in green. On each monomer, these are backbone hydrogen bonds between these residue pairs: Ala28–Arg87, Asp29–Asn88, Gly49–Gly52, and Gly16–Gln18. The cavity associated with substrate binding contains the first three dehydrons, with dehydron 49–52 located in the flap and dehydrons 28–87 and 29–88 adjacent to the catalytic site (Asp25), to anchor the substrate. This "sticky track" determined by dehydrons 28–87 and 29–88 is required to align the substrate peptide across the cavity as needed for nucleophilic attack by the Asp25. The flap, on the other hand, needs an exposed and hence labile hydrogen bond to confer the flexibility associated with the gating mechanism. The lack of protection on the flap (49–52) hydrogen bond becomes the reason for its stickiness, as the bond can be strengthened by the exogenous removal of surrounding water (Fig. 7.2b). The positioning of all three dehydrons in the cavity (six in the dimer) promotes inhibitor association (Fig. 7.2a).

Indinavir is a wrapper of packing defects in the enzymatic cavity: It contributes 12 desolvating groups to the 49–52 hydrogen bond (Fig. 7.2b), 10 to the 28–87 hydrogen bond, and 8 to the 29–88 hydrogen bond. All functionally relevant residues are either polar or expose the polarity of the peptide backbone (Asp25, Thr26, Gly27, Ala28, Asp29, Arg87, Asn88, Gly49, and Gly52) and thus are not *themselves* promoters of association with hydrophobic groups of the ligand. The strategic position of dehydrons involving these residues in their microenvironments becomes a decisive factor to drive water removal or charge de-screening required in facilitating the enzymatic nucleophilic attack.

Central to drug design is the minimization of toxic side effects and the enhancement of safety. Achieving such goals is proving to be extremely challenging and

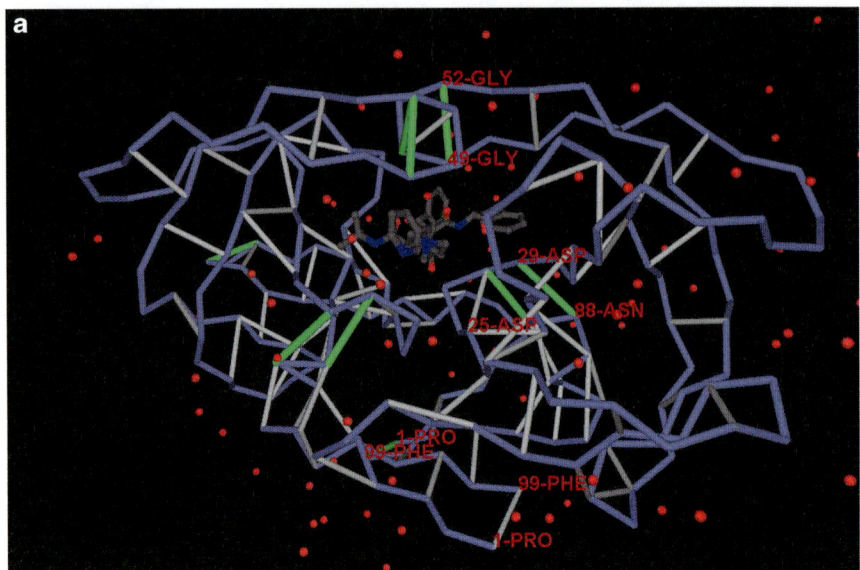

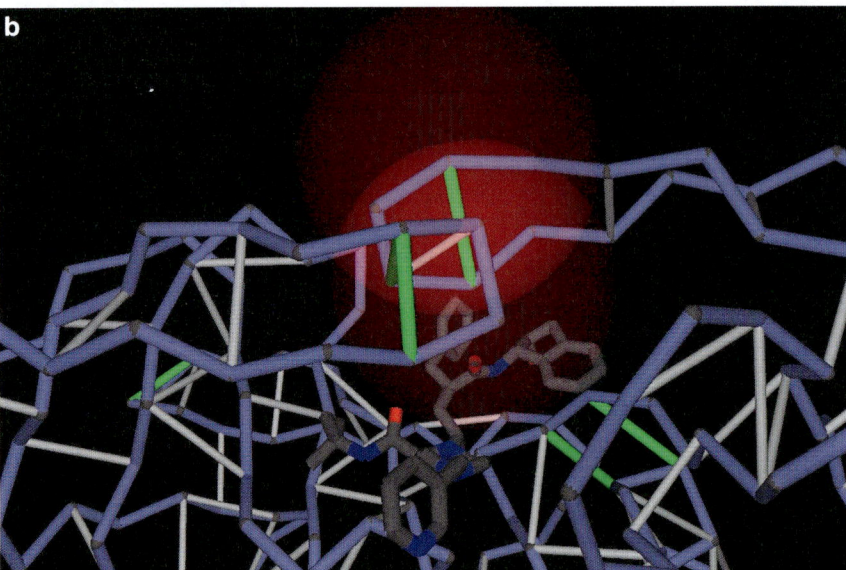

Fig. 7.2 HIV-1 protease in complex with inhibitor that acts as dehydron wrapper. (**a**) Indinavir (Crixivan) inhibitor crystallized in complex with HIV-1 protease (PDB.2BPX). The packing defects in the dimeric HIV-1 protease are displayed jointly with their spatial relation to the inhibitor position. The chain backbone is represented by *blue* virtual bonds joining α-carbons, well-wrapped backbone hydrogen bonds are shown as *light gray* segments joining the α-carbons of the paired residues, and dehydrons are shown as *green* segments. (**b**) Desolvation spheres for flap Gly49–Gly52 dehydron containing nonpolar groups of the wrapping inhibitor. Reprinted from [1], copyright 2005 with permission from Elsevier

yet a more modest yet crucially related aim, the control of specificity, seems clearly within reach, as advocated in this section and proved in the subsequent chapters. Because paralog proteins are likely to share a common structure [31], the possibility of multiple binding partners for a given protein–inhibitor arises, unless the inhibitor specifically targets nonconserved features. Thus, the problem of controlling specificity may in principle be circumvented by targeting dehydrons because, in contrast to the fold, the wrapping is generally *not* conserved across homologs [3]. This evolutionary fact makes the dehydron pattern a possible selectivity filter for drug design. The usefulness of this filter of course depends on our ability to design drugs that may contribute intermolecularly to a better wrapping of the target.

To determine whether dehydron targeting is likely to enable a control of specificity, we first investigated the extent of conservation of dehydrons across human paralogs in PDB. The paralogs for every crystallized protein–inhibitor complex were identified and dehydron patterns at binding cavities were compared. A 30% minimal sequence alignment was required for paralog identification. Packing defects were found to be a differentiating marker in paralogs of 527 of the investigated 631 proteins crystallized in complex with inhibitors. Protein chains are often reported in complexes with different inhibitors.

The PDB contains 440 redundancy-free pairs of human paralogs. Of these, 308 involve some of the 527 proteins containing binding dehydrons. In 269 pairs, the intramolecular wrapping at the binding cavity differs in the location or presence of at least one dehydron, and in 203 pairs, the difference extends to two or more dehydrons. Thus, *the probability of avoiding side effects by selectively wrapping packing defects is estimated at 88%* (269 pairs out of 308).

7.3 Poor Dehydron Wrappers Make Poor Drugs

The natural product *staurosporine* has been singled out as the ligand inhibitor most promiscuously interactive with the human kinome [32]. Its complexes are extensively reported in the PDB and high-throughput screening reveals nanomolar inhibitory impact over ~80% of the screened kinases. We have examined all 96 staurosporine–protein PDB complexes and found no instance of the inhibitor acting as dehydron wrapper. All interactions were hydrophobic in nature, involving nonpolar groups on the protein surface. An illustration of such type of binding is given in Fig. 7.3, where the Syk tyrosine kinase–staurosporine complex (PDB.1XBC) is shown to hinge on three pivotal hydrophobic interactions engaging Phe382, Val385, and Leu377. It should be noticed that staurosporine does not wrap the single dehydron Gly380–Gly382 closest to the binding site. Thus, since surface hydrophobic residues tend to be conserved as markers for interactivity and staurosporine does not discriminate among packing differences in the kinome, its promiscuity becomes entirely expected, and so is its uselessness as a therapeutic agent.

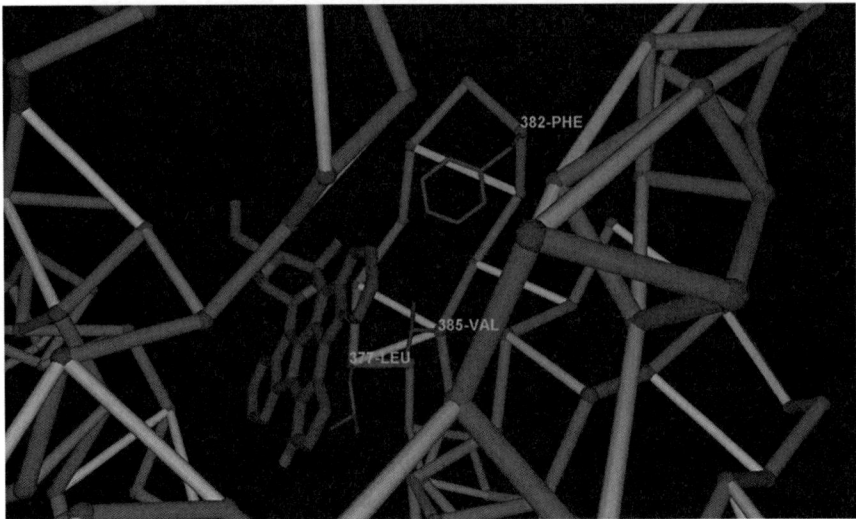

Fig. 7.3 Staurosporine does not wrap dehydrons and thus, it does not discriminate proteins with different packing, which makes it highly promiscuous within the kinome. The staurosporine–Syk complex (PDB.1XBC) reveals a binding mode determined exclusively by hydrophobic interactions engaging residues Leu377, Phe382, Val385. The closest Gly380–Gly383 dehydron is not wrapped by staurosporine

7.4 Wrapping as a Selectivity Filter

The subtle structural differences among paralogs of a kinase target may be quantified by introducing a "wrapping homology." Assessing wrapping similarity across paralogs requires comparing the microenvironments of backbone hydrogen bonds [8] that can be effectively carried out following the structural alignment [33]. This "wrapping homology" becomes a useful tool to control specificity by targeting a packing defect. For a protein chain of length N, a matrix of dehydrons $D_{ij}, i, j = 1, 2, \ldots, N$ may be constructed by choosing $D_{ij} = 1$ if residues i and j are paired by a dehydron and $D_{ij} = 0$, otherwise. This representation is illustrated in Fig. 7.4a. When the dehydron matrices for two proteins m and n are aligned, they are trimmed by restricting them to those residues that have been structurally aligned in a one-to-one correspondence. Then a Hamming distance $M_{\mathrm{H}}(m, n) = \Sigma_{i<j} |D_{ij}(m) - D_{ij}(n)|$ may be computed to serve as an indicator of the wrapping similarity between proteins m, n. Given a collection of paralogs $m, n = 1, \ldots, F$, and its corresponding packing-similarity matrix $\mathbf{M}_{\mathrm{H}} = [M_{\mathrm{H}}(m, n)]$, a wrapping similarity tree (WST) may be easily constructed by adopting the premise that the minimal pathway between protein nodes m and n crosses a number of nodes equal to $M_{\mathrm{H}}(m, n)$.

A comparison of dehydron matrices for proteins with aligned structure enables an automated digitalization of the drug selectivity problem. The alignment of dehydron matrices reveals the pattern of a priori cross-reactivities: Thus, targeting a

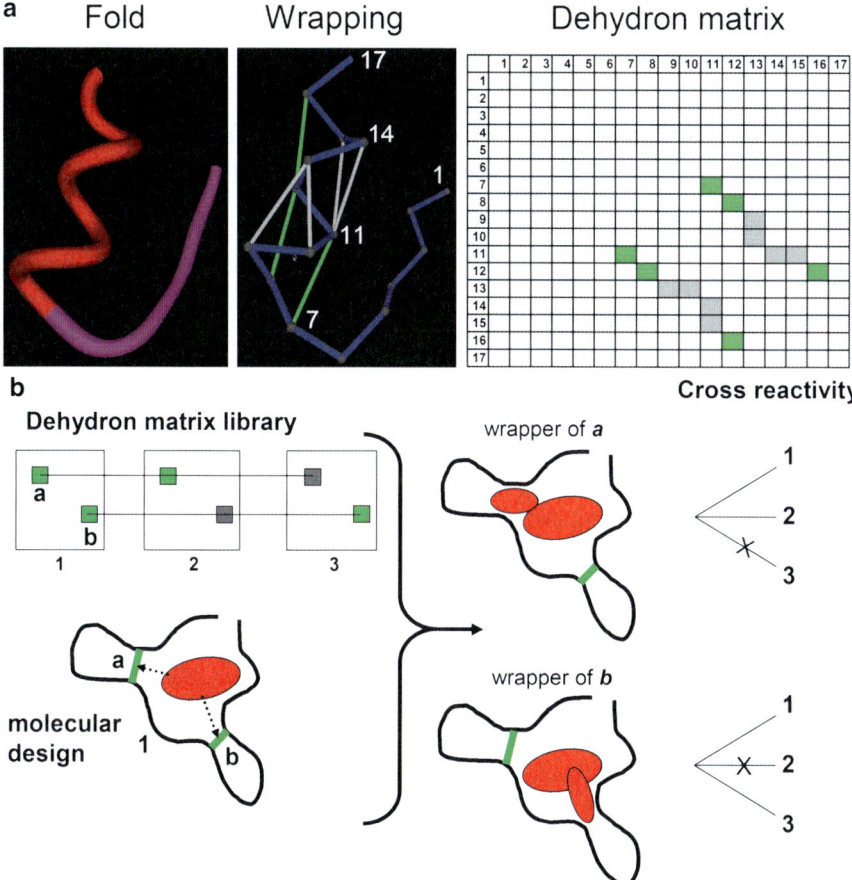

Fig. 7.4 Dehydron matrices and a priori cross-reactivities derived from their alignment. (**a**) Three different representations of a folded peptide chain of length $N = 17$: fold, wrapping, and dehydron matrix. The *gray* entries in the matrix indicate well-wrapped hydrogen bonds and the *green* entries represent dehydrons. (**b**) A priori cross-reactivities across three proteins (**1, 2, 3**) of drugs/ligands designed to wrap specific dehydrons (**a, b**) in protein **1**. The cross-reactivities are inferred from the alignment of the dehydron matrices

specific dehydron in one particular protein is likely to promote cross-reactivities with other targets where this dehydron is found at the aligned position and preclude inhibitory activity whenever the dehydron aligns with a well-wrapped hydrogen bond (Fig. 7.4b).

7.5   Wrapping as a Selectivity Filter: An Exercise in Drug Design

We now present an exercise on controlling selectivity based on the wrapping design concept. Because of the evolutionary proximity of kinases, side effects arising from off-target ligand binding often arise with kinase inhibitors, especially in cancer therapy [32]. Thus, a need arises to sharpen the binding affinity within the pharmacokinome associated with a specific drug. For example, the selective inhibition of the constitutively active or deregulated Brc-Abl (Abl: Abelson tyrosine kinase), the fusion product of an aberrant chromosomal translocation, has proven decisive to treat chronic myeloid leukemia (CML) [34]. Brc-Abl has been proven to be a primary target for the potent inhibitor imatinib (Gleevec) [34] but is not its only target [32]. Of the alternative targets with reported structure, the C-Kit tyrosine kinase has been recognized as a binding partner with clinical relevance, making Gleevec a therapeutic agent to treat certain solid tumors like gastrointestinal stromal tumor [35, 36]. In addition Gleevec binds tightly to the lymphocyte-associated kinase (Lck) [37], an immunosuppression-related target probably undesirable, and to the platelet-derived growth factor receptor (PDGFR) kinase, a close homolog of Kit. Thus, we sought to modify Gleevec to improve its selectivity for Brc-Abl by targeting dehydrons not conserved across paralogs.

The protein–inhibitor complex PDB.1FPU (Fig. 7.5a) reveals three electrostatic interactions in Brc-Abl, the dehydrons Gly249–Gln252, Gln300–Glu316, and the salt bridge Lys271–Glu286 that can be better wrapped by methylating Gleevec at the positions indicated. Thus, methylation at positions I, II, and III would presumably improve the wrapping of dehydrons 249–252, 300–316, and salt bridge 271–286, respectively. A structural alignment of the paralogs of Brc-Abl was performed using the program Cn3D [33] to investigate the microenvironment conservation for these intramolecular interactions. The alignment of six paralog proteins is shown in Fig. 7.5b. The dehydron 249–252 (crankshaft-like kink marked in yellow on Fig. 7.5b) is not conserved in *any* of the six paralogs of Brc-Abl, while the dehydron 300–316 becomes well wrapped in the paralogs. On the other hand, the microenvironment of the salt bridge 271–286 is conserved.

The WST for Abl and its paralogs is shown in Fig. 7.5c. We have also restricted the WST computation to the hydrogen bonds/dehydrons in Brc-Abl which become wrapped by Gleevec upon association. *Adopting this restricted measure, the closest proteins to Brc-Abl become Lck, C-Kit, and PDGFR, precisely the alternative Gleevec targets with reported structure.*

In order to enhance affinity and selectivity for Brc-Abl, we modified the inhibitor methylating at positions I and II (Fig. 7.5d). The synthesis of the wrapping prototype recapitulates imatinib synthesis [38], as described in [39]. To test whether the specificity and affinity for Brc-Abl improved, we conducted a spectrophotometric kinetic assay to measure the phosphorylation rate of peptide substrates in the presence of the kinase inhibitor at different concentrations. This assay couples production of adenosine diphosphate (ADP), the byproduct of downstream phosphorylation, with the concurrent detectable oxidation of reduced nicotinamide adenosine dinucleotide (NADH). The oxidation results upon transfer of phosphate from PEP (phosphoenolpyruvate) to ADP followed by the NADH-mediated reduction of PEP to lactate. Thus, phosphorylation activity is monitored by the decrease in 340 nm absorbance due to the oxidative conversion $NADH \rightarrow NAD^+$ [34, 39].

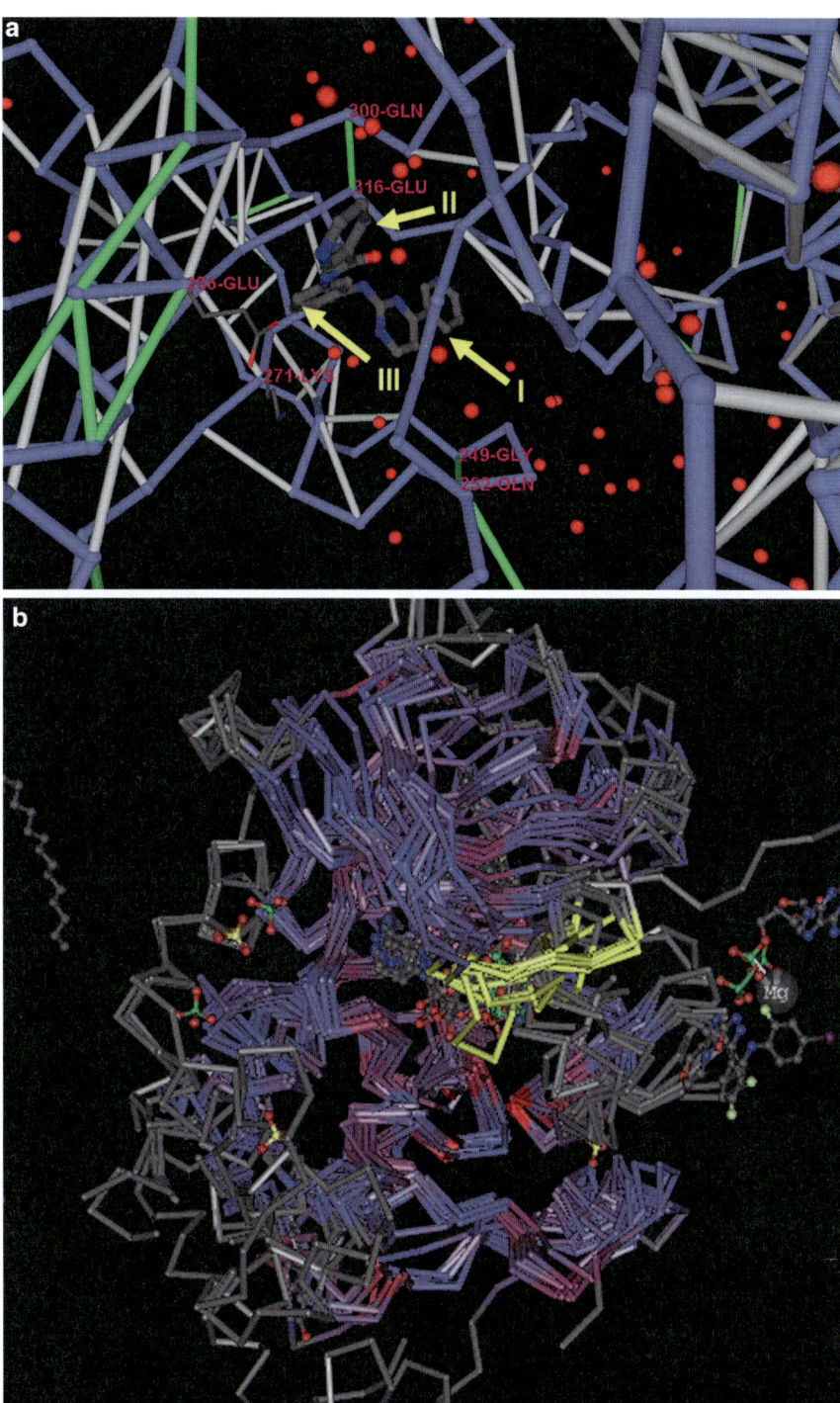

Fig. 7.5 (continued)

c

Homology of gleevec-wrapped region

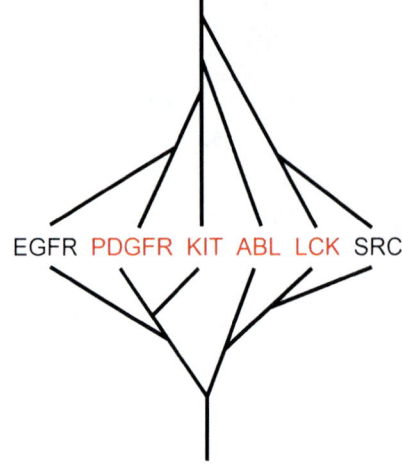

EGFR PDGFR KIT ABL LCK SRC

Wrapping homology

d

Fig. 7.5 (continued)

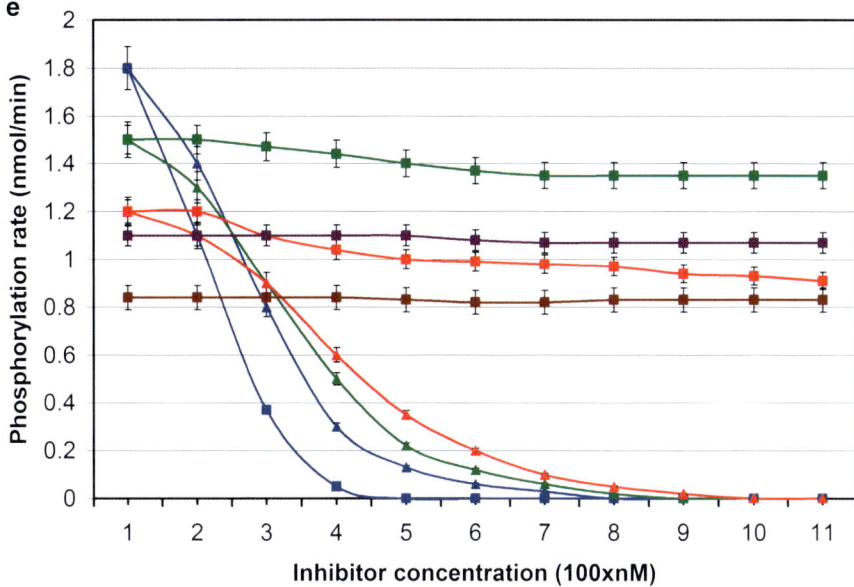

Fig. 7.5 Wrapping as selectivity filter in inhibitor drug design: Modifications of Gleevec geared at improving selectivity and affinity for Brc-Abl. (**a**) Three possible sites for Gleevec methylation (I–III) aimed at selectively improving the wrapping of packing defects of Brc-Abl (PDB.1FPU). (**b**) Structural alignment of Brc-Abl and its six paralogs using the program Cn3D. The *yellow* region corresponds to residues 247–257 in Abl. (**c**) Wrapping similarity tree (WST) for the six structurally aligned paralogs including Brc-Abl. The WST restricted to the alignments of the Gleevec-wrapped region in Brc-Abl is shown above. The paralogs in *red* have the most similar wrapping in the region that aligns with the Gleevec-wrapped region in Brc-Abl. (**d**) The modified Gleevec-based molecule methylated at sites I and II and assayed in vitro. (**e**) Phosphorylation rates of Brc-Abl (*blue*), C-Kit (*green*), Lck (*red*), PDGFR (*purple*) and Src (*brown*) in the presence of Gleevec (*triangles*) and in the presence of the I- and II-methylated modified Gleevec (*squares*). The latter compound was designed to better wrap the nonconserved dehydrons in Brc-Abl. To measure the rate of phosphorylation due to kinase activity in the presence of inhibitors, a standard spectrophotometric assay has been adopted in which the adenosine diphosphate (ADP) production, a marker of phosphorylation activity, is coupled to NADH (reduced nicotinamide adenosine dinucleotide) oxidation and determined by absorbance reduction at 340 nm, as described in [34, 39]. Basically, ADP production is coupled to the final steps of glycolysis, so that the enzymatic reaction of ADP with reactant phosphoenol pyruvate generates pyruvate whose production can be measured by reducing it to lactate using redoxagent NADH. Reprinted from [1], copyright 2005 with permission from Elsevier

As indicated in Fig. 7.5e, the spectrophotometric assay revealed the inhibition of the unphosphorylated Brc-Abl by the wrapper of the 249–252 and 300–316 dehydrons (the I- and II-methylation product) improved over Gleevec levels. Furthermore, the inhibitory impact of the dehydron wrapper became selective for Brc-Abl vis-à-vis C-Kit and Lck. Dehydrons 249–252 and 300–316 are absent in the latter kinases and, consistently, the drug designed to better wrap them has very low inhibitory impact against C-Kit and Lck.

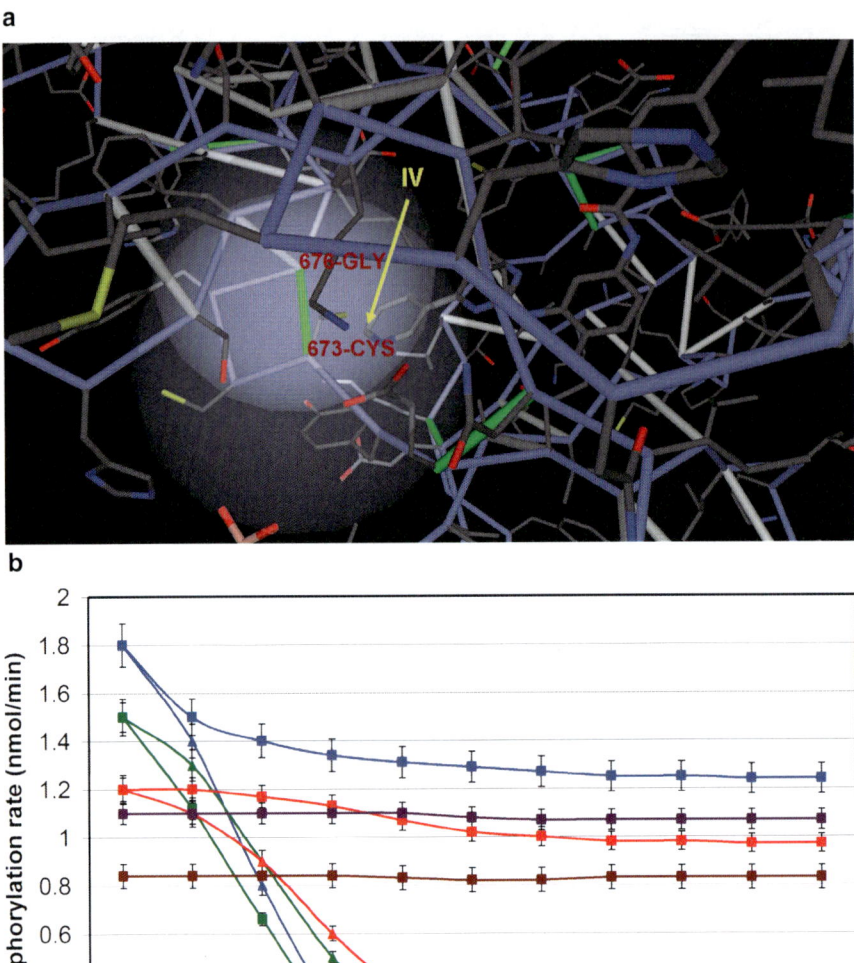

Fig. 7.6 Selectivity switch determined by packing differences across paralog proteins. The structural alignment (Fig. 7.5b), reveals that dehydron Gly249–Gln252 in Brc-Abl maps onto the spatially displaced dehydron Cys673–Gly676 in C-Kit. (**a**) Relative position of 673–676 dehydron of C-Kit with respect to Gleevec in protein–inhibitor complex (PDB.1T46). The desolvation domain of the dehydron is displayed by two intersecting spheres. By methylating Gleevec at the *para*-position (IV) rather than the *ortho*-position (I, Fig. 7.5a), the inhibitor becomes a selective wrapper of the packing defect in C-Kit. (**b**) Phosphorylation rates from spectrophotometric assay on the kinases indicated in Fig. 7.5e inhibited by Gleevec methylated at positions II and IV. The *squares* correspond to the Gleevec variant methylated at positions II and IV. Notice the selective and enhanced inhibition of C-Kit. Reprinted from [1], copyright 2005 with permission from Elsevier

7.6 Wrapping-Based Selectivity Switch

Can we really control imatinib specificity through a reworking of the parent compound as suggested in Fig. 7.4b? If this is indeed the case, we should be capable of re-funneling the inhibitory impact toward any of the alternative primary targets of imatinib. A structural alignment of the Abl and Kit kinases reveals that dehydron Gly249–Gln252 in Abl (Fig. 7.5a) maps onto dehydron Cys673–Gly676 in Kit (Fig. 7.6a), spatially displaced with respect to the former. Given the relative position of the 673–676 dehydron of Kit with respect to Gleevec in the protein–inhibitor complex (PDB.1T46), we may determine which modification of Gleevec would selectively direct its impact toward C-Kit. By methylating Gleevec at the *para*-position (IV) in the terminal ring, rather than at the *ortho*-position (I, Fig. 7.5a), the inhibitor becomes a selective wrapper of the packing defect in C-Kit, while it no longer wraps the 249–252 dehydron in Brc-Abl. The inhibitory impact of Gleevec methylated at positions II and IV becomes apparent in the phosphorylation rates obtained from the spectrophotometric assay on the same kinases studied previously (Figs. 7.5e and 7.6b). *The inhibitory impact of the II and IV Gleevec variant is now concentrated on C-Kit, rather than Brc-Abl. Thus, by redesigning Gleevec according to the wrapping differences between Brc-Abl and C-Kit, we redirected the inhibitory impact from one kinase to the other.* Taken together, both design exercises reveal that wrapping technology may enable us to introduce a selectivity switch in inhibitor design.

References

1. Fernández A (2005) Incomplete protein packing as a selectivity filter in drug design. Structure 13:1829–1836
2. Fernández A, Scott R, Berry RS (2004) The nonconserved wrapping of conserved folds reveals a trend towards increasing connectivity in proteomic networks. Proc Natl Acad Sci USA 101:2823–2827
3. Fernández A, Berry RS (2004) Molecular dimension explored in evolution to promote proteomic complexity. Proc Natl Acad Sci USA 101:13460–13465
4. Fernández A, Berry RS (2002) Extent of hydrogen-bond protection in folded proteins: A constraint on packing architectures. Biophys J 83:2475–2481
5. Fernández A, Berry RS (2003) Proteins with hydrogen-bond packing defects are highly interactive with lipid bilayers: Implications for amyloidogenesis. Proc Natl Acad Sci USA 100:2391–2396
6. Fernández A, Rogale K, Scott LR, Scheraga HA (2004) Inhibitor design by wrapping packing defects in HIV-1 proteins. Proc Natl Acad Sci USA 101:11640–11645
7. Deremble C, Lavery R (2005) Macromolecular recognition. Curr Opin Str Biol 15:171–175
8. Fernández A, Scheraga HA (2003) Insufficiently dehydrated hydrogen bonds as determinants of protein interactions. Proc Natl Acad Sci USA 100:113–118
9. Ban Y-E, Edelsbrunner H, Rudolph J (2006) Interface surfaces for protein-protein complexes. J Assoc Comp Mach 53:361–378
10. Adesokan AA, Roberts VA, Lee KW, Linz RD, Briggs JM (2004) Prediction of HIV-1 integrase/viral DNA interactions in the catalytic domain by fast molecular docking. J Med Chem 47:821–828

11. Kortemme T, Baker D (2004) Computational design of protein-protein interactions. Curr Opin Chem Bio 8:91097
12. Kortvelyesi T, Silberstein M, Dennis S, Vajda S (2003) Improved mapping of protein binding sites. J Comput Aided Mol Des 17:173–186
13. Lu L, Lu H, Skolnick J (2002) Multiprospector: An algorithm for the prediction of protein-protein interactions by multimeric threading. Proteins Struct Funct Genet 49:350–364
14. Sen TZ, Kloczkowski A, Jernigan RL et al (2004) Predicting binding sites of hydrolase-inhibitor complexes by combining several methods. BMC Bioinform 5:205–215.
15. Huber R (1979) Conformational flexibility in protein molecules. Nature 280:538–539
16. Dunker KA, Brown C, Obradovic Z (2002) Identification and function of usefully disordered proteins. Adv Protein Chem 62:25–49
17. Verkhivker G, Bouzida D, Gehehaar D, Rejto P, Freer ST, Rose P (2003) Simulating disorder-order transition in molecular recognition of unstructured proteins: Where folding meets binding. Proc Natl Acad Sci USA 100:5148–5153
18. Fernández A, Crespo A (2008) Protein wrapping: A molecular marker for association, aggregation and drug design. Chem Soc Rev 37:2372–2382
19. Stevens RC (2004) Long live structural biology. Nat Struct Mol Biol 11:293–295
20. Fauman EB, Hopkins A, Groom C (2003) Structural bioinformatics in drug discovery. In: Structural Bioinformatics, Bourne P, Weissig H, eds, Wiley-Liss, New York
21. Wlodawer A, Vondrasek J (1998) Inhibitors of HIV-1 protease: A major success of structure-assisted drug design. Annu Rev Biophys Biomol Struct 27:249–284
22. Arkin MR, Wells JA (2004) Small-molecule inhibitors of protein-protein interactions: progressing towards the dream. Nat Rev Drug Discov 3:301–317
23. Katz BA (2000) Structural basis for selectivity of a small molecule, S1-Binding, submicro-molar inhibitor of urokinase-type plasminogen activator. Chem Biol 7:299–307
24. Steinmetzer T, Hauptmann J, Sturzebecher J (2001) Expert Opin Invest Drugs 10:845–864
25. Nicholls A, Sharp KA, Honig B (1991) Protein folding and association: Insights from the interfacial and thermodynamic properties of hydrocarbons. Proteins Struct Funct Genet 11: 281–196
26. Liang J, Edelsbrunner H, Woodward C (1998) Anatomy of protein pockets and cavities: Measurement of binding site geometry and implications for ligand design. Protein Sci 7:1884–1897
27. Lee B, Richards FM (1971) The interpretation of protein structures: Estimation of static accessibility. J Mol Biol 55:379–400
28. Fernández A, Scott LR (2003) Adherence of packing defects in soluble proteins. Phys Rev Lett 91:18102
29. Fernández A, Scott LR (2003) Dehydron: A structure encoded signal for protein interactions. Biophys J 85:1914–1928
30. Munshi S, Chen Z, Li Y et al (1998) Rapid X-ray diffraction analysis of HIV-1 protease-inhibitor complexes: Inhibitor exchange in single crystals of the bound enzyme. Acta Crystallogr D 54:1053–1060
31. Mount DW (2001) Bioinformatics, Cold Spring Harbor Laboratory Press, New York
32. Karaman MW, Herrgard S, Treiber DK et al (2008) A quantitative analysis of kinase inhibitor selectivity. Nat Biotechnol 26:127–132
33. NIH/NCBI website for the structural alignment program Cn3D. http://www.ncbi.nlm.nih.gov/structure/CN3D/cn3d.shtml
34. Schindler T, Bornmann W, Pellicena P et al (2000) Structural mechanism for STI-571 inhibition of Abelson tyrosine kinase. Science 289:1938–1942
35. Attoub S, Rivat C, Rodrigues S et al (2002) The C-kit tyrosine kinase inhibitor STI-571 for colorectal cancer therapy. Cancer Res 62:4879–4883
36. Skene RJ, Kraus ML, Scheibe DN (2004) Structural basis for autoinhibition and STI-571 inhibition of C-kit tyrosine kinase. J Biol Chem 279:31655–31663

37. Perlmutter RM, Marth JD, Lewis DB et al (1988) Structure and expression of lck transcripts in human lymphoid cells. J Cell Biochem 38:117–126
38. Li J, Johnson D, Sliskovic D, Roth B (2004) Contemporary Drug Synthesis. Wiley Interscience, Hoboken
39. Fernández A, Sanguino A, Peng Z et al (2007) An anticancer C-kit kinase inhibitor is re-engineered to make it more active and less cardiotoxic. J Clin Invest 117:4044–4054

Chapter 8
Re-engineering an Anticancer Drug to Make It Safer: Modifying Imatinib to Curb Its Side Effects

As indicated in the previous chapter, molecular therapeutic approaches to cancer are often geared at interfering with signaling pathways that govern cell fate or proliferation. These strategies target kinases, the quintessential signal transducers in the cell. This approach remains challenging because kinases are evolutionarily and therefore structurally related and thus kinase inhibitors often lack specificity or possess uncontrolled cross-reactivities, which may lead to toxic side effects. This chapter illustrates the power of the wrapping concept in developing safer kinase inhibitors, thus unraveling a rational approach to fulfill this therapeutic imperative. The focus of this chapter is the redesign of the anticancer drug imatinib (Gleevec) used to treat chronic myeloid leukemia, where its primary target is the chimeric Bcr-Abl kinase, as well as certain solid tumors based on its impact on the C-Kit kinase. Imatinib also has potentially cardiotoxic effects traceable to its impact on the Abl kinase in off-target cells. This chapter describes a wrapping modification of imatinib purposely designed to prevent inhibitory impact against Abl (and Bcr-Abl), to re-focus the impact toward C-Kit and to promote inhibition of an additional target, JNK, required to reinforce the prevention of cardiotoxicity. The molecular blueprint for target discrimination, already suggested in Chaps. 4 and 6, is now established through assays that span a vast testing ground, from in silico assessment and test-tube probes to animal models. The findings surveyed in this chapter identify the wrapping-based re-engineered imatinib as an agent to treat gastrointestinal stromal tumors with reduced side effects and hence illustrate the potentiality of the wrapping approach.

8.1 Rational Control of Specificity: Toward a Safer *Imatinib*

The possibility of molecular interference with signaling pathways that control cell fate and cell proliferation opened up many therapeutic opportunities for the treatment of malignancy. Thus, protein kinases, the signal transducers of the cell, have become paradigmatic targets for drug-based cancer therapy [1–13]. However, their evolutionary – and hence structural – relatedness often leads to unforeseen and uncontrollable cross-reactivities [13–16]. Although the relationship between specificity, anticancer activity, and therapeutic efficacy remains opaque to our understanding, a lack of controlled specificity toward targets of therapeutic relevance typically underlies toxic side effects [5, 14]. Undesired and even health-threatening side effects can even be traced to inhibitory impact on a primary molecular target in off-target (i.e., normal) cells. This scenario is superbly illustrated by the reported potential cardiotoxicity of imatinib (STI571, Gleevec) attributed to its impact on Bcr-Abl [15], the chimeric Abelson (Abl) kinase target in the treatment of CML (chronic myeloid leukemia) already introduced in the preceding chapter.

In this chapter we survey a rational drug redesign strategy to control and re-focus the inhibitory impact but adopt a more sophisticated and physically rigorous blueprint than the dehydron pattern to guide molecular design [14]. Furthermore, for validation of the design strategy, we present a broader testing ground, from in silico assay to animal model, than that presented in the preceding chapter.

While wrapping remains a structure-based target discriminator, we now exploit its link to the physically more tangible and unique hydration structure of the target protein (Chap. 4, [16]) to guide our molecular design. Thus, exploiting the target dehydration or de-wetting hot spots as a blueprint for drug design, we modified imatinib to significantly reduce its impact on Bcr-Abl, prima facie curbing its car-diotoxicity, while retaining the anticancer activity associated with its other primary targets.

In concrete terms, by sculpting into the ligand the de-wetting differences across imatinib targets, we are able to focus the inhibitory impact specifically on the C-Kit kinase, a target for GIST (gastrointestinal stromal tumor) [9–11] while suppressing Bcr-Abl inhibition. Finally, we promote JNK inhibition, as needed to reinforce the prevention of cardiotoxicity [15]. The discriminating molecular design is reflected in selective anticancer activity on GIST cell lines and a GIST animal model, and a prima facie reduction of cardiotoxicity. The program for imatinib redesign is sketched in Fig. 8.1.

Imatinib wrapping-based re-engineering

Imatinib Targets	Therapeutic impact	wrapping drug desired targets
C-KIT	GIST, KIT-dependent melanomas?	KIT
Bcr-ABL	CML / cardiotoxicity	
PDGFR	Anti-Angiogenic	PDGFR
LCK	Immunodeficiencies	
		JNK

Fig. 8.1 Imatinib targets and specific goals of its wrapping modification

8.2 Unique De-wetting Hot Spots in the Target Protein Provide a Blueprint for Drug Design

As noted in Chap. 4, dehydrons generate de-wetting hot spots, that is, defects in the hydration layer of the protein. This physical property clearly suggests a molecular design strategy where, upon association with the target, the drug/ligand contributes to the removal of the hydrating molecules at the de-wetting hot spot. This design strategy is often but not always equivalent to the dehydron-wrapping approach. The latter is based on an arithmetic counting of side-chain nonpolar groups in protein microenvironments disregarding their relative positions, while the former is rooted in the physical properties of interfacial biological water, providing a more rigorous approach.

To engineer an affinity discriminator for C-Kit following the de-wetting blueprint, we compare the patterns of residence times of water molecules that solvate the aligned interfacial regions of PDB-reported imatinib targets Bcr-Abl (PDB.1FPU), C-Kit (PDB.1T46), and Lck (lymphocyte-specific tyrosine kinase, PDB.3LCK) (Fig. 8.2a–c, generated as indicated in Chap. 4. Thus, we introduce a descriptor of hydration tightness for targeted soluble proteins defined as the mean residence time of hydrating molecules within a domain around each residue on the protein surface. Hydrating molecules with short residence times constitute our blueprint for ligand re-engineering since they signal a local propensity for water removal. *The crux of the redesign strategy is then the sculpting in the ligand of nonconserved local de-wetting propensities in the aligned targets: the ligand is engineered to remove interfacial water upon association according to weaknesses in the target hydration shell* (see Fig. 4.5). This blueprint is typically not conserved across drug–targets (cf. Fig. 8.2a), enabling us to modulate the inhibitory impact of the drug and funnel it on the realm of therapeutic relevance. The key de-wetting hot spots correspond to amide–carbonyl backbone hydrogen bonds pairing backbone-exposed residues [13]. As noted in Chap. 4, such bonds become energetically enhanced and stabilized upon removal of surrounding water and thus constitute de-wetting sites.

A de-wetting hot spot in C-Kit not conserved in the alternative targets corresponds to the residue pair C673–G676, which is mapped into the M318–G321 pair in Bcr-Abl and the M319–G322 pair in Lck (Fig. 8.2a–c). Thus, this local difference in de-wetting propensity prompted us to conceive and synthesize a methylated variant of imatinib, named WBZ_4 [14] (Fig. 8.2d, e). To achieve a higher level of favorable and selective interactivity required an exogenous modulation of the target microenvironment, in turn, requiring a modification of the parental compound. As dictated by the de-wetting blueprint, the added methyl group would promote the favored dehydration of the pair C673–G676 upon binding to C-Kit, while it would hamper the association with Bcr-Abl and Lck, since the latter kinases favor hydration of the nucleotide-binding loop residues aligned with the targeted C-Kit residues C673–G676.

A comparison of the aligned dehydron matrices (see also Chap. 7) of the three imatinib-targeted kinases restricted to the imatinib-binding region singles out the C673–C676 dehydron in C-Kit as the selectivity-promoting feature (red circles in

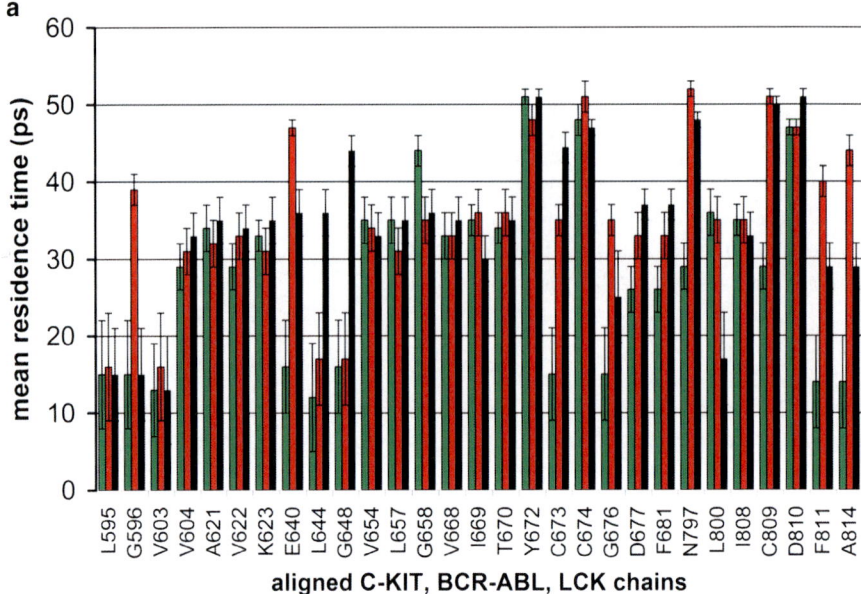

a

y-axis: mean residence time (ps)

x-axis: aligned C-KIT, BCR-ABL, LCK chains

(residue labels: L595, G596, V603, V604, A621, V622, K623, E640, L644, G648, V654, L657, G658, V668, I669, T670, Y672, C673, C674, G676, D677, F681, N797, L800, I808, C809, D810, F811, A814)

Fig. 8.2 (**a**) De-wetting propensities of C-Kit residues in contact with imatinib (PDB.1T46, *green bars*) and of aligned residues in Bcr-Abl kinase (PDB.1FPU, *red bars*) and Lck (PDB.3LCK, *black bars*). Residue *i* is in contact with the ligand if an atom of the latter lies within its domain $D(i)$. The de-wetting propensity is quantified by the mean lifetime of solvating water molecules. Error bars denote Gaussian dispersion over 5 MD (molecular dynamics) runs. (**b**) Pattern of de-wetting hot spots that arise as backbone hydrogen-bonded residues in C-Kit kinase (ribbon backbone representation in *light blue* and de-wetting hot spots in *green*), aligned with Bcr-Abl kinase (*magenta* backbone and de-wetting hot spots in *red*). The imatinib methylation site leading to the expulsion of water from C-Kit C673–G676 de-wetting hot spot is highlighted (*yellow rectangle*). Imatinib is thus modified to favorably expel interfacial water molecules from the C-Kit microenvironment, a feature not conserved in Bcr-Abl. Hydrogen bonds are represented as segments joining α-carbons of the paired residues and those lacking the propensity for dehydration are shown as *light gray* segments. Residues from C-Kit chain are labeled in *green* and those from Bcr-Abl are labeled in *white*. (**c**) Differences in de-wetting hot spots upon alignment of C-Kit (*light blue* backbone and hot spots in *green*) and Lck kinase (*gold backbone* and *black hot spots*). Labels for Lck residues are shown in *white*. (**d**) Prototype molecule WBZ_4 (*N*-{5-[4-(4-methyl piperazine methyl)-benzoylamido]-2-methylphenyl}-4-[3-(4-methyl)-pyridyl]-2-pyrimidine amine). The added methyl group is indicated in *red*. (**e**) Total synthesis of WBZ_4 [14]. The wrapping modification (methylation) of imatinib is indicated by the *thick arrow*. (**f–h**) Dehydron matrices of Kit, Abl, and Lck kinases, respectively, derived from PDB entries 1T46, 1FPU, and 3LCK and restricted to the residues in contact with imatinib. Rows and columns are labeled by residue numbers, well-wrapped bonds are displayed in *gray*, and dehydrons are shown in *green, red,* and *black* for Kit, Abl, and Lck, respectively. The entry denoting a selectivity-promoting feature for the wrapping design is circled in *red*. (**a–e**) reprinted from the [14], copyright 2007 with permission from the American Society for Clinical Investigation

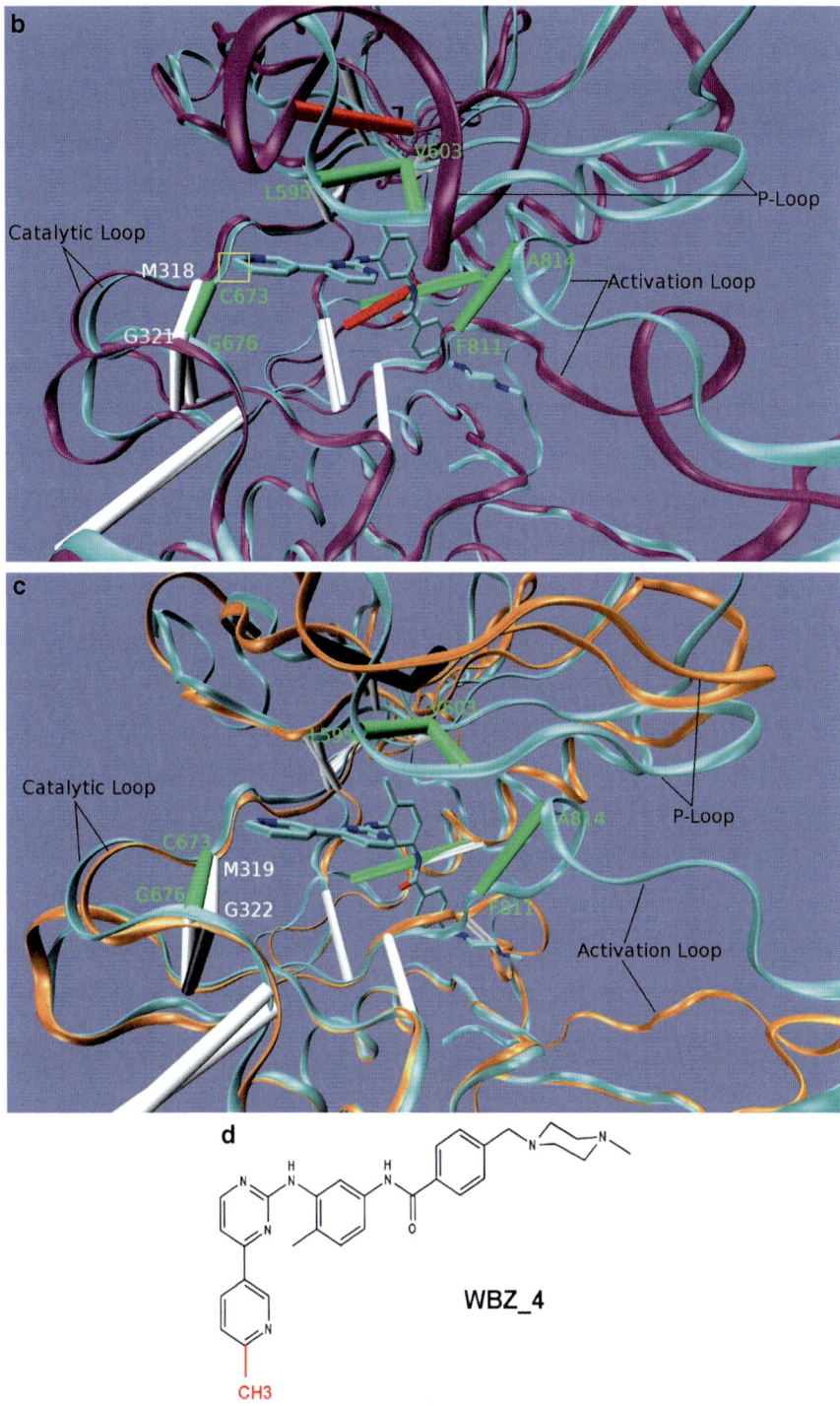

Fig. 8.2 (continued)

e

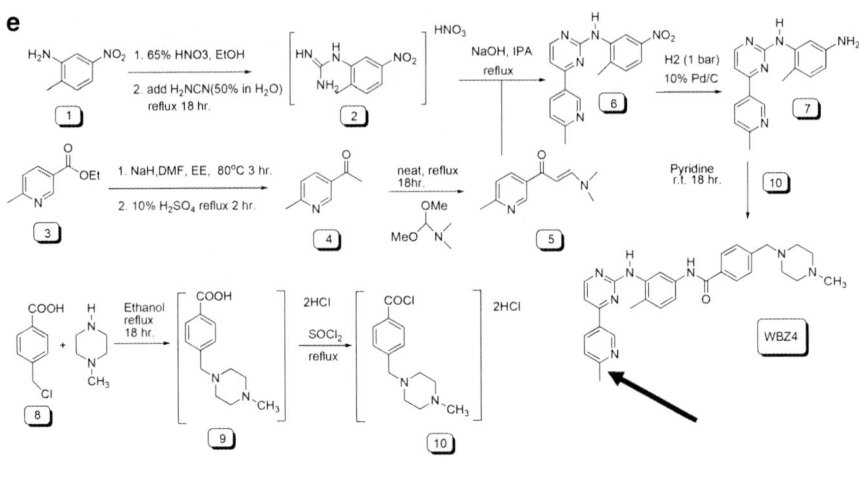

f

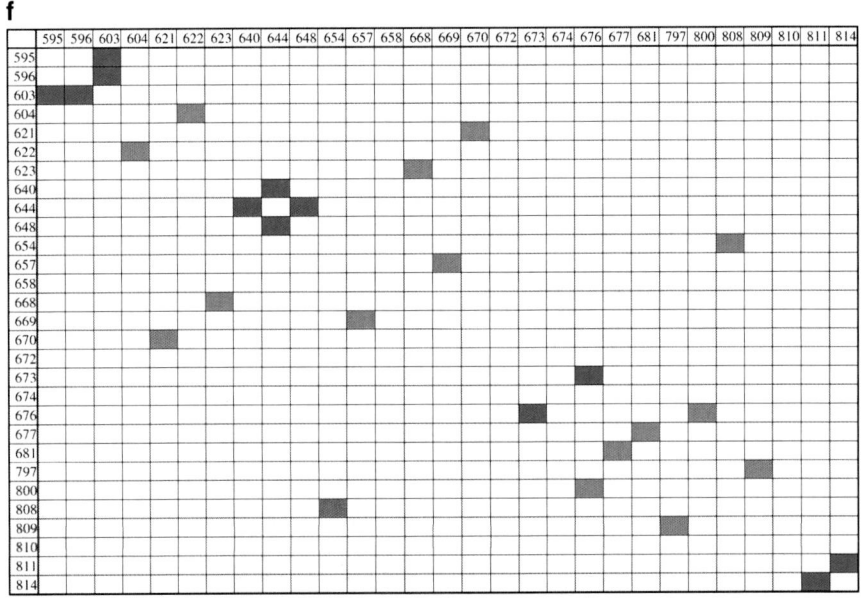

Fig. 8.2 (continued)

g

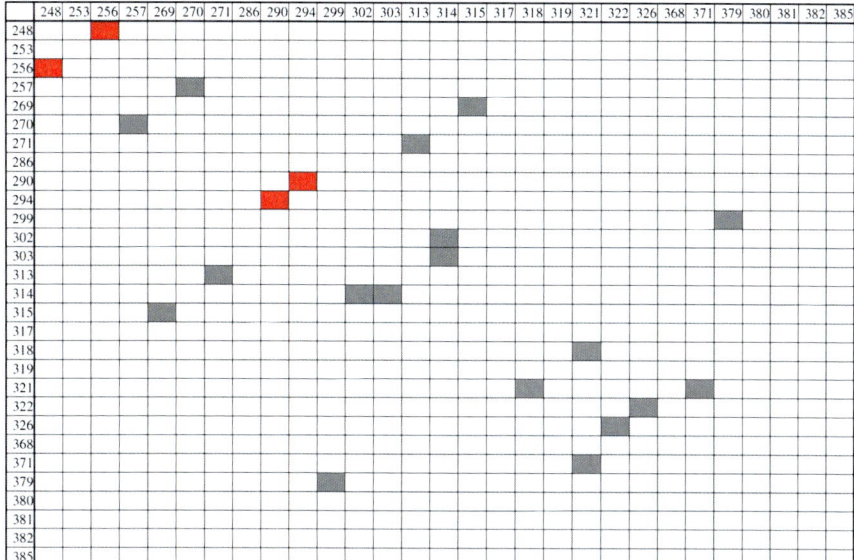

h

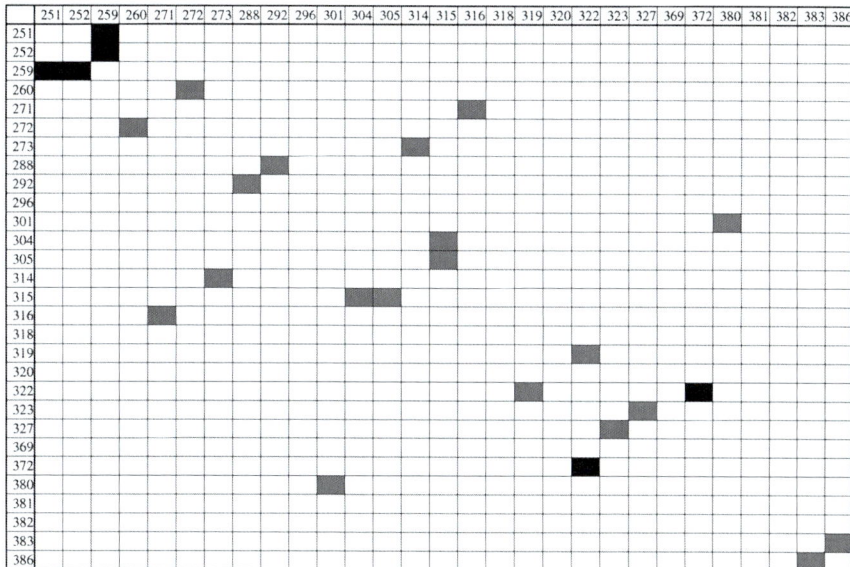

Fig. 8.2 (continued)

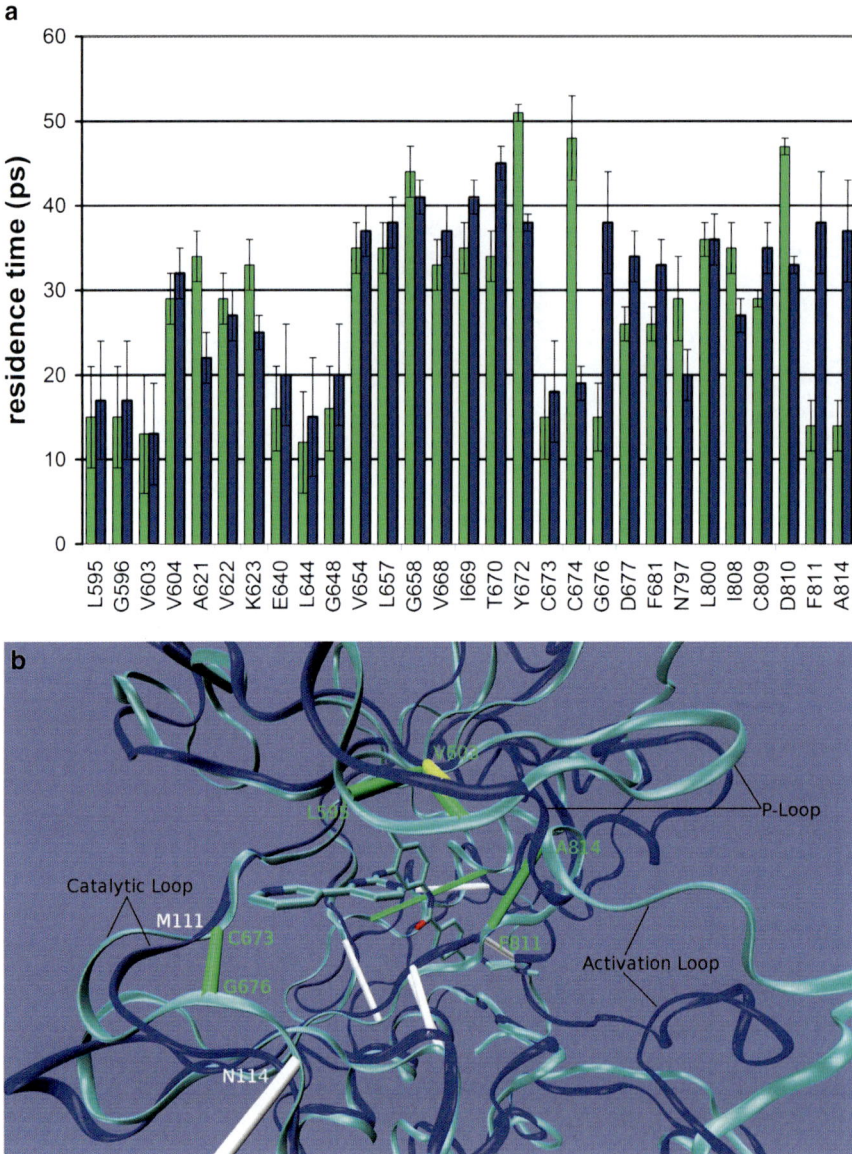

Fig. 8.3 (**a**) Aligned de-wetting patterns for C-Kit kinase (*green*) and JNK1 (*blue*) restricted to the C-Kit residues in contact with imatinib. (**b**) De-wetting hot spots arising as backbone hydrogen-bonded residues in C-Kit kinase (backbone in *light blue* and de-wetting hot spots in *green*), aligned with JNK (*blue* backbone and de-wetting hot spots in *yellow*). The JNK residues M111 and N114, aligned with the de-wetting hot spot C673–G676 in C-Kit, are not paired by a hydrogen bond. Yet, M111 is a de-wetting hot spot for JNK (**a**), hence a harnessing spot for the designed imatininb modification. Reprinted from the [14], copyright 2007 with permission from the American Society for Clinical Investigation

Fig. 8.2f–h), in good agreement with the de-wetting analysis. The dehydron matrix is precisely the representational tool that enables the simplest and fastest in silico assessment of drug cross-reactivity across the entire human kinome, as described in Chap. 9.

Reported assays on imatinib-promoted cardiomyocyte damage suggest that JNK (JNK1) inhibition may be beneficial to reduce the induced collapse of mitochondrial membrane potential and preserve sarcolemmal integrity [15]. For this reason, we tested the binding of WBZ_4 as it anchors on the de-wetting pattern of JNK1 (PDB.2G01). This analysis required alignment of JNK1 with the imatinib–C-Kit complex (PDB.1T46) (Fig. 8.3a, b). The high de-wetting propensity at JNK1 residue M111, the residue that aligns with C673 in C-Kit, instills confidence in the affinity of WBZ_4 for JNK.

8.3 In Silico Assays of the Water-Displacing Efficacy of a Wrapping Drug

As described in Chap. 4, molecular dynamics of the target-wetting patterns has proven to be a useful tool to assess the efficacy of drugs or ligands in displacing interfacial water from the target. In this regard, we have probed WBZ_4 in silico by monitoring the spatio-temporal effect of the specific methylation on the kinase–inhibitor association through molecular dynamics (Fig. 8.4a, b). As C-Kit associates with imatinib, the loop facing the *para*-position on the terminal ring (Figs. 8.2b and 8.4a) was found to be unstable. In all five simulation trajectories, the backbone hydrogen bond between Cys673 and Gly676 was irreversibly replaced by water-mediated interactions within 1 ns (Fig. 8.4a). When WBZ_4 replaced imatinib within the C-Kit complex, the same loop was found to be stabilized due to the improved dehydration of the hydrogen bond by the added methyl group on the ligand (Fig. 8.4b). Thus, for C-Kit, WBZ_4 introduces favorable interactions engaging the solvent-exposed pre-formed hydrogen bond and consequently stabilizes the inhibitor binding. The effect of WBZ_4 is the opposite on Bcr-Abl. The simulations demonstrate that in this case, the imatinib complex has a better stability in the same loop region, while WBZ_4 significantly destabilizes the loop. This is due to the fact that the Met318–Gly321 backbone hydrogen bond on the Bcr-Abl loop is well dehydrated intramolecularly and positioned closer to imatinib; thus this bond becomes well protected [17] by imatinib and the addition of the methyl group clashes sterically with the loop. Thus, molecular dynamics provides a rationale for the discriminatory power of WBZ_4 relative to the parent drug.

8.4 High-Throughput Screening: Test-Tube Validation of the Engineered Specificity

Screening methodologies for kinase inhibitors have been revolutionized in their breadth with the advent of bacteriophage libraries of kinase display [6], enabling

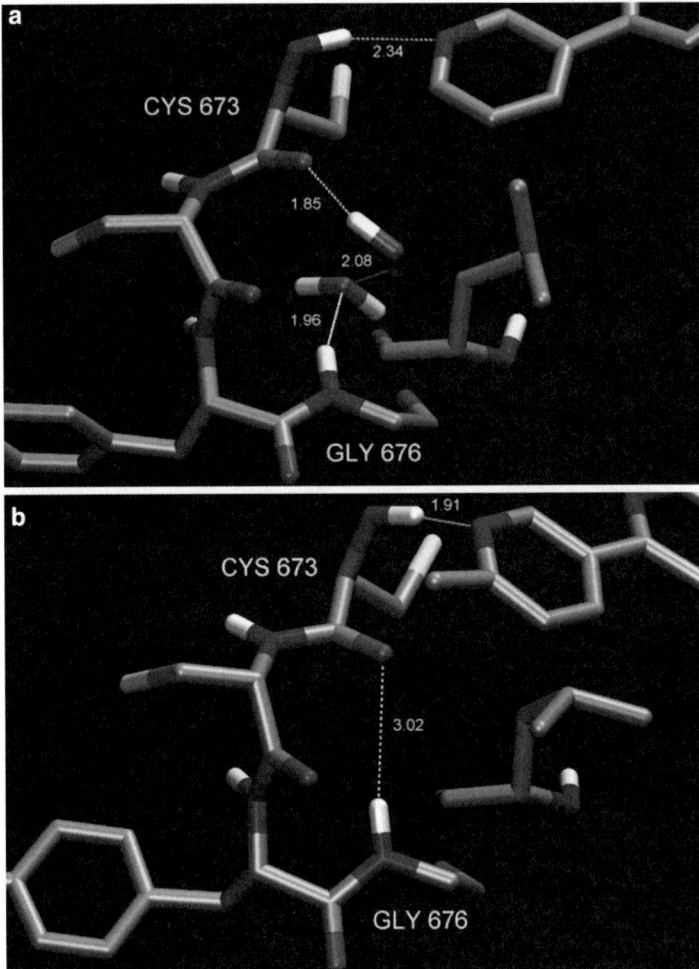

Fig. 8.4 (a) Snapshot of the molecular dynamics simulation for C-Kit kinase bound to imatinib at 1 ns. The main chain hydrogen bond between Cys673 and Gly676 is competitively and irreversibly replaced by hydrogen bonding to a water molecule, revealing the instability of the intramolecular interaction. (b) Snapshot of the simulation of C-Kit kinase in complex with WBZ_4 at 1 ns. The main chain hydrogen bond between Cys673 and Gly676 is stabilized by the water expulsion promoted by the added methyl on the inhibitor. Reprinted from the [14], copyright 2007 with permission from the American Society for Clinical Investigation

a huge amplification of the affinity signal of the test compound. With panels now covering 442 of the 519 kinases that comprise the human kinome, this screening methodology has become the benchmark of high-throughput assays in the test tube. Thus, kinases are expressed on the capside of bacteriophages and this display construct is subject to a competitive binding assay. In the assay, the test compound and the kinase display construct are in solution and the high-affinity and promiscuous

ATP-mimicking ligand is in a fixed phase within a chromatographic column. Thus, the affinity of the test compound is measured relative to that of the fixed phase ligand through its ability to displace the latter and hence enable the kinase display bacteriophage to elute from the column.

To test whether wrapping-based design truly enables the control of specificity, WBZ_4 was screened for affinity against the T7-bacteriophage kinase display library (Fig. 8.5), using imatinib screening as control [14]. The selective affinity of WBZ_4 for C-Kit kinase and JNK1 is noteworthy. Predictably, the affinity of WBZ_4 for ABL1 is reduced by 75% and by 95% or more on all other Abl variants, while, in contrast with imatinib, WBZ_4 shows no detectable affinity for Lck. The impact of WBZ_4 on all additional imatinib targets was comparable, while its controlled specificity is apparent.

8.5 In Vitro Assays: Selectively Modulating Imatinib Impact

As shown in [14], the specificity of WBZ_4 toward C-Kit was established first through in vitro assays that probe the phosphorylation kinetics of competitive inhibition. These spectrophotometric assays couple ADP production, the byproduct of phosphorylation activity, with oxidative steps following the glycolysis cycle [18, 19], as described in Chap. 7. The inhibitory impact of imatinib (triangles) and WBZ_4 (squares) on the rate of phosphorylation was determined by spectrophotometry, assaying for the activity of C-Kit and Bcr-Abl (blue and red plots, respectively, Fig. 8.6). These kinetic assays revealed a high specificity of WBZ_4 for C-Kit, in contrast with imatinib. WBZ_4 enhances the inhibition of C-Kit activity even beyond imatinib levels, revealing a higher competitive affinity of the prototype compound for the adenosine tri phosphate (ATP)-binding pocket (K_I (imatinib) $\approx 55 \pm 7$ nM; K_I (WBZ_4) $\approx 43 \pm 5$ nM). On the other hand, the pattern of inhibition for Bcr-Abl is dependent on the phosphorylation state of this kinase [19, 20]. At 1 μM concentration, imatinib reduces 66% of the activity of Tyr412-phosphorylated kinase [8] ($K_I \approx 5 \pm 1$ μM) and ~100% of the activity of the unphosphorylated state ($K_I \approx 50 \pm 5$ nM). By contrast, 1 μM WBZ_4 reduces by less than 20% the activity of both states of Bcr-Abl. The prototype compound has reciprocal affinity constants $K_I \approx 18 \pm 3$ μM and $K_I \approx 11 \pm 2$ μM, for phosphorylated and unphosphorylated Bcr-Abl, respectively.

In vitro colorimetric assays were performed over a 1 pM to 100 μM range in ligand concentration to assess the inhibition of phosphorylating activity by antibody recognition of phosphorylated peptide substrates [14]. The IC50 (50% inhibition concentration) for imatinib/Abl is ~1 μM, while the WBZ_4/Abl value is above 100 μM (Fig. 8.7a). The active recombinant Abl kinase and its substrate (Abl-tide) were incubated in the presence of various WBZ_4 or imatinib concentrations and ATP (100 nM). Phosphorylation of Abl-tide peptide was detected by spectrophotometry following incubation with phospho-Abl-tide antibodies.

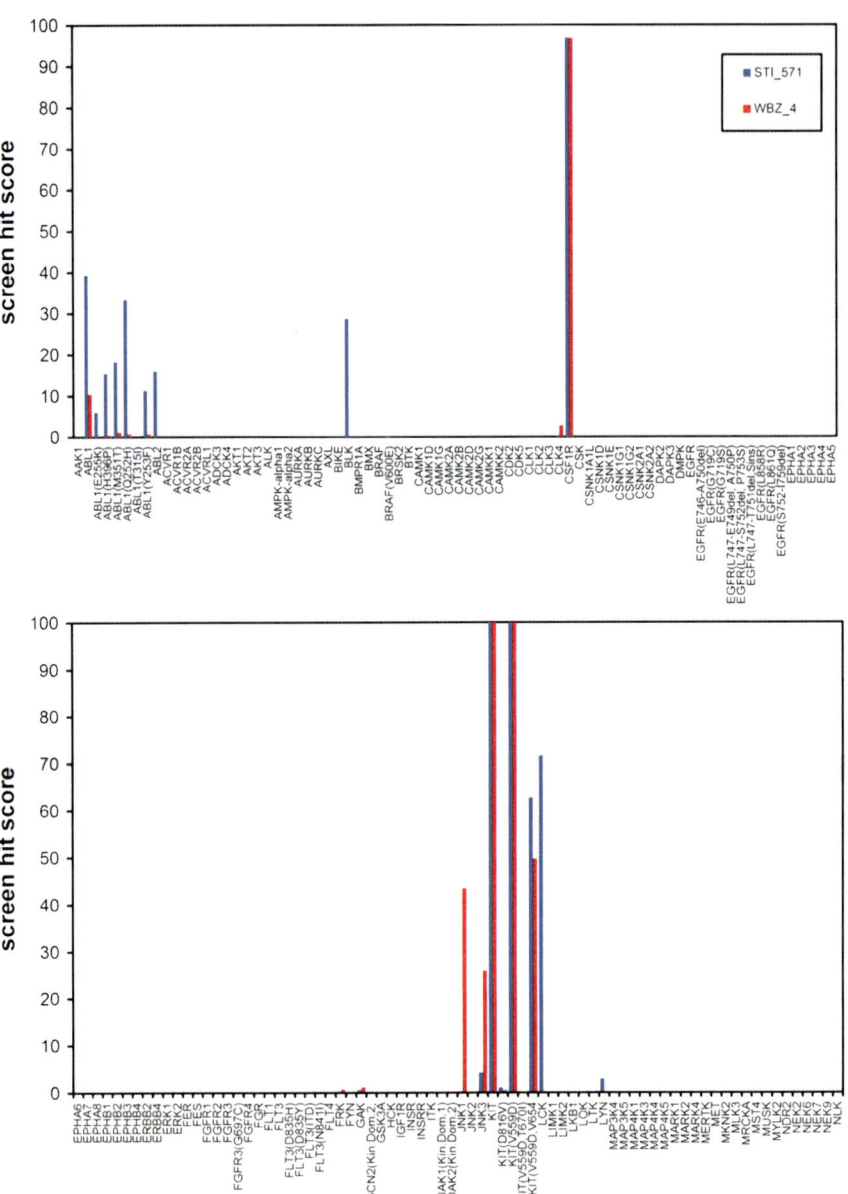

Fig. 8.5 High-throughput screening at 10 mM for WBZ_4 (*red*) and imatinib (STI_571, *blue*, control) over a battery of human kinases displayed in a T7-bacteriophage library (Ambit Biosciences, San Diego, CA). Hit values are reported as percentage-bound kinase. Reprinted from the [14], copyright 2007 with permission from the American Society for Clinical Investigation

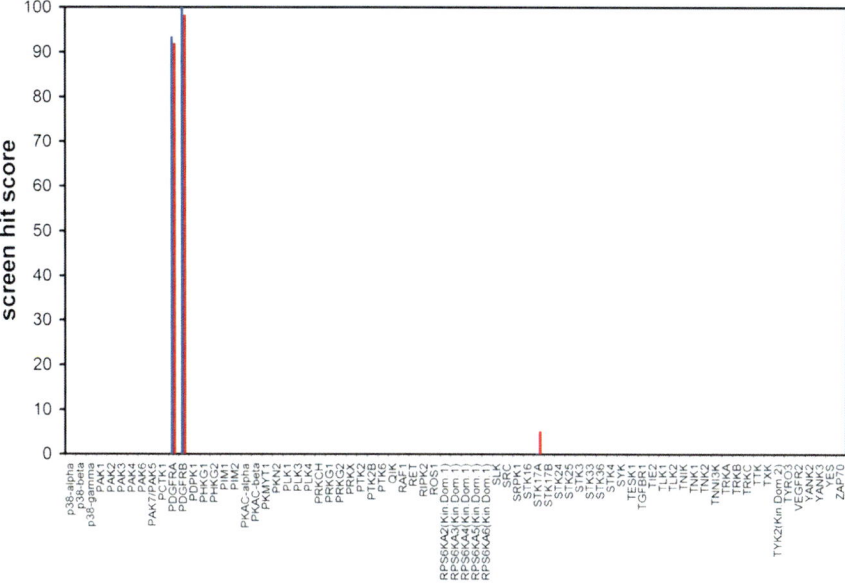

Fig. 8.5   (continued)

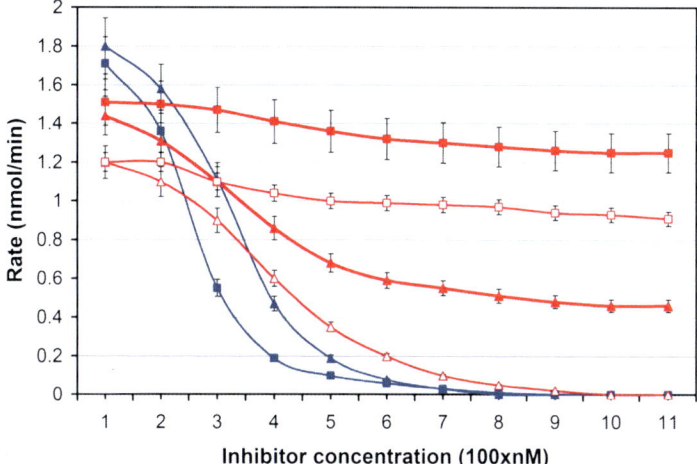

Fig. 8.6   Kinetic inhibitory impact of compounds WBZ_4 and imatinib determined by measuring phosphorylation rates through spectrophotometric assays of C-Kit and Bcr-Abl kinase activity. The kinases are inhibited by WBZ_4 (*squares*) and by the parental compound (*triangles*). Phosphorylation rate plots are given for Bcr-Abl (*red*) and C-Kit (*blue*). The *open red symbols* correspond to inhibition of unphosphorylated Bcr-Abl, while the *full red symbols* correspond to the Tyr412-phosphorylated form. Error bars represent dispersion over five runs for each kinetic assay. Reprinted from the [14], copyright 2007 with permission from the American Society for Clinical Investigation

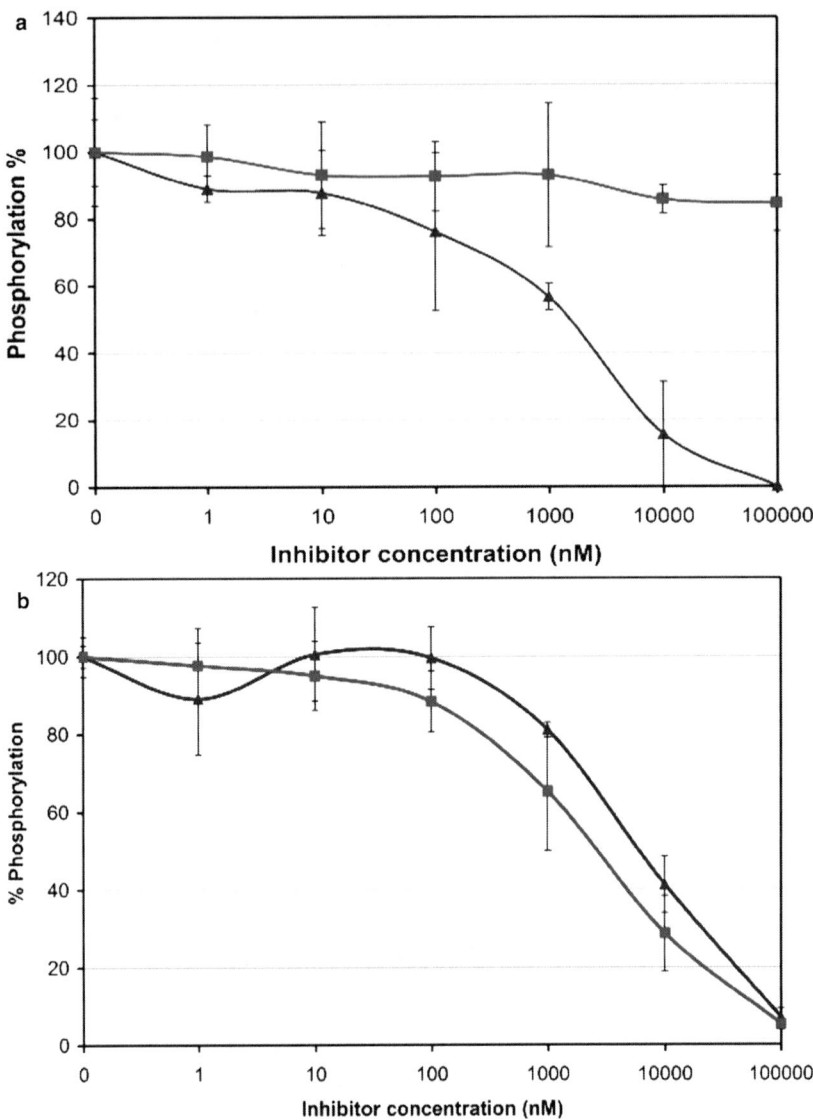

Fig. 8.7 (**a**) In vitro phosphorylation inhibition assay for Abl enzyme in the presence of WBZ_4 (*squares*) or imatinib (*triangles*). Active recombinant Abl enzyme (1 μg/ml) and its substrate (Abl-tide, 1 μg/ml) were incubated for 1 h at 37°C in the presence of various WBZ_4 or imatinib concentrations. ATP (100 nM) was added to the reaction mixture. Phosphorylation of Abl-tide peptide was detected by incubation in consecutive order with antirabbit phospho-Abl-tide antibody and antirabbit horseradish peroxidase (HRP) antibody. Phosphorylation of the substrate was quantified as absorbance units (AU) by spectrophotometry at 450 nm. Values obtained with the enzyme without the inhibitors (WBZ_4 or imatinib) were assumed to be 100% phosphorylation and were compared to the values obtained with the addition of the inhibitors. (**b**) In vitro phospho-rylation inhibition assay for C-Kit in the presence of WBZ_4 (*squares*) or imatinib (*triangles*). Active recombinant C-Kit kinase (25 ng/ml) and its substrate poly (Glu4–Tyr, 150 nM) were

The specificity of the inhibitory impact on C-Kit is significantly enhanced as WBZ_4 substitutes imatinib (Fig. 8.7a, b), while C-Kit inhibition is $23 \pm 12\%$ greater for WBZ_4 (Fig. 8.7b). The impact of the wrapping prototype ligand on alternative imatinib targets, such as the platelet-dependent growth factor receptor (PDGFR) kinase, could not be modulated since such proteins are not reported in PDB and thus, no de-wetting pattern can be reliably identified (cf. high-throughput screening of WBZ_4 versus imatinib in Fig. 8.5). On the other hand, as shown in Chap. 7, the PDGFR kinase is a close paralog of the Kit kinase and hence it is largely expected that any Kit kinase inhibitor would be cross-reactive with the PDGFR kinase.

8.6   In Vitro Assay of the Selective Anticancer Activity of the Wrapping Design

The controlled inhibitory impact of wrapper prototype WBZ_4 was further tested through in vitro assays on cancer cell lines representative of CML and GIST [14]. The initial tests entailed treating the GIST882 [12] cell line and the CML K562 [7] cell line with WBZ_4 and contrasting its inhibitory impact on cell proliferation and in-cell specificity with those associated with imatinib. Because of its relative insolubility, WBZ_4 was incorporated into liposomes to promote cellular delivery. The proliferation of C-Kit-expressing GIST882 cells treated with WBZ_4 was significantly inhibited in a quantitative dose-dependent manner similar to imatinib, with maximum impact at $1~\mu M$ bulk concentration (Fig. 8.8a). By contrast, the inhibitory impact of WBZ_4 on CML K562 cells at the same bulk physiological dose of $1~\mu M$ is almost negligible (cell proliferation less than 10% lower than the proliferation of untreated cells), while $1~\mu M$ imatinib promotes a decrease in cell proliferation of approximately 66% (Fig. 8.8b). These results demonstrate the higher specificity in anticancer activity of WBZ_4.

A western blot assay on treated cancer-derived cell lines (Fig. 8.9a, b) was performed to determine the *in-cell* specificity of WBZ_4. The immunoblots revealed specificity toward C-Kit consistent with the selective anticancer activity on the GIST882 cell line that expresses C-Kit. Thus, the activating phosphorylation of C-Kit at sites Tyr703 and Tyr721 in ST882 cells is inhibited by WBZ_4 in a dose-sensitive manner similar to imatinib (Fig. 8.9a). By contrast, phosphorylation of Bcr-Abl at Tyr 412 (22) in K562 cells was not significantly inhibited (<15%) by WBZ_4, while densitometry revealed an imatinib-induced inhibition of ~85% (Fig. 8.9b).

Fig. 8.7 (continued) incubated for 1 h at 37°C in the presence of various WBZ_4 or imatinib concentrations. ATP (100 nM) was added to the reaction mixture. Phosphorylation of poly (Glu4–Tyr) peptide was detected by incubation in consecutive order with antiphosphotyrosine antibody and antirabbit horseradish peroxidase (HRP) antibody. Reprinted from the [14], copyright 2007 with permission from the American Society for Clinical Investigation

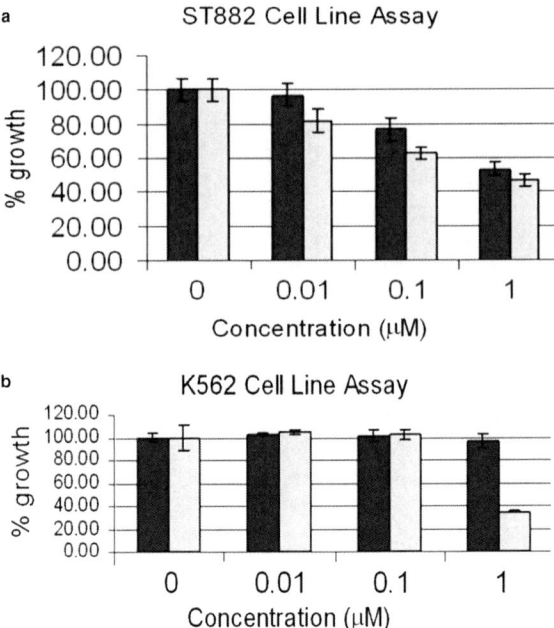

Fig. 8.8 (a) Cell proliferation assay for GIST 882 cells. WBZ_4 inhibits cell proliferation of C-Kit-positive ST-882 cells. GIST cancer cells ST882 (GIST 882) were seeded in 96 wells plates at a density of 8×10^3 cells per well. The cells were treated with various concentrations of WBZ_4 (*dark gray*) and imatinib (*light gray*) for an additional 48 h. Cell proliferation is expressed as the percentage of proliferating cells relative to untreated cells. (**b**) Cell proliferation assay for K562 cells. WBZ_4 does not significantly inhibit cell proliferation of Bcr-Abl-expressing K562 cells. K562 cells were seeded in 96-well plates at a density of 1×10^4 cells per well in 50 μl of medium. Two hours later, 50 μl medium containing different concentrations 0.01, 0.1, 1 μM of liposome-encapsulated WBZ_4 or soluble imatinib were added to the wells to reach a final volume of 100 μl per well. Following 48 h of exposure, the Alamar blue assay was performed. Plates were read at dual wavelength (570–595 nm) in an Elisa plate reader. Reprinted from the [14], copyright 2007 with permission from the American Society for Clinical Investigation

Since imatinib is a micromolar inhibitor of phosphorylated Bcr-Abl (Fig. 8.6), the attack on CML cells at 1 μM bulk concentration (growth decrease 63%, Fig. 8.8b) must be attributed to the inhibition of phosphorylated Bcr-Abl (Fig. 8.9b) combined with the effective nanomolar inhibition of the unphosphorylated form (Fig. 8.6). By contrast, WBZ_4 hinders the phosphorylation of Bcr-Abl only partially (Fig. 8.9b) and in any case, it is an ineffective (micromolar) inhibitor of *both* forms of Bcr-Abl (Fig. 8.6). Hence, its antitumor activity is predictably minimal on CML cells (Fig. 8.8b). In the case of GIST cells, comparable antitumor activity of both compounds (Fig. 8.8a) is likely to arise from comparable inhibitory impact (~85%, Fig. 8.9a) on C-Kit phosphorylation, and the fact that both compounds are nanomolar-affinity inhibitors of C-Kit (Fig. 8.6).

a   ST 882 Cell Line Assay

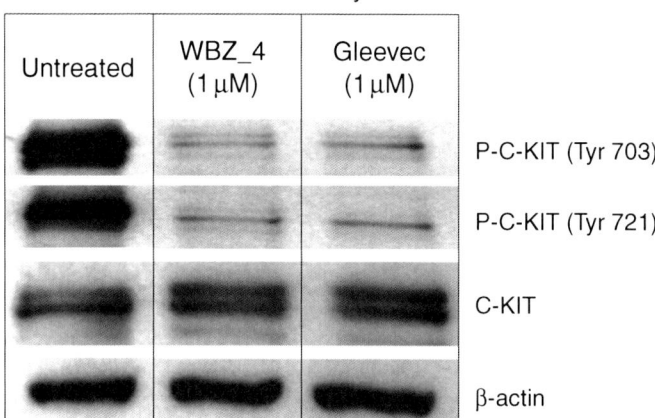

b   K562 Cell line assay

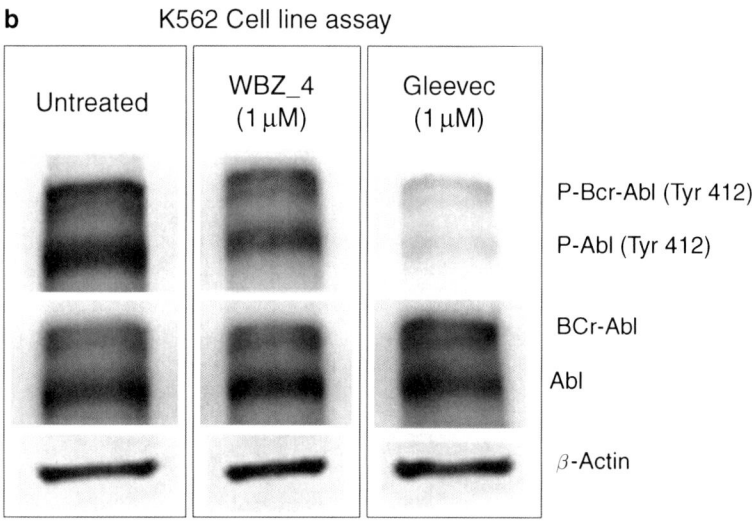

Fig. 8.9 (**a**) Western blot of C-Kit inhibition. WBZ_4 inhibits phosphorylation of C-Kit kinase in ST-882 GIST cells. Gel bands from the western blot assays of C-Kit and its phosphorylated (P) form in GIST cells treated with WBZ_4 and imatinib. The β-actin assay was adopted as control. (**b**) Western blot of Bcr-Abl inhibition. Phosphorylation of Bcr-Abl kinase is not significantly inhibited by WBZ_4 in K562 CML cells. Electrophoretic gel bands for western blots for Bcr-Abl kinase and its phosphorylated (P) form in CML cells treated with WBZ_4 and imatinib. Reprinted from the [14], copyright 2007 with permission from the American Society for Clinical Investigation

8.7 Enhanced Safety of the Wrapping Redesign in Animal Models of Gastrointestinal Stromal Tumor

To assay WBZ_4 for anticancer activity in vivo, an animal model for GIST growth based on female C.B-17/IcrHsd-$Prkdc^{SCID}$ mice was developed de novo [14]. The model involved the subcutaneous injection of GIST882 cells [21]. The efficacy of WBZ_4 was found to be comparable to that of imatinib, as determined by the decrease in tumor volume and weight (Fig. 8.10a, b). No obvious toxicities were observed in the animals during treatment. Selectivity of WBZ_4 in the animal model was corroborated by assaying on the xenograft induced by CML cells K562 (Fig. 8.10c). While imatinib is shown to significantly impair tumor growth ($p < 0.01$), the prototype WBZ_4 has virtually no effect, in accord with its engineered specificity and hence lack of inhibitory impact on Bcr-Abl kinase.

Previous studies of imatinib-induced cardiotoxicity identified a protective JNK (JNK1) inhibition as a means to reduce the collapse of mitochondrial membrane

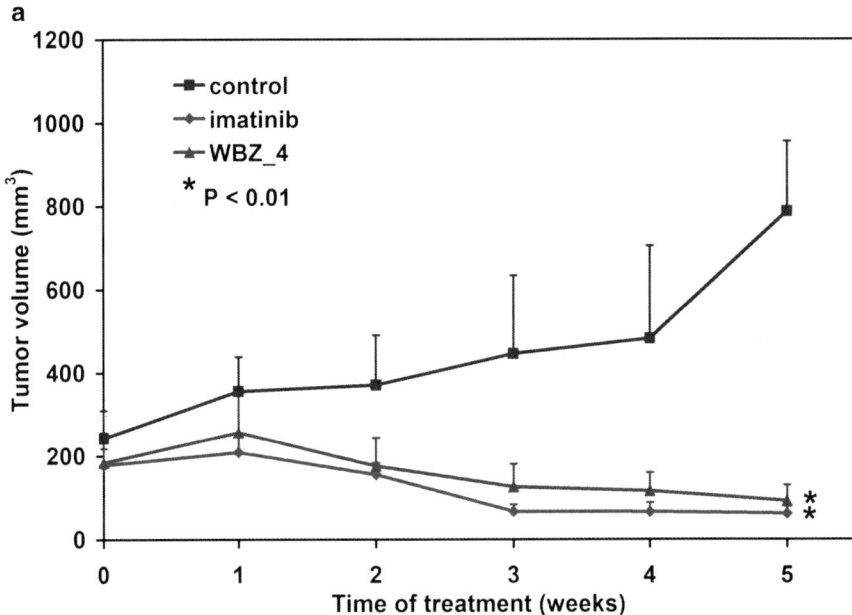

Fig. 8.10 Xenograft models of anticancer activity. (**a**) Effect of WBZ_4 or imatinib therapy on in vivo GIST growth determined by longitudinal tumor volume measurements. Mice were randomized to treatment with either control (normal PBS and empty liposomes give indistinguishable results within experimental uncertainty), imatinib, or liposome-formulated WBZ_4. (**b**) Effect of WBZ_4 or imatinib therapy on in vivo GIST growth determined by weight measurements. Animals from all groups were sacrificed after 6 weeks of therapy, tumors were excised, and the weight was recorded. (**c**) Effect of WBZ_4 or imatinib therapy on in vivo CML growth induced through a xenograft of K562 tumor cells, determined by longitudinal tumor volume measurement. The WBZ_4 selectivity is hereby corroborated in vivo. Reprinted from the [14], copyright 2007 with permission from the American Society for Clinical Investigation

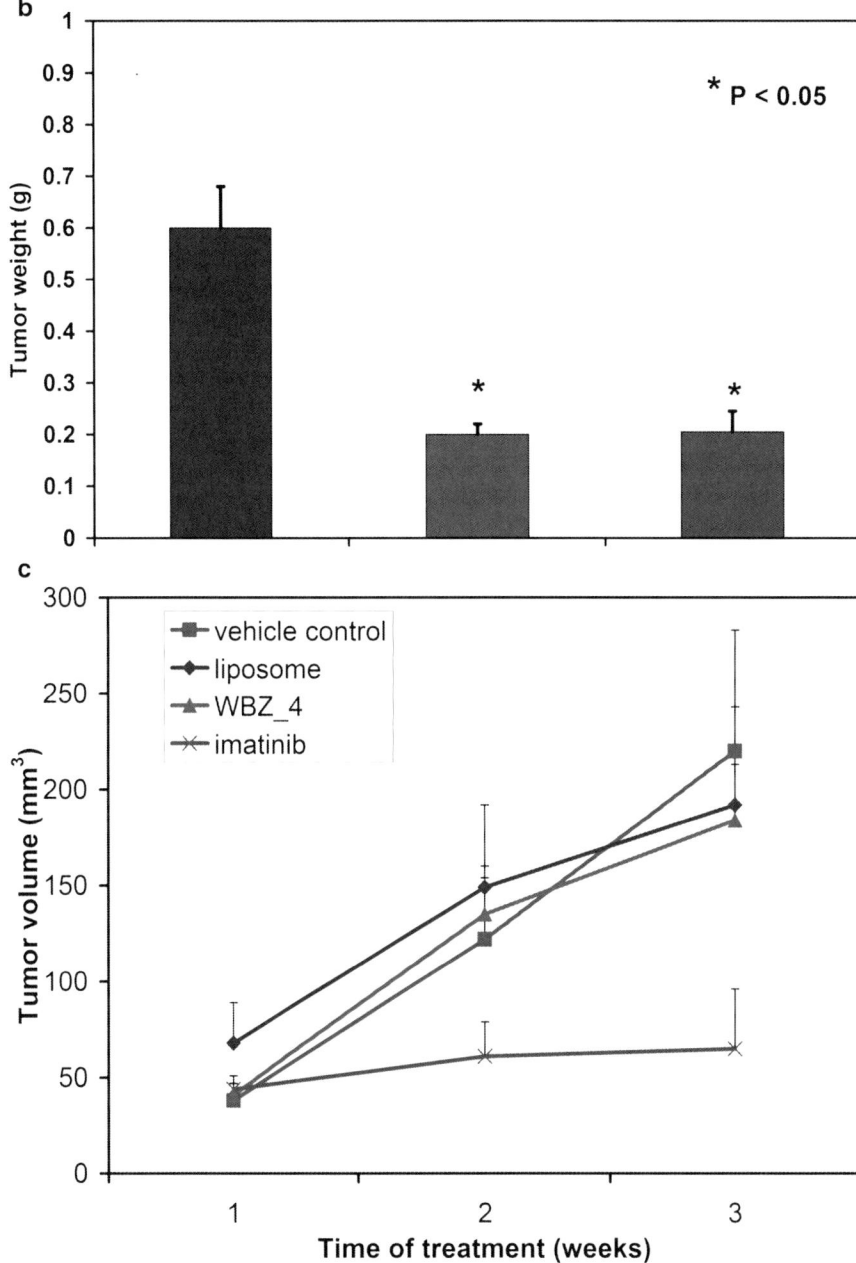

Fig. 8.10   (continued)

potential [15]. Accordingly, WBZ_4 was engineered to also inhibit JNK, a feature preliminarily corroborated in vitro (Fig. 8.5). This protective role of WBZ_4
is also validated in vivo. Western blotting of cell extracts derived from the same
imatinib- and WBZ_4-treated cardiomyocytes (Fig. 8.11a) revealed that both drugs
increased the phosphorylation state of ERK1/2 and p38^{MAPK} as compared to
untreated cultures. In contrast, cells treated with WBZ_4 showed reduced levels
of JNK activation at each concentration tested, as compared to untreated cells, or
cells treated with imatinib. These results confirm other observations [15] demon-
strating that imatinib induces the endoplasmic reticulum (ER) stress response and
activates JNKs in neonatal rat ventricular myocytes (NRVM). JNK activation has
been clearly associated with cardiomyocyte apoptosis via collapse of the mitochondrial membrane potential and release of cytochrome c [22]. It was further confirmed
that imatinib (and WBZ_4) had little effect on p38^{MAPK} activation at concentrations below 50 μM and actually increased ERK activation in these cultured cells.
In contrast, WBZ_4 appeared to reduce levels of pJNK at each dose tested, suggesting that upstream activation of JNKs was inhibited by WBZ_4. Thus, a direct
inhibitory effect on JNK activity would predict that WBZ_4 should have reduced
cardiotoxicity in vivo.

This expectation is supported by other observations using the surrogate
marker brain natriuretic peptide (*BNP*), a most sensitive indicator of myocardial
hypertrophy and cardiac impairment [23]. The expected curbing of cardiotoxicity in WBZ_4 anticancer therapy was confirmed by examining the mRNA levels
of BNP in the left ventricle of mice from the same groups assayed for anticancer
activity. The *BNP* mRNA levels were about 58% higher in the ventricles from
imatinib-treated animals (Fig. 8.11b), while no significant difference was detected

---

Fig. 8.11 Cardiotoxicity assays. (**a**) Western blot of JNK inhibition in cardiomyocytes. Western
blots were probed with primary antibodies specific for the phosphorylated forms of ERKs, JNKs,
and p38^{MAPK}. The position of molecular weight standards is indicated to the left of each blot. (**b**)
Effect of WBZ_4 or imatinib therapy on mouse heart–brain natriuretic peptide (*BNP*). The mRNA
levels of *BNP* (a sensitive marker of myocardial hypertrophy and cardiac impairment) were examined in the left ventricle of mice from the groups in Fig. 8.10a. The *BNP* mRNA levels were about
58% higher in the ventricles from imatinib-treated animals ($p = 0.02$), but no significant difference
was noted in the WBZ_4-treated animals. (**c**) Comparison of left ventricular ejection fraction after
6 weeks of control (groups treated with either PBS or empty liposomes), imatinib, or WBZ_4 therapy in mice. (**d**) Magnetic resonance imaging of cardiac function. Assessment of left ventricular
function (*red circle, lower cavity*) was performed as follows: representative in vivo axial images
of the left ventricle in diastole and systole of control and test mice were acquired in a 4.7 T MR
scanner cardiac gating using a magnetization-prepared spoiled gradient echo sequence. To assess
cardiac function, short axis cardiac cine images were acquired using a magnetization-prepared,
cardiac-gated spoiled gradient echo sequence (TE/TR 2.1 ms/~23.5 ms; 16 phases covering 1.5
R–R cycles; in-plane resolution 312 μm × 312 μm; 1.25 mm slice thickness). For assessing EF,
regions of interest encompassing the cavity of the left ventricle were drawn. For each animal, the
average region of interest of four central slices in systole was divided by the average region of
interest of four central slices in diastole taken at the same location in the left ventricle to derive the
EF. (**a–c**) reprinted from the [14], copyright 2007 with permission from the American Society for
Clinical Investigation

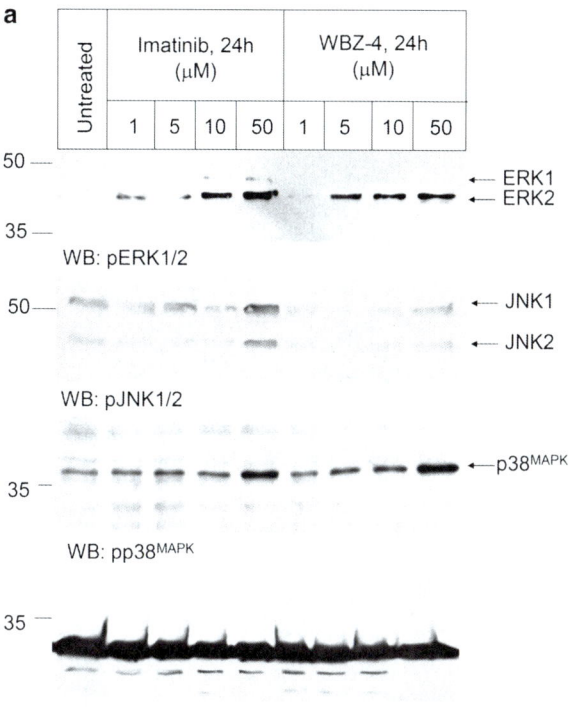

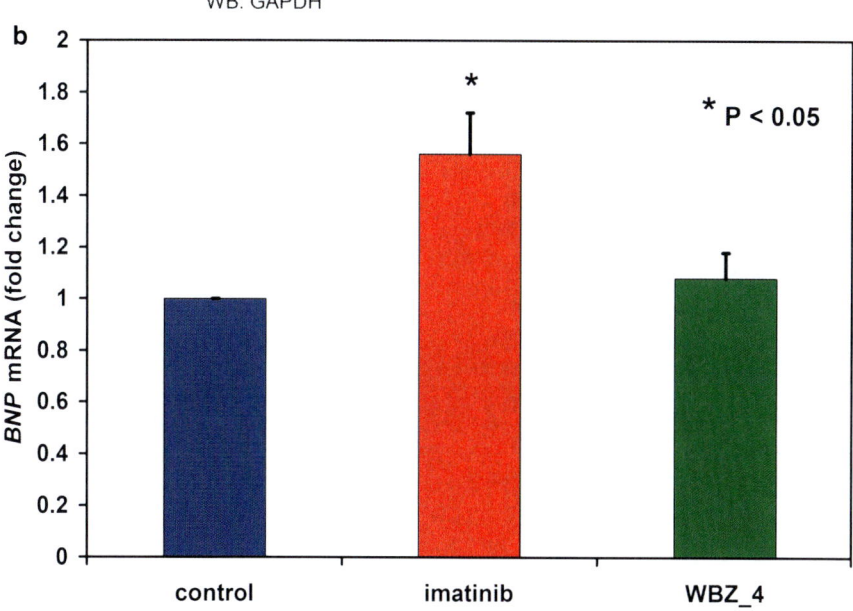

Fig. 8.11 (continued)

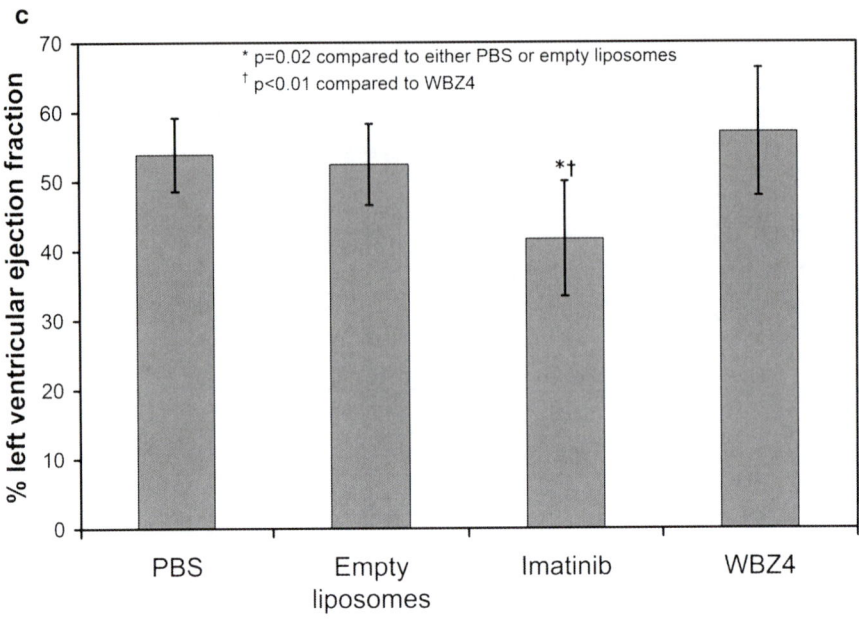

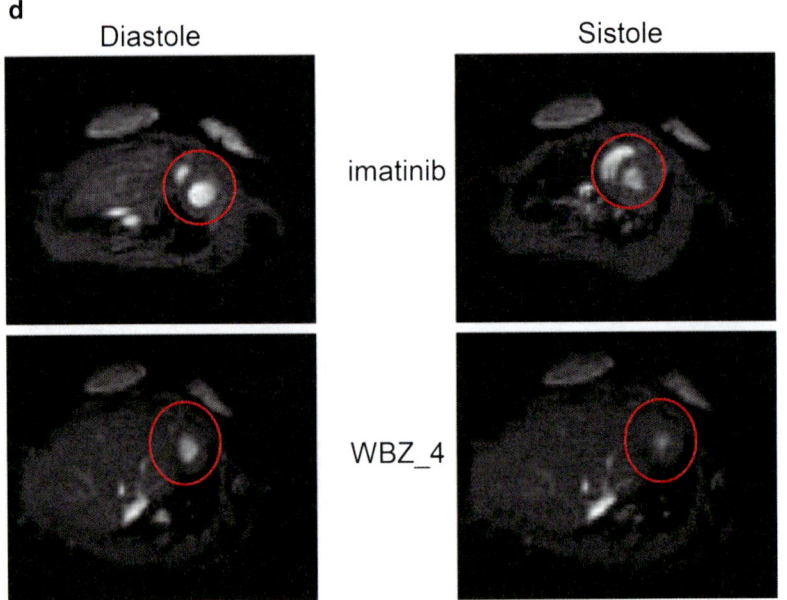

Fig. 8.11 (continued)

in the WBZ_4-treated animals when compared with untreated mice. Finally, the reduced cardiotoxicity of WBZ_4 was directly tested in mice by determination of the percentage ejection fraction (EF) in the left ventricle by magnetic resonance (MR) imaging of treated mice subject to imatinib or WBZ_4 therapy (Fig. 8.11c, d). Following 6 weeks of therapy, the cardiac EF was significantly lower in the imatinib group compared to controls ($p = 0.02$), which is consistent with previous findings [15]. Remarkably, WBZ_4 treatment had no effect on cardiac EF despite prolonged therapy (Fig. 8.11c, d).

8.8 Controlled Specificity Engineered Through Rational Design: Concluding Remarks

The top-down approach presented in this chapter is centered in a rational ligand re-design sculpting target-specific molecular features. The approach harnesses fundamental concepts pertaining to the discrimination of proteins with a common ancestry, sketched in Chaps. 4 and 6. These discriminatory features are associated with unique dehydration or de-wetting hot spots found on the protein/water interface of the purported target. The de-wetting propensities are in turn promoted by dehydrons in the protein structure, and hence the hydration blueprint of the target becomes a design tool complementing the wrapping concept. In this particular illustration, the re-design had three goals: (a) to re-focus the primary impact on C-Kit kinase; (b) to reduce the inhibitory impact on Bcr-Abl kinase; and (c) to promote JNK inhibition in order to reduce cardiotoxicity. These three goals have been achieved [14], as attested by the in vitro and in vivo results described in this chapter, clearly revealing that *drug specificity can be rationally controlled*. The imatinib re-design turned the wrapping ligand into a selective proto-therapeutic agent to treat GIST with a prima facie reduction of cardiotoxicity.

References

1. Dancey J, Sausville EA (2003) Issues and progress with protein kinase inhibitors for cancer treatment. Nat Rev Drug Discov 2:296–313
2. Levitski A, Gazit A (1995) Tyrosine kinase inhibition: An approach to drug development. Science 267:1782–1788
3. Tibes R, Trent J, Kurzrock R (2005) Tyrosine kinase inhibitors and the dawn of molecular cancer therapeutics. Annu Rev Pharmacol Toxicol 45:357–84
4. Gibbs J, Oliff A (1994) Pharmaceutical research in molecular oncology. Cell 79: 193–198
5. Donato NJ, Talpaz M (2000) Clinical use of tyrosine kinase inhibitors: Therapy for chronic myelogenous leukemia and other cancers. Clin Cancer Res 6:2965–2966
6. Fabian MA, Biggs WH, Treiber DK et al (2005) A small molecule kinase interaction map for clinical kinase inhibitors. Nat Biotechnol 23:329–336
7. Gambacorti-Passerini C, le Coutre P, Mologni L et al (1997) Inhibition of the ABL kinase activity blocks the proliferation of BCR/ABL+ leukemic cells and induces apoptosis. Blood Cells Mol Dis 23:380–394

8. Schindler T, Bornmann W, Pellicena P et al (2000) Structural mechanism for STI-571 inhibition of Abelson tyrosine kinase. Science 289:1938–1942
9. Attoub S, Rivat C, Rodrigues S et al (2002) The C-kit tyrosine kinase inhibitor STI-571 for colorectal cancer therapy. Cancer Res 62:4879–4883
10. DeMatteo RP (2002) The GIST of targeted cancer therapy: A tumor (gastrointestinal stromal tumor), a mutated gene (c-kit), and a molecular inhibitor (STI571). Ann Surg Oncol 9: 831–839
11. Skene RJ, Kraus ML, Scheibe DN et al (2004) Structural basis for autoinhibition and STI-571 inhibition of C-kit tyrosine kinase. J Biol Chem 279:31655–31663
12. Tuveson DA, Willis NA, Jacks T et al (2001) STI571 inactivation of the gastrointestinal stromal tumor C-kit oncoprotein: Biological and clinical implications. Oncogene 20:5054–5058
13. Chen JP, Zhang X, Fernández A (2007) Molecular basis for specificity in the druggable kinome: Sequence-based analysis. Bioinformatics 23:563–572
14. Fernández A, Sanguino A, Peng Z et al (2007) An anticancer C-kit kinase inhibitor is re-engineered to make it more active and less cardiotoxic. J Clin Invest 117:4044–4054
15. Kerkela R, Grazette L, Yacobi R et al (2006) Cardiotoxicity of the cancer therapeutic agent imatinib mesylate. Nat Med 12:908–916
16. Druker BJ (2004) Molecularly targeted therapy: Have the floodgates opened? Oncologist 9:357–360
17. Fernández A, Scheraga HA (2003) Insufficiently dehydrated hydrogen bonds as determinants of protein interactions. Proc Natl Acad Sci USA 100:113–118
18. Barker S, Kassel D, Weigl D et al (1995) Characterization of pp[60c-src] tyrosine kinase activities using a continuous assay: Autoactivation of the enzyme is an intermolecular autophosphorylation process. Biochemistry 34:14843–14851
19. Songyang Z, Carraway KL, Eck M et al (1995) Catalytic specificity of protein-tyrosine kinases is critical for selective signalling. Nature 373:536–539
20. Clarkson B, Strife A, Wisniewski D, Lambek CL, Liu C (2003) Chronic myelogenous leukemia as a paradigm of early cancer and possible curative strategy. Leukemia 17: 1211–1262
21. Prenen H, Deroose C, Vermaelen P, Sciot R, Debiec-Rychter M (2006) Establishment of a mouse gastrointestinal stromal tumour model and evaluation of response to Imatinib by small animal positron emission tomography. Anticancer Res 26:1247–1252
22. Baines CP, Molkentin JD (2005) Stress signaling pathways that modulate cardiac myocyte apoptosis. J Mol Cell Cardiol 38:47–62
23. Scheuermann-Freestone M, Simon Freestone N, Langenickel T et al (2001) A new model of congestive heart failure in the mouse due to chronic volume overload. Eur J Heart Fail 3:535–543

Chapter 9
Wrapping Patterns as Universal Markers for Specificity in the Therapeutic Interference with Signaling Pathways

Chapter 8 illustrated the power of the wrapping concept as a guidance to engineer specificity and enhance safety in a kinase inhibitor. Yet, only a universal selectivity filter, applicable to the entire human kinome – even to kinases with unreported structure – would be truly useful for the drug designer. This chapter thematically belongs to the bioinformatics realm and addresses this issue at the broadest possible level. The surveyed findings reveal that targeting the polar dehydration patterns defined by protein dehydrons heralds a new generation of drugs that enable a tighter control of specificity. The universality of this selectivity filter in the field of therapeutic interference with cell signaling is thus established.

9.1 The Need for a Universal Selectivity Filter for Rationally Designed Kinase Inhibitors

As indicated in Chaps. 7 and 8, kinase targeting is a central theme of molecular cancer therapy [1–5]. Be as it may, a structural basis for the rational design of kinase inhibitors appears to be sorely absent [1, 3]. In practice, most leads and scaffolds for drug development are actually discovered through screening techniques and are generally transformed into therapeutic tools through a painstaking trial-and-error process that often renders a final product plagued with clinical uncertainties.

While the paradigm of target specificity may be shifting to controlled multi-target drugs [3], the structural factors determining cross-reactivities are not yet fully understood, in spite of some progress. For instance, the solvent-accessible nonpolar surface on the surface of a protein, frequently invoked as a major determinant of protein associations [6, 7], actually fosters promiscuity in the context of kinase inhibition [3, 8]. This is so because nonpolar groups on the surface of proteins tend to be highly conserved across paralog proteins and even within the entire kinase superfamily [9, 10]. Progress in controlling specificity is also hampered by the lack of structural information. For instance, only about 22% of the human kinome is represented in the PDB.

In this chapter we focus on assessing the universality of wrapping as the molecular feature controlling nonpromiscuous drug targeting. To fulfill this objective an extensive bioinformatics exploration of drug profiles is required. Recent high-throughput screenings of commercially available kinase inhibitors lead us to an operational definition of promiscuity as a significant cross-reactivity (dissociation constant K_d < 100 nM) extended over >30% of the sampled kinome. The problem is complicated due to the scarcity of affinity-profiled kinases with reported structure: To date, only about 24–25% of the human kinome (123 out of the ~519 discovered protein kinases) is reported in the PDB. Furthermore, kinase homology models are quite unreliable to make inferences on such detailed features as wrapping patterns since the level of pairwise sequence identity across the kinase superfamily is typically low (<30%) [11].

To mitigate these problems, we take advantage of the high degree of topological conservation of the kinase folds, a known consequence of their common ancestry. Thus, reliable sequence-based attributes such as disorder propensity [12] are here mapped onto structurally threaded models to make inferences about dehydron patterns and ultimately about drug specificity/promiscuity patterns. As described in Chap. 5, these cross-validations of dehydron inferences in proteins with unreported structure take advantage of the crucial link between disorder propensity and the degree of exposure of the protein backbone in the native fold [13]. The latter is itself a measure of the propensity of backbone hydration, the main factor competing with the hydrophobic collapse of the polypeptide chain.

Thus we seek a sequence-based attribute that enables a classification of kinase space in a way that accurately reproduces similarities/differences in affinity profiling against a background of drugs [14, 15]. The classification introduces a partition into disjoint sets of kinases within which pharmacological differences are accurately predicted from molecular differences. This methodology could be implemented because the relevant classification of kinases was performed by comparing the regions deprived of adequate packing or intramolecular dehydration, i.e., the flexible regions, which are precisely the markers differentiating kinases at the structural level [4].

The chapter is organized as follows: First, we determine the type of kinase molecular similarity that promotes promiscuity whenever targeted. Thus, we demonstrate that solvent-exposed nonpolar regions promoting ligand associations foster promiscuity. To draw statistics at a kinomic scale, we formally define a region, the nonpolar hull, enabling comparison of targeted exposed nonpolar regions across different kinases. Then, we examine other targeted features in molecular design aimed at fostering pairwise interactions between ligand and kinase and show that the high degree of conservation of the partner groups on the protein surface does not enable a suitable control of specificity. Subsequently, we introduce a means of calculating dehydration propensities on polar-paired regions on the protein surface and formally define a region, the so-called environmental (wrapping) hull, enabling a comparison of polar dehydration propensities across targets. We demonstrate that this molecular attribute is responsible for controlling specificity. To carry out the analysis at a sequence level, a technique named *environmental alignment* is implemented. This technique enables the identification of residues whose microenvironments are likely to be perturbed by ligand association. Such residues are identified by aligning the kinase sequence against a background of sequences of homologous kinase–ligand complexes reported in the PDB.

9.2   Computational Tool Box for Comparative Analysis of Molecular Attributes Across the Human Kinome

9.2.1   Wrapping Inferences on Proteins with Unreported Structure

To establish the universal validity of wrapping as a selectivity filter, the comparative analysis of molecular attributes in target paralogs cannot be constrained to PDB-reported kinases (32 out of the 119 assayed in [14]). Homology-based predictions of native structures [16, 17] become plausible given the extent of PDB representation ($\sim$24%) of paralogs within the kinase superfamily and also because dehydrons may be directly inferred from sequence (Chap. 5, [13]) and ultimately contrasted with the homology-based predictions for mutual validation. Thus, the disorder-based prediction of dehydrons hinges on the fact that native disorder arises essentially from the impossibility to dehydrate intramolecular hydrogen bonds, as noted in Chap. 5. The choice of disorder score to benchmark homology predictions is an obvious one given the particularly low accuracy of the latter on loopy or flexible regions [10].

The disorder propensity is given by a score determined by the program PONDR® (predictor of native disorder), a neural-network predictor of native disorder. Only 0.4% of more than 900 nonhomologous PDB proteins give false-positive predictions in regions with 40 or more consecutive sites of predicted disorder. Even this 0.4% of false positives is an overestimation, as many disordered regions in monomeric chains become ordered upon ligand binding or in crystal contacts [12]. The false-negative error rate ($\sim$11% for regions of 40 or more consecutive predicted ordered residues) is also compelling in regard to the predictor quality.

The correlation between solvent exposure of hydrogen bonds and disorder propensity implies that it is possible to predict dehydrons directly from sequence [13]: It suffices to determine the PONDR-generated pattern associated with the desired feature. The correlation implies that the propensity to adopt a natively disordered state becomes pronounced for proteins which, due to a chain composition reflecting high hydrophilicity, cannot protect the backbone hydrogen bonds even minimally. Thus, we can infer the existence of dehydrons from the PONDR score (f_d) with 92% accuracy in regions with $f_d > 0.35$ provided such regions are flanked by well-protected regions ($f_d < 0.35$), to ensure the existence of structure. The accuracy of this sequence-based dehydron predictor was established by inferring the location of dehydrons in proteins with reported structure, for which the microenvironment of each hydrogen bond can be determined unambiguously [13]. The false negatives constitute 368 of the 8,215 dehydrons in a PDB database of 1,466 proteins free from structural redundancy and less than 25% sequence identity in pairwise alignment. The false positives correspond to 2,721 of the 133,623 backbone hydrogen bonds examined.

9.2.2 Alignment of Targetable Molecular Features Across the Human Kinome

In order to assess differences in the exposed nonpolar regions of kinase targets that interact with different ligands we define a common region, named the *nonpolar hull*. A residue a is defined as making contact with a ligand L within a PDB-reported complex if a side chain heavy atom (H excluded) is found to be within 3.6 Å (upper bound for any bond length) of a heavy atom in the ligand. The nonpolar hull for protein chain i, $H_{np}(i)$, is dependent on a "structural background set" of chains, $S(i)$, which includes all homolog chains that align with chain i [18] for which there are protein–ligand complexes with reported structure. Thus, the nonpolar hull is defined as $H_{np}(i) = \cup_{j \in S(i)} \Phi_i(R_{np}(j))$, where $\Phi_i(a)$, with $a \in$ chain j, is the residue in chain i that aligns with residue a in chain j, and $R_{np}(j)$ is the set of nonpolar residues in chain j in contact with its respective ligand L_j. For any pair i, j, the following relation holds: $\Phi_i(H_{np}(j)) = H_{np}(i)$, enabling a comparison of kinases by examining differences in nonpolar hulls.

To assess differences in the dehydration propensies of polar regions for purported kinase targets, we introduce a common region for comparison named *environmental hull*. First, the set $R_{env}(j)$ is defined for protein chain j as the set of residues paired by dehydrons in chain j within a protein–ligand complex with reported structure subject to the following condition: ligand L_j contributes to the dehydration of the dehydron, that is, it has some carbonaceous nonpolar group within the dehydration domain of the dehydron. Then, the set $E_{env}(j)$ is defined as the set of residues from chain j that contribute to the dehydration microenvironment of a dehydron contained in $R_{env}(j)$, that is, they are either paired by the dehydron or they contain a side chain nonpolar group within the microenvironment of the dehydron. Then, the environmental hull for chain j, $H_{env}(j)$, is defined as the union of the residues in chain j that align with residues framing the environments of dehydrons that in turn are environmentally affected by ligands in PDB-reported complexes: $H_{env}(j) = \cup_{i \in S(j)} \Phi_j(E_{env}(i))$ (notation has been followed consistently, the structural background $S(j)$ is defined above). As with nonpolar hulls, for any pair i, j, the following relation also holds: $\Phi_i(H_{env}(j)) = H_{env}(i)$. This property is needed to actually compare environments of different proteins.

9.3 Is Wrapping Pharmacologically Relevant? A Bioinformatics Analysis

In this section we report on previous attempts to correlate different molecular attributes of the protein targets with available drug-screening data for a sizable set of kinase targets. Essentially, the scope of this section is to contrast differences in the pharmacological profiles of kinases with structure-based dissimilarities that take into account specific molecular attributes. The large drug/kinase panel experimentally assayed [14] and exploited in our analysis is highly underreported in the

PDB. Consequently, a reliable sequence-based predictor of the relevant molecular attributes of kinases had to be implemented as shown in the previous sections.

We first define a pharmacological distance, d_{phar}, that quantifies differences in the affinity profiling of kinases against a background of available drugs. This metric is the Euclidean distance between affinity vectors with entries given in negative logarithmic or $\Delta G/RT$ units (ΔG = free energy change for protein–ligand association, R = universal gas constant, T = absolute temperature). A positive cutoff value $\Delta G/RT = ln10 \approx 2.3$ is adopted for affinities reported as "no hit" in the screening ($K_{\text{d}} > 10 \ \mu\text{M}$). Figure 9.1a displays the matrix $\mathbf{D}_{\text{phar}} = \left[d_{\text{phar}}(i,j)\right]$ for all pairs (i,j) from the 119 assayed kinases. The affinity profiling adopted included 19 of the 20 drugs originally screened [14]: only the promiscuous staurosporine was excluded since it does not belong to the pharmacology realm.

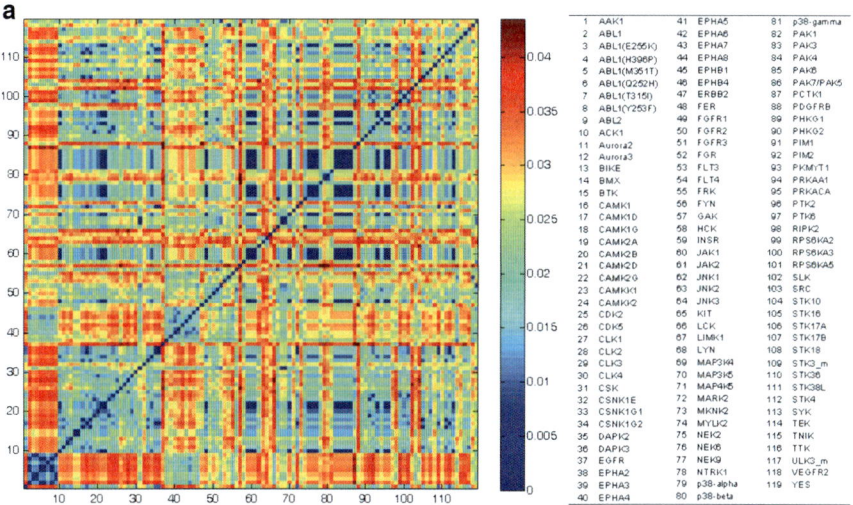

Fig. 9.1 (a) Pharmacological distance matrix $\mathbf{D}_{\text{phar}} = \left[d_{\text{phar}}(i,j)\right]$ for all pairs (i, j) from the 119 kinases assayed through affinity profiling against a background of 19 drugs [14]: SB202190; SB203580; sp600125; imatinib (Gleevec); VX-745; BIRB 796; BAY-43-9006; GW-2016; gefitinib; erlotinib; CI-1033; EKB-569; ZD-6474; Vatalanib; SU11248; MLN-518; LY-333531; roscovitine/CYC202; and flavopiridol. The distance is given by $d_{\text{phar}}(i,j) = \left[\sum_{m \in \text{inhibitors}} (K(i,m) - K(j,m))^2\right]^{1/2}$, where $K(i,m)$, $K(j,m)$ represent, respectively, the negative logarithm of equilibrium constants for complexation of kinase i and kinase j with drug inhibitor m. (b) Aligned backbones $(\text{RMSD} \approx 0.33\text{\AA})$ for paralog kinases PDK1 (*blue*) and CHK1 (*lilac*) in their active folds. The structures were reported in complex with ligands BIM8 (PDB.1UVR) and 3A3 (PDB.2GCU), respectively. The nonpolar hulls are depicted in *yellow*. (c) Nonpolar distance matrix $\mathbf{D}_{\text{np}} = \left[d_{\text{np}}(i,j)\right]$ over the 119 assayed kinases. (d) Plot of nonpolar distance versus pharmacological distance. *Each circle* represents a kinase pair. No correlation is observed, while there is some bimodality in each dimension. (e) Correlation between pseudopharmacological distance (including staurosporine in the drug-screening background) and nonpolar distance between kinases. The sole outliers are pairs involving the EGFR kinase, the kinase whose affinity vector is *not* dominated by staurosporine [14]. Reprinted from [10] with permission from Oxford University press

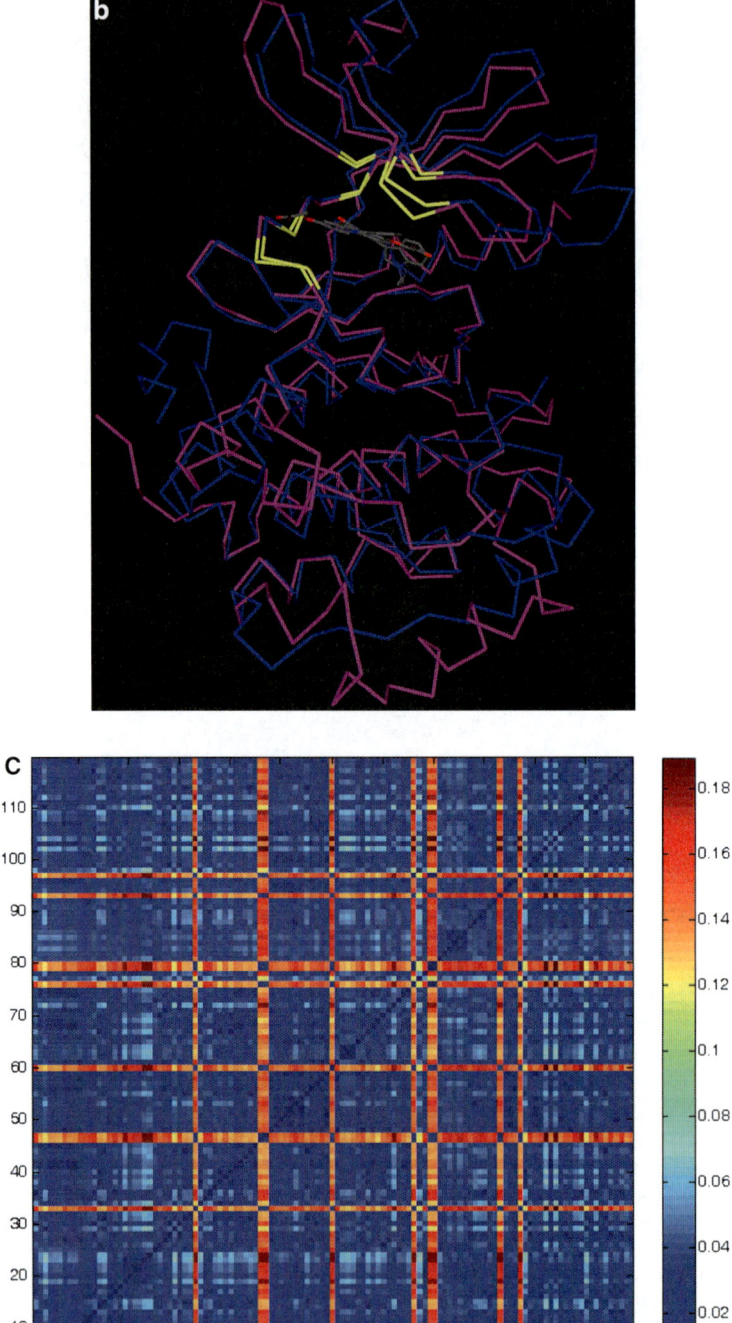

Fig. 9.1 (continued)

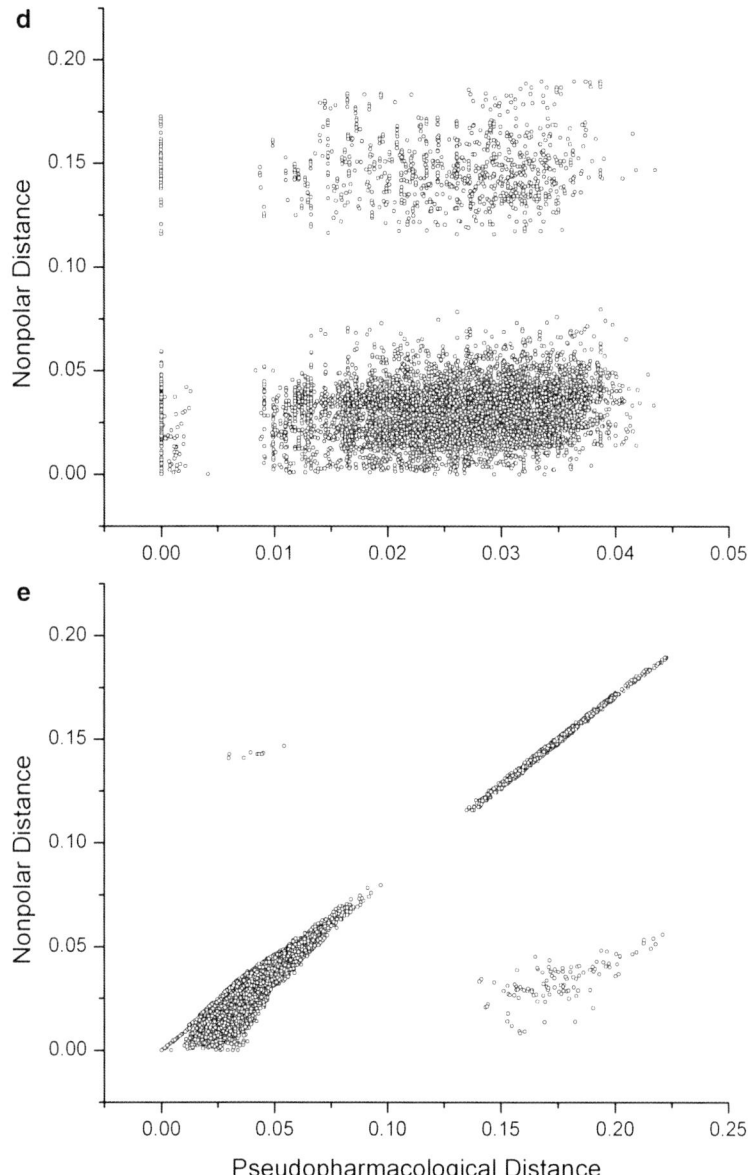

Fig. 9.1 (continued)

To determine whether pharmacological differences are dictated by differences in nonpolar accessible surfaces of the targets within ligand-binding sites, a nonpolar distance, $d_{np}(i,j)$, between the affinity-assayed kinases i, j is introduced. The $d_{np}(i,j)$ is determined by differences in accessible nonpolar surface areas of the respective nonpolar hulls, $H_{np}(i)$, $H_{np}(j)$. As rigorously defined in the previous section, the nonpolar hull of a kinase is comprised of the nonpolar residues of the kinase in contact with its drug ligands, whenever the complexes are reported in PDB, and residues in the kinase (not necessarily nonpolar) that align with nonpolar residues in contact with ligands in homolog PDB-reported ligand–kinase complexes (Fig. 9.1b). The latter constitute the structural background for the comparison. Introducing the hull becomes necessary to compare kinases at this level. Thus, the nonpolar distance is defined as

$$d_{np}(i,j) = \left[A\left(H_{np}(i)\right)\right]^{-1} \sum_{a \in Hnp(i)} \left|A(a) - A\left(\Phi_j(a)\right)\right|, \qquad (9.1)$$

with A = nonpolar accessible area [19, 20]; a = generic residue in the nonpolar hull $H_{np}(i)$ of chain i; $A(a) = 0$ if a is polar; and $\Phi_j(a)$ = residue in chain j that aligns with a.

Only 32 of the 119 assayed kinases are reported in PDB complexes, yet, structural inferences can be made with confidence given the significant extent of structural alignment (RMSD <1 Å) across reported structures [10]. The prediction accuracy decreases somewhat but in a quantifiable manner ($\sim$12%) on loopy regions [10]. The matrix $\mathbf{D}_{np} = \left[d_{np}(i,j)\right]$ (Fig. 9.1c) reveals remarkable nonpolar similarity across kinases: $< d_{np} > /(\text{maximum}(d_{np}) \approx 11\% (<>= \text{average over}$ kinase pairs)$; < \left[d_{np} - < d_{np} >\right]^2 >^{1/2} / < d_{np} > \approx 16\%$. *This similarity implies a high conservation of accessible nonpolar surface, an indication that ligands whose affinity is dominated by hydrophobic interactions should be highly promiscuous.*

The plot d_{np} versus d_{phar} for all (119 × 118)/2 kinase pairs (i, j) shows no correlation between the two metrics (Fig. 9.1d). However, when the highly promiscuous affinity-dominant staurosporine is incorporated to the affinity profile [14] and the affinity-based distance matrix is recalculated $\left(d_{phar} \rightarrow d_{ps} = \text{pseudopharmacological distance}\right)$, a good correlation $\left(R^2 = 0.875\right)$ between d_{ps} and d_{np} is obtained (Fig. 9.1e). *This correlation reveals that promiscuity, the dominant affinity trait when staurosporine is incorporated, is fostered by targeting accessible nonpolar moieties, in turn, a highly conserved feature of protein interfaces* [21]. The strong correlation shown in Fig. 1e implies that staurosporine should bind mainly through hydrophobic contacts, as it is indeed the case in its PDB complexes [22].

With typical ligand dissociation constants in the nanomolar range, the coordinate for staurosporine dominates the affinity vector for each kinase to such an extent that most affinity profiles look similar when this compound is included in the drug background. This similarity is reflected in the low rank ($\sim$7) of the pseudopharmacological distance matrix. The nonpolar matrix possess a comparable rank ($\sim$6) because the basis for comparison, i.e., exposed nonpolar residues, is highly conserved, leading to high levels of target similarity.

Sequence alignment including the 32 assayed kinases with PDB-reported complexes reveals that residues in ATP-binding sites engaged in hydrogen bonding with ligands are highly conserved, with $0 \leq \sigma(n) \leq 0.21$ (n = chain position for hydrogen bonding residue; σ = information entropy reflecting amino acid variability after sequence alignment; average σ in $H_{np} = 0.87$; maximum ($\sigma = \ln 20 \approx 4.2$). As expected, differences in hydrogen bonding capabilities do not appreciably correlate with d_{ps} ($R^2 \approx 0.19$), and there is no correlation with d_{phar}.

These observations lead to the question: Is wrapping the *universal* feature to be targeted to promote specificity? *In other words, is the dehydron pattern a molecular feature with sufficient variability across homologs and capable of significantly influencing ligand affinity?* To answer this question, we focus on "environmental residues," i.e., those framing the dehydron microenvironments. As we recall from Chaps. 1, 3, and 5, dehydrons promote protein associations because the enhancement and stabilization of the underlying electrostatic interactions overcome the thermodynamic cost associated with removing the surrounding water molecules that hydrate amide or carbonyl groups [23, 24]. Environmental residues include the hydrogen-bonded residues themselves. To compare environments of different kinases, we define the environmental hull of a kinase, H_{env}, as the reunion of all environmental residues in the chain and residues aligning with environmental residues from other chains (Fig. 9.2a, b). Thus, an environmental distance $d_{env}(i,j)$ between kinases i and j is obtained by comparing the aligned hydrogen bond microenvironments within $H_{env}(i)$ and $H_{env}(j)$: $d_{env}(i,j) = M(i,j)^{-1} \left[\sum_{n=1,...,M(i,j)} \Delta_n(i,j) \right]$, where $M(i,j) =$ number of residue pairs in $H_{env}(i)$ corresponding to dehydrons in kinase i or to hydrogen bonds or nonbonded residue pairs that align with dehydrons in $H_{env}(j)$, n = dummy index denoting residue pair, and $\Delta_n(i,j) = 0$ if residue pair n corresponds to a dehydron in $H_{env}(i)$ that aligns with a dehydron in $H_{env}(j)$, and $\Delta_n(i,j) = 1$, otherwise. Thus defined, the environmental distance compares local dehydration propensities associated with dehydron patterns in kinases. The validity of the relation $\Phi_i(H_{env}(j)) = H_{env}(i)$ enables a rigorous comparison between environments of two different proteins.

The matrix $\mathbf{D}_{env} = [d_{env}(i,j)]$ for the 119 kinases assayed (Fig. 9.2c) is obtained after inference of the dehydrons for the 87 kinases unreported in PDB from direct structure prediction. The predictions are validated through a correlation with an independent and accurate sequence-based prediction of the disorder propensity. This attribute was chosen because loopy regions, in the twilight between order and disorder (Chap. 5), compromise somewhat the accuracy of a structure prediction.

Fig. 9.2 (**a**) Environmental hull (*light blue*) for CHK1 (obtained from alignment with PDK1). Dehydrons are indicated as *green* segments joining the α-carbons of the paired residues. The virtual bonds are shown as *blue segments*. The three dehydrons perturbed by the ligand (named 3A3) are C87-G90, G90-L138, and G16-V23. (**b**) Aligned backbones for PDK1 (*blue*) in complex with BIM8 (PDB.1UVR) and CHK1 (*lilac*) in complex with 3A3 (PDB.2GCU), with the environmental hulls depicted in *light blue*. (**c**) Environmental distance matrix $\mathbf{D}_{env} = [d_{env}(i,j)]$ for the 119 kinases assayed. (**d**) Correlation of environmental versus pharmacological distance. The *line* indicates the optimal linear fit obtained from linear regression. The *red diamonds* correspond to the six pairs including ABL1, the primary target for imatinib, and its six mutants (listed in Fig. 9.1a) that confer drug resistance. Reprinted from [10] with permission from Oxford University press

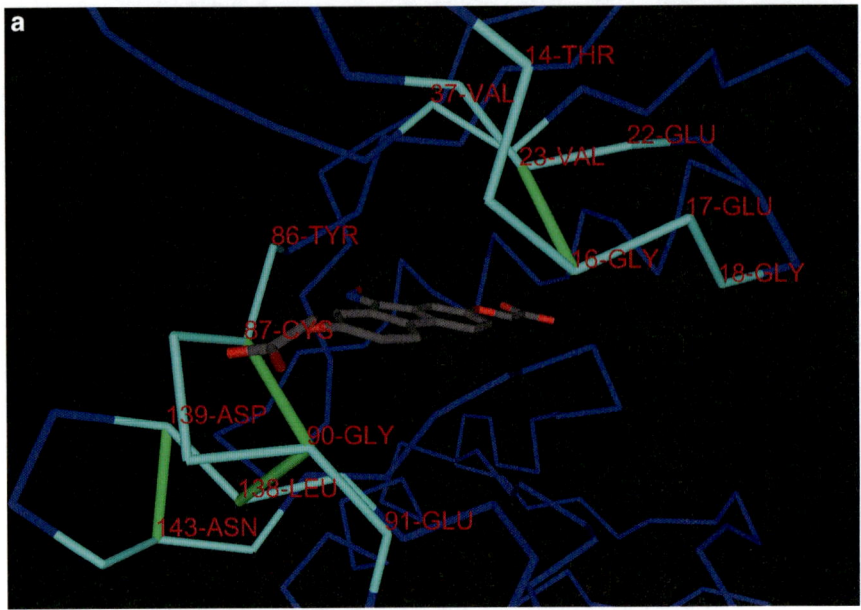

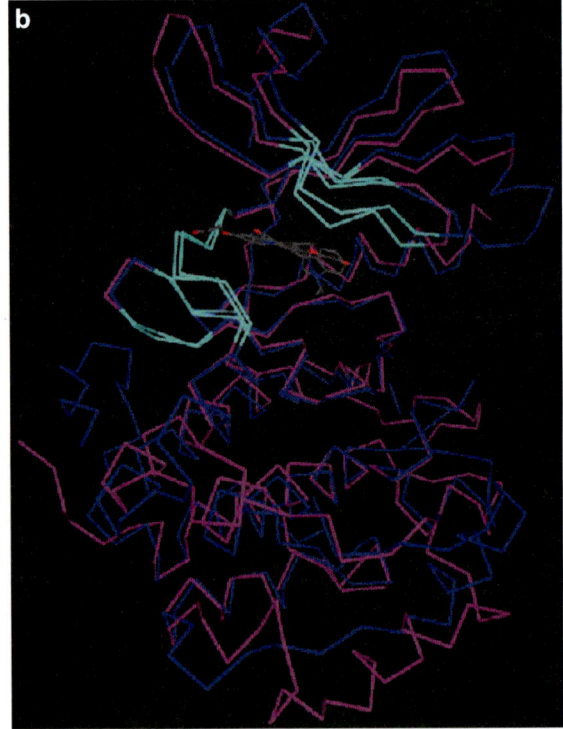

Fig. 9.2 (continued)

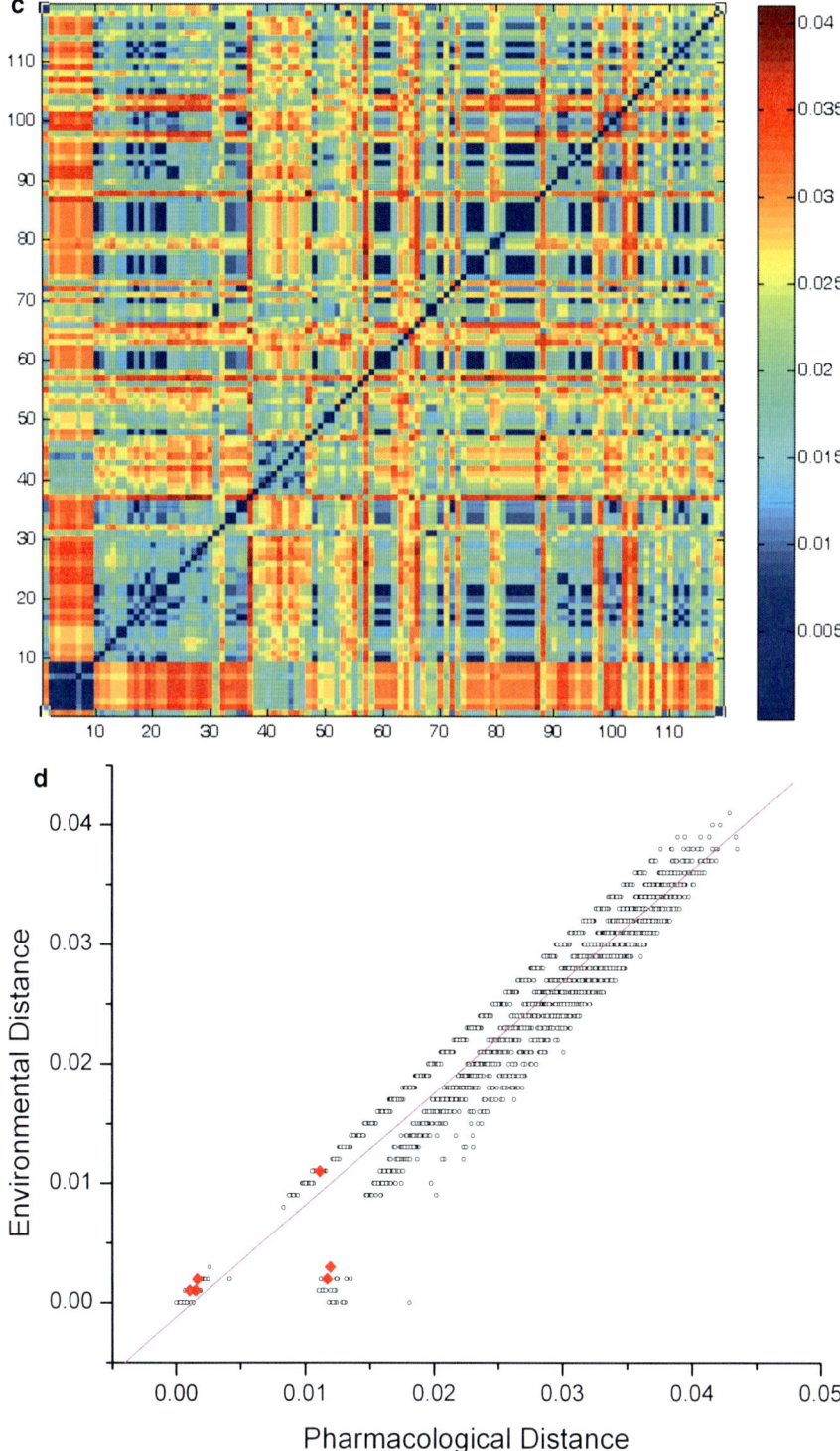

Fig. 9.2 (continued)

The strong correlation between d_{env} and d_{phar} $(R^2 \approx 0.917)$ (Fig. 9.2d) reveals that the impact of drugs on the human kinome is dictated by differences in hydration microenvironments or dehydron patterns across the ligand-binding regions of the kinases. The diversity in hydration microenvironments needed to yield specificity across paralog kinases results from the variability ($<\sigma> \approx 1.38$) of environmental residues ($<\sigma> \approx 0.21$ when the average is restricted to residues paired by hydrogen bonds that are environmentally affected by the ligands).

The strong correlation shown in Fig. 9.2d provides compelling evidence in support of the universality of protein wrapping as the selectivity filter for molecular therapeutic approaches targeting the human kinome.

9.4 A Target Library for the Human Kinome: Broadening the Technological Basis of Drug Discovery

As shown in the previous section, the wrapping pattern of a target kinase provides a universal selectivity filter for the drug designer. This fact begets the obvious question: How do we take advantage of our selectivity filter to actually broaden the technological basis of drug discovery? To address this question we envision the construction of a dehydron library in drug-targeted proteins and then exploit this bioinformatics library to generate a map of controlled cross-reactivity that results when such specialized patterns are targeted. The controlled cross-reactivity map will reflect the superior level of specificity that may be achieved by targeting packing patterns. This resource will exploit the matrix representation of the wrapping pattern of a protein that enables comparative alignment across an entire target superfamily and will digitalize information on subtle packing differences across protein targets. The long-term application of this resource is the control of side effects in molecular-targeted therapy adopting the new drug design paradigm, *the drug as dehydron wrapper.*

As shown in Chaps. 7 and 8, specific drugs may be chemically modified to turn them into wrappers of dehydrons, protecting them from water attack. This type of ligand optimization has been shown in specific cases to retain anticancer activity while reducing potential life-threatening side effects. Thus, the molecular basis for target discrimination was delineated in vitro and through in vivo assays for antitumor activity, albeit over limited sets of purported targets.

Against this background, it becomes essential to introduce bioinformatics results of wide applicability that precisely map out the optimizing chemical modifications for all drug chemotypes and all kinase targets with the goal of controlling and enhancing inhibitor specificity across the entire human kinome. Thus, in principle at least, the technological basis of drug design can indeed be broadened contingent on our ability to construct a comparative and annotated dehydron library. This library should compile the cross-reactivity map that will contain "raw" information on wrapping defects as selectivity-promoting features and contain the annotated information to guide the drug designer to develop superior drugs with controlled specificity. The raw data should include the following:

- Identification of all packing defects – *dehydrons* – targetable by ATP competing inhibitors for each human kinase.
- Comparative alignment of the dehydron pattern of one kinase against the respective patterns in all the others to obtain a map of a priori cross-reactivities for dehydron-wrapping drugs.
- Identification of dehydrons with the lowest levels of conservation for each kinase together with the kinases where the dehydron is replaced by a well-wrapped hydrogen bond upon alignment. Thus, the optimal specificity-controlling dehydrons will be determined through a "conservation score."
- Determination of accessibility of optimal dehydrons vis-à-vis wrapping/targeting within a given drug chemotype and the set of kinases that is discriminated as the optimal dehydrons are targeted. Obviously, a specificity-promoting dehydron would be of no practical use if its exogenous wrapping is not accessible from the particular lead chemotype considered.

Thus, the results expounded in this chapter pave the way for the construction of the matrix-based library of dehydrons for human kinases enabling direct comparative alignment across the structure-reported kinome. This library is destined to be used to decompose PDB-reported target space into clusters of cross-reactive kinases, enabling direct assessment of the full potential of the novel selectivity filter on targets with reported structure. The results can be validated by examining compounds independently screened using a phage display kinase library and contrasting their affinity (K_D) profile against predicted cross-reactivity clusters. The results should also be cast in usable format to guide the engineering of safer inhibitors with controlled specificity. Dehydrons should be prioritized according to their decreasing levels of conservation across homologs in the human kinome, hence serving as the most reliable molecular features to be targeted in order to control and reduce cross-reactivity.

9.5 Useful Annotations of a Library of Specificity-Promoting Target Features

In the first stages of the resource building, we envision a comparative analysis confined to kinases with PDB-reported structures, thus guaranteeing unambiguous dehydron identification. The structural alignment required for comparative analysis will be largely facilitated by the kinase fold conservation resulting from common ancestry. Targetable dehydrons should be reported for each kinase jointly with several annotations instrumental to the drug optimization strategy: (a) extent of dehydron conservation across the superfamily and expected cross-reactivity arising if targeted by a wrapping ligand; (b) accessibility of the dehydron vis-à-vis a specific drug chemotype; (c) activation state of the kinases under comparison; (d) variations in dehydron patterns due to different induced fits of the same target; and (e) indication on whether a dehydron identified as targetable belongs to an induced fit region within a kinase–ligand complex (in which case the lead optimization should preserve the ligand chemotype) or is an inherent feature of the kinase.

Such a digital library of packing defects enabling a comparative kinome-wide alignment will represent a powerful tool to guide lead optimization geared at controlling specificity and reducing side effects. Thus, the results of this chapter call for a superfamily-wide bioinformatics platform to safely and reliably control cross-reactivity of kinase inhibitors based on the dehydron concept.

We also envision extensions of the library to targets with unreported structure. Specifically, we envision a homology-based extension of the dehydron-matrix library to the entire human kinome and concurrent extension of the map of controlled cross-reactivity enabling a comprehensive assessment and validation of the novel packing-based design paradigm. As described in the preceding sections, only about 25% of the human kinome has PDB representation and hence many kinase targets of clinical interest become intractable when adopting structure-based strategies for drug design/optimization. Thus, homology-based approaches become essential to turn the dehydron library into a useful tool to guide the optimization of therapeutic agents. Our adopted homology-based predictors can generate kinase 3D structures by satisfying spatial restraints imposed by threading with the template kinase structure and applying a force field with refinement protocols based on simulated annealing. The predicted dehydron patterns should be corroborated by two independent routes: (I) through comparison with sequence-based inferences of disorder propensity, harnessing on a known correlation between poor packing and disorder propensity (see Chap. 5) and (II) by contrasting inferred dehydron-based cross-reactivities with experimental high-throughput screening of kinase inhibitors. The annotations for homology-based dehydron inference will include those indicated previously (a–e) and should incorporate additional annotations pertaining to inference reliability: (f) extent of template probe sequence identity (s.i.); (g) quality of template probe alignment defined by number of sequence insertions and deletions.

The construction of the dehydron library and its validation as resource of kinome-wide applicability for the drug developer involves nine interrelated milestones described in Fig. 9.3. As indicated above, such a comparative dehydron library should provide "raw" information on packing defects as selectivity-promoting features in drug design. This information ought to be supplemented with annotations that should aid the drug designer in the interpretation and exploitation of the raw data and in addressing the problems pertaining to the concrete application of the concept of "drug as wrapper of packing defects." These annotations are now described in detail:

(A) *Extent of Conservation of Each Dehydron Within the Kinase Superfamily and Expected Cross-Reactivity Arising when Targeted by a Wrapping Ligand:* In general, the dehydron pattern of a kinase may not enable a *complete* discrimination against all other ~518 kinases. For example, the wrapping distances across pairs from the set of 119 kinases originally screened for drug affinity by AMBIT [14, 15] reveal that the focal adhesion kinase FAK (synonym: PTK2) has the same dehydron

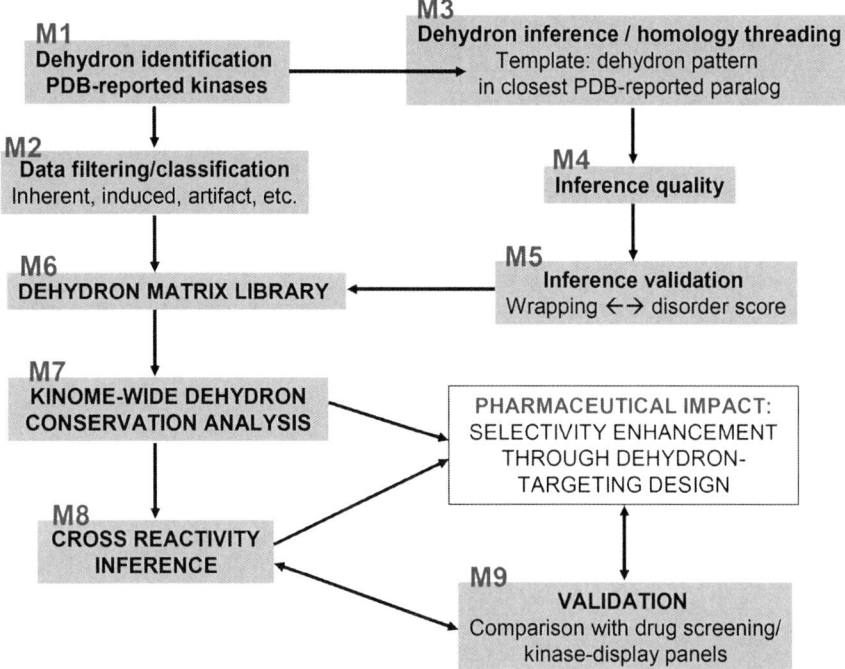

Fig. 9.3 Nine milestones (M1–M9) for the construction and validation of the kinome-wide dehydron library as a resource to guide drug design

pattern as members of the CAMK, JAK, JNK, NEK, and PAK families (Figs. 9.1a and 9.2c). This example illustrates the fact that *certain degree of cross-reactivity may be unavoidable when adopting the wrapping technology, pointing at the very limitations of this approach. The whole issue then becomes whether the particular cross-reactivity is clinically tolerable or whether it provides the right compromise between activity and toxicity.*

Besides giving the level of conservation of each packing defect, the library should provide an assessment of the expected cross-reactivity for a drug that wraps the defect. As a concrete example, let us determine the level of control of cross-reactivity that may be achieved when wrapping the KIT dehydron C673–G676, as described in Chap. 8. Table 9.1 provides a comparative analysis of the conservation of the KIT dehydron across KIT paralogs and other members of the kinase superfamily.

Thus, a drug inhibitor that wraps the KIT dehydron C673–G676 is a priori likely to also hit the following kinases: CSF1R, FLT3, TIE2, MET, MUSK, IGFR1, and INSR, where the KIT dehydron aligns with a dehydron. However, the actual levels of cross-reactivity may be smaller depending on the drug chemotype and its associated steric clashes with the target. For instance, of this group of kinases, the imatinib modification that wraps KIT dehydron C673–G676 binds only to CSF1R, as shown in Chap. 8.

Table 9.1 Conservation of KIT dehydron C673–G676 within an assortment of PDB-reported kinases. (DH, dehydron; HB, well-wrapped hydrogen bond; "–", no bond in alignment)

Group	Protein	PDB accession code	Sequence identity	Wrapping analysis of HB aligning with KIT dehydron C673–G676
TK	KIT	1T46	100	**DH**
		1PKG	99	DH
		1T45	95	DH
	CSF1R	2I1M	70	DH
	FLT3	1RJB	65	DH
	VEGFR2/KDR	2YWN	59	–
	FGFR2	1GJO	54	HB
	FGFR1	1FGI	52	HB
	RET	2IVS	49	HB
	TIE2	2OO8	44	DH
	HSK	1QCF	44	HB
	MET	1R0P	42	DH
	MUSK	1LUF	42	DH
	ABL	2HZ0	42	HB
		2HYY	41	HB
	LCK	2OG8	42	HB
		2PL0	41	HB
		2OFV	41	HB
	INSR	1IRK	39	DH
	IGF1R	1M7N	36	DH
TKL	BRAF	1UWH	36	–
CMCG	CDK6	1BLX	31	–
CAMK	MARK1	2HAK	31	HB
	TTN	1TKI	29	–
STE	PAK1	1F3M	28	–
AGC	AKT2	1MRV	28	–
	MSK1	1VZO	27	–

(B) *Accessibility of a Dehydron as a Targetable Feature Vis-à-vis a Specific Drug Chemotype:* The mere identification of a specificity-promoting feature in a kinase is of limited practical use unless it can be targeted through a chemical modification within a specific chemotype. For example, the selectivity-promoting E282–E286 dehydron in the Abelson (Abl) kinase may be targeted through a wrapping chemical modification (methylation) of imatinib (Fig. 9.4a). However, the same dehydron is not accessible through wrapping methylation of the compound PD173955 (Fig. 9.4b) since the incorporated nonpolar group lies outside the microenvironment of the E282–E286 dehydron.

(C) *Variations in Dehydron Patterns Due to Different Induced Fits:* The identification of a targetable feature to promote drug specificity is not independent of the chemotype of the drug to be re-designed. This is not only due to the fact that the chemotype determines the accessibility of the dehydron (point B) but also because a kinase may present different induced fits depending on the ligand to which it is associated, and hence different ligand-dependent dehydron patterns. Thus, if a dehydron in a kinase is identified as targetable but only occurs in a particular induced fit with

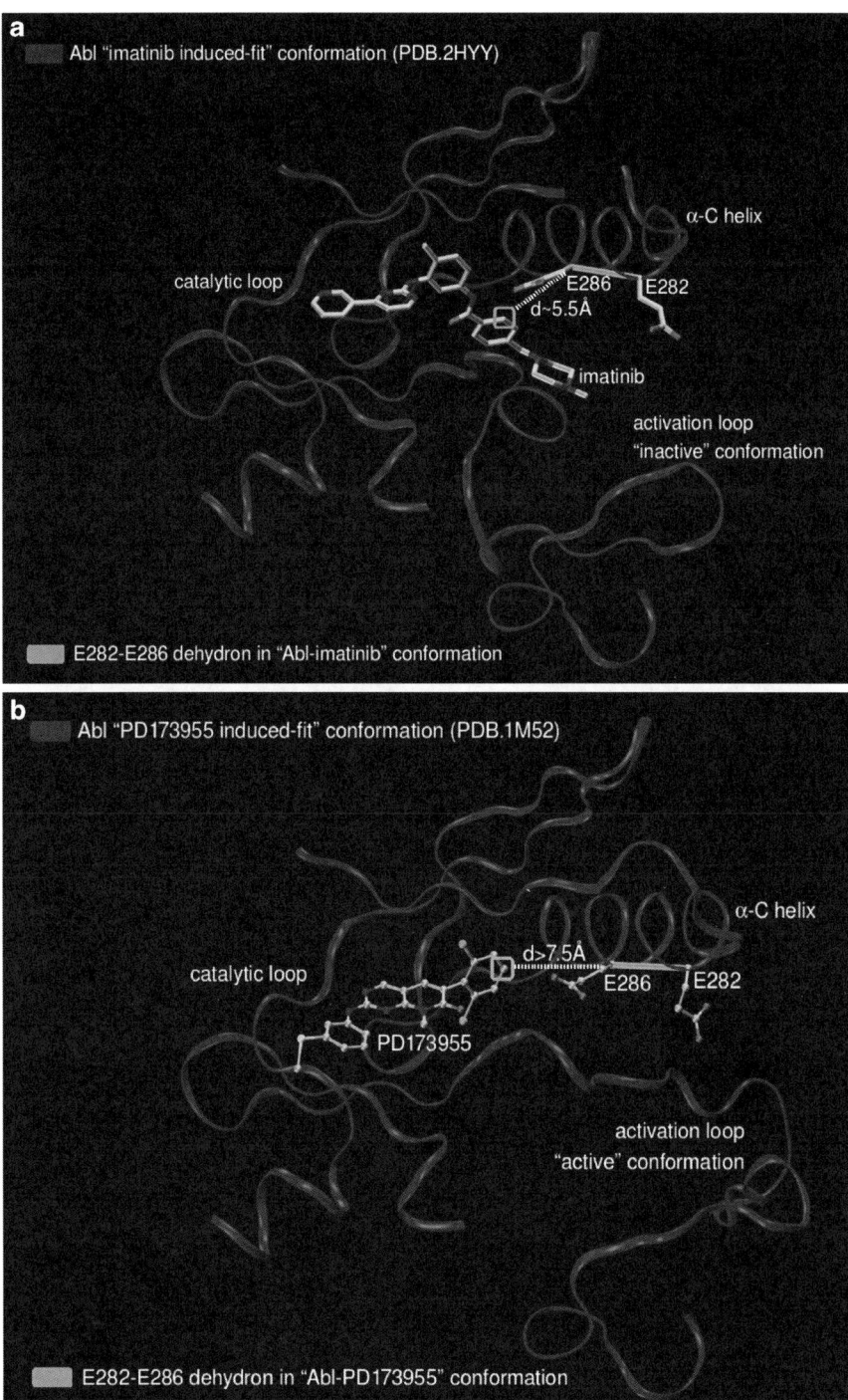

Fig. 9.4 Differences in accessibility of ABL dehydron E282–E286 according to drug chemotype. (**a**) Dehydron-wrapping methylation of imatinib and (**b**) inaccessibility of dehydron to wrapping methylation in compound PD173955

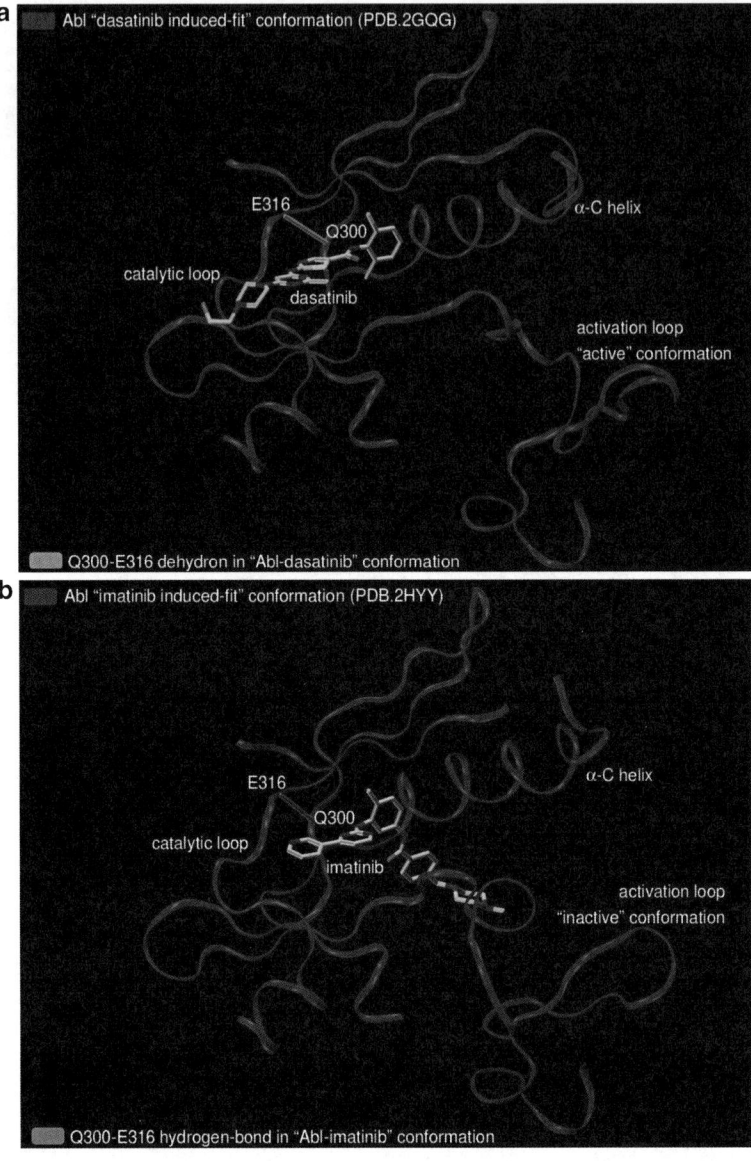

Fig. 9.5 Variations in dehydron patterns for the ABL kinase due to different induced fits. (**a**) ABL/*dasatinib* complex and (**b**) ABL/*imatinib* complex

a particular drug ligand, it may be most advantageous to respect the chemotype of the ligand when targeting/wrapping the dehydron in order to enhance speci-ficity. To address this problem, when identifying a particular selectivity-promoting dehydron to be targeted, a dehydron library should always include the ligand and PDB accession of the complex where the dehydron has been identified.

For example, the Q300–E316 dehydron in Abl kinase may be a good target to enhance the specificity of *dasatinib* and drugs within its chemotype since this dehydron is present in the induced fit of the Abl upon complexation with dasatinib (Fig. 9.5a). However, that dehydron becomes a well-wrapped hydrogen bond in the induced fit resulting upon Abl complexation with imatinib (Fig. 9.5b). Thus, the library should contain annotation guarding *against* adopting the imatinib chemotype when targeting the Q300–E316 dehydron in Abl, since this dehydron is not present upon imatinib association.

9.6 The Dehydron Library as a Technological Resource

The dehydron library will generate chemotype-dependent decompositions of target space (human kinome) into cross-reactivity clusters. These results are crucial to assess the full potential of the packing-based drugs as therapeutic agents of superior controllable selectivity. Cross-reactivity clusters include only kinases that are inhibited by the same wrapping modification within a fixed lead chemotype. The wrapping modification, in turn, targets a dehydron in a particular kinase, and all other kinases presenting a dehydron at the aligned position will belong to the same chemotype-dependent cross-reactivity cluster. The cartoon in Fig. 9.6 represents the

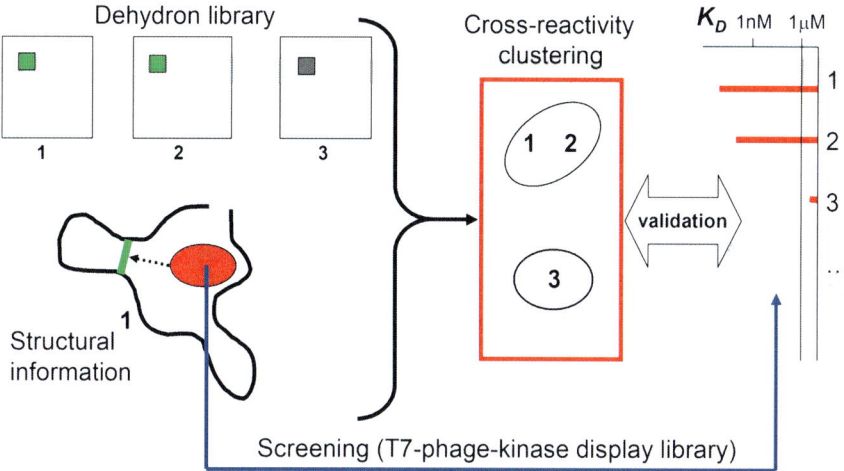

Fig. 9.6 Schematic representation of the generation of a chemotype-dependent cross-reactivity map. A dehydron (*green*) in a particular protein (**1**) that may be wrapped (*dashed-line arrow*) by a chemical substitution of a lead within a specific chemotype (*red*) will be singled out for comparison across the kinome and to define the cross-reactivity cluster to which this protein belongs. The extent of conservation of the dehydron will then be established by comparative alignment of dehydron matrices within the bioinformatics library. Conservation of the dehydron (across proteins **1** and **2** in the scheme) promotes cross-reactivity of the wrapping compound, defining the cluster for protein **1** within the chemotype-dependent decomposition of target space (*red rectangle*). Proteins that lack drug-accessible dehydrons or those that contain unique (not conserved) drug-accessible dehydrons constitute singleton clusters. Validation is readily obtained by examining compounds independently screened using AMBIT's high-throughput kinase panels [14, 15] and contrasting the predicted cross-reactive clusters with the reported affinity (K_D) values

construction/validation of the cross-reactivity map combining the dehydron library and structural information on a single kinase/inhibitor complex for the particular drug chemotype.

The cross-reactivity map is thus obtained by a straightforward procedure that combines structural input from a kinase/inhibitor complex with a kinome-wide comparative alignment of packing defects provided by the bioinformatics library. The results will be validated by examining compounds independently screened using the T7-phage display kinase library ($\sim$360 kinases in latest panels) [15] and contrasting their affinity (K_D) profile with their predicted cross-reactivity cluster: if a compound is identified to wrap a particular dehydron in a kinase, then this compound should be cross-reactive with all other kinases for which the dehydron is present at the aligned position except for kinases where structural alignment reveals steric clashes. This target space decomposition will assess the full potential of the wrapping-based design motif.

References

1. Bain J, McLauchlan H, Eliott M, Cohen P (2003) The specificities of protein kinase inhibitors: An update. Biochem J 371:199–204
2. Druker BJ (2004) Molecularly targeted therapy: Have the floodgates opened? Oncologist 9:357–360
3. Hopkins AL, Mason JS, Overington JP (2006) Can we rationally design promiscuous drugs? Curr Opin Str Biol 16:127–136
4. Huse M, Kuriyan J (2002) The conformational plasticity of protein kinases. Cell 109:275–282
5. Knight ZA, Shokat KM (2005) Features of selective kinase inhibitors. Chem Biol 12: 621–637
6. Chothia C (1974) Hydrophobic bonding and accessible surface area in proteins. Nature 248:338–339
7. Whittle PJ, Blundell TL (1994) Protein structure-based drug design. Annu Rev Biophys Biomol Struct 23:349–375
8. Feng BY, Shelat A, Doman TN, Guy RK, Shoichet BK (2005) High throughput assays for promiscuous inhibitors. Nat Chem Biol 1:146–148
9. Zhang X, Crespo A, Fernández A (2008) Trung promiscuous kinase inhibitors into safer drugs. Trends Biotechnol 26:295–300
10. Chen JP, Zhang X, Fernández A (2007) Molecular basis for specificity in the druggable kinome: sequence-based analysis. Bioinformatics 23:563–572
11. Manning G, Whyte DB, Martinez R, Hunter T, Sudarsanam S (2002) The protein kinase complement of the human genome. Science 298:1912–1934
12. Braken C, Iakoucheva LM, Romero PR, Dunker AK (2004) Combining prediction, computation and experiment for the characterization of protein disorder. Curr Opin Struct Biol 14:570–576
13. Fernández A, Berry RS (2004) Molecular dimension explored in evolution to promote proteomic complexity. Proc Natl Acad Sci USA 101:13460–13465
14. Fabian MA, Biggs WH, Treiber DK et al (2005) A small molecule kinase interaction map for clinical kinase inhibitors. Nat Biotechnol 23:329–336
15. Karaman MW, Herrgard S, Treiber DK et al (2008) A quantitative analysis of kinase inhibitor selectivity. Nat Biotechnol 26:127–132
16. Bonneau R, Straus CE, Rohl CA et al (2002) De novo prediction of three-dimensional structures for major protein families. J Mol Biol 322:65–78

17. Chivian D, Kim DE, Malmstrom L et al (2005) Prediction of CASP6 structures using automated Robetta protocols. Proteins 61 (Suppl 7):157–166
18. Higgins DG, Thompson JD, Gibson TJ (1996) Using CLUSTAL for multiple sequence alignments. Method Enzymol 266:383–402
19. Fraczkiewicz R, Braun W (1998) Exact and efficient analytical calculation of the accessible surface areas and their gradient for macromolecules. J Comput Chem 19: 319–333
20. Ooi T, Oobatake M, Nemethy G, Scheraga HA (1987) Accessible surface area as a measure of the thermodynamic parameters of hydration of peptides. Proc Natl Acad Sci USA 84: 3086–3090
21. Ma B, Elkayam T, Wolfson T, Nussinov R (2003) Protein-protein interactions structurally conserved residues distinguish between binding sites and exposed protein surfaces. Proc Natl Acad Sci USA 100:5772–5777
22. Fernández A, Maddipati S (2006) A priori inference of cross reactivity for drug-targeted kinases. J Med Chem 49:3092–3100
23. Fernández A, Sosnick TR, Colubri A (2002) Dynamics of hydrogen bond desolvation in protein folding. J Mol Biol 321:659–675
24. Fernández A (2004) Keeping dry and crossing membranes. Nat Biotechnol 22:1081–1084

Chapter 10
Fulfilling a Therapeutic Imperative in Cancer Treatment: Control of Multi-target Drug Impact

The causal heterogeneity and complexity of cancer have inspired drug developers to seek for multi-pronged strategies of therapeutic attack. Fuelled by systems biology insights, the burgeoning interest in multi-target drugs is motivating a reassessment of the therapeutic value of promiscuity. While drug efficacy may not correlate with specificity, it would be risky to take the opposite stance and welcome promiscuous compounds without a rational strategy to funnel their therapeutic impact. In this chapter we survey approaches to control the therapeutic impact of multi-target kinase inhibitors to fulfill the therapeutic imperatives of human cancer. Thus, we advocate for the application of the wrapping design concept to turn multi-target kinase inhibitors into clinical opportunities through judicious chemical modification.

10.1 Is There Really a Case for Promiscuous Drugs in Anticancer Therapy?

The need to reduce toxic side effects obviously makes specificity of small-molecule drugs a desired attribute in molecularly targeted therapy [1–5]. The side effects, arising from drug cross-reactivity and from the diverse roles of the target protein in off-target cellular contexts, prompted researchers to advocate for the "magic bullet" paradigm [6], epitomized by compounds with high binding specificity. However, this paradigm faces serious challenges since there is no obvious correlation between drug specificity and therapeutic index (lethal dose (LD50) over therapeutically effective dose (ED50)) [6–11]. Thus, the causal complexity of cancer, with its manifold and heterogeneous scenarios for the unfolding of the disease, clearly advocates for a multi-pronged treatment often translated into a multi-target molecular therapy.

Be as it may, welcoming dirty drugs as a means to fulfill the therapeutic imperatives of human cancer is obviously not the paradigm shift that most researchers in the field have in mind. In principle, multi-target drugs are more prone to trigger side effects than their more specific counterparts. Thus, while there is a perception that a shift from the magic bullet paradigm is needed, there is no clue on what the right replacing concept should be.

A. Fernández, *Transformative Concepts for Drug Design: Target Wrapping*,
DOI 10.1007/978-3-642-11792-3_10, © Springer-Verlag Berlin Heidelberg 2010

Uncertain clinical outcome resulting from health-threatening side effects is a central issue for therapies based on multi-target kinase inhibitors since kinases play different signal transduction roles in different cellular contexts and a favorable inhibition in one scenario may prove fatal in another [12]. For instance, a specific inhibitor of the Abelson (Abl) kinase may be assumed to be most efficacious in treating chronic myeloid leukemia (CML), since an aberrantly deregulated Abl kinase is a recognized primary target for treating this malignancy [5]. However, as extensively described in Chaps. 7 and 8, a systems biology assessment [12] has recently revealed that Abl inhibition is also a likely culprit of cardiotoxicity: Abl inhibition recruits a signaling cascade that leads to mitochondrial depolarization, ATP depletion, and ultimately apoptosis in cardiomyocytes [12]. Thus, a more cross-reactive drug, i.e., one that *also* inhibits another kinase along the pro-apoptotic pathway triggered by Abl inhibition (hence suppressing it) is expected to have a higher therapeutic efficacy in treating CML [13]. In this regard, an inhibitor that also impacts JNK would be highly desirable, as described in Chap. 8. The rationale for this assertion is that higher therapeutic doses may be tolerable due to the removal of cardiotoxicity, albeit at the expense of sacrificing specificity [13–15]. One possible downturn of such an approach is that JNK inhibition may introduce uncertain clinical outcomes in off-target cells (i.e., cells other than cardiomyocytes). Given the downstream position of JNK in most signal transduction cascades, this is not likely to be the case, as JNK may be replaced by alternative and redundant signal transducers in other cellular contexts.

The causal complexity of cancer and other diseases has motivated much effort directed at reassessing the therapeutic value of promiscuity. This conceptual shift is in part motivated by telling cases. For instance, multi-target kinase inhibitors such as *Sutent* (sunitinib) or *Nexavar* (sorafenib) have recently received FDA approval, albeit with important caveats highlighting their side effects, especially their cardiotoxicity [7, 12]. Thus, the possibility of exploiting drug promiscuity as a therapeutic opportunity is being considered in novel approaches to treat complex disorders such as cancer, depression, and cardiovascular disease [7]. The simultaneous modulation of multiple targets is often required to alter a clinical phenotype as biological redundancies and alternative pathways can often bypass the inhibition of a single target or of multiple targets along a single pathway [16]. Thus, in contrast with the magic bullet, a "magic shotgun" targeting multiple proteins may in some instances possess a higher therapeutic index than a specific drug [6].

Another argument favoring promiscuity arises with the observation that highly cross-reactive drugs may be more resilient against drug-resistant mutations [17]. Cross-reactivity arises because, upon association with the target protein, such drugs typically make interactions with evolutionarily conserved residues and with backbone groups, and less interactions with the residues that mutate to confer drug resistance [17, 18]. Targeting conserved residues or unspecific parts of the chain (backbone) begets promiscuity.

The safety of less selective drugs, and especially that of promiscuous kinase inhibitors, is often addressed by assessing the full extent of their cross-reactivity [19]. This assessment has been facilitated by novel high-throughput screening assays such as (a) the kinase assay using the T7-phage expression panels from Ambit Biosciences [20, 21]; (b) a thermal stability shift assay using a 60-Ser/Thr kinase panel [22]; and (c) the living-cell assays of pathway inhibition that assess the impact of the drug on the protein-recruiting capability of the target [23].

In vitro assays of cross-reactivity are affected by the complexities of tissue distribution and, generally, by pharmacodynamic issues [24]. Yet, target affinity, being governed by thermodynamics of ligand binding, represents a telling parameter independent of in vivo heterogeneities, except for allosteric antagonism. On the other hand, crowding, membrane adsorption, and other effects can modulate in vivo drug concentrations, increasing local levels in different spatial locations. Thus, certain cross-reactivities undetectable in vitro may surface in vivo, causing unexpected side effects [24, 25]. Conversely, tissue and subcellular distribution may prevent a ligand from binding an in vitro-established target, introducing another caveat in the interpretation of high-throughput screening results.

In spite of timely efforts to establish a paradigm shift, promiscuous drugs obviously carry a higher risk of inducing side effects than their more specific counterparts. Even the successful anticancer drug *Gleevec* (imatinib), with a reduced spectrum of primary targets, i.e., the kinases Bcr-Abl, C-Kit, Lck, PDGFR, and CSF1R [20, 21], has been recently shown to be potentially cardiotoxic [12, 26] and labeled as such by the FDA (see: http://www.fda.gov/medwatch/safety/2006/Gleevec_DHCP_10-19-2006.htm). Not surprisingly, the more promiscuous anticancer kinase inhibitors sunitinib and sorafenib are also cardiotoxic, to a larger extent than imatinib [12, 26]. In Ambit screenings [20] sunitinib was shown to bind 79 kinases out of 119 assayed, while sorafenib binds to 41.

Hence, it is forbiddingly risky to welcome promiscuous compounds into the therapeutic arena without a rational strategy to control their specificity and therapeutic index. Such a control may be achieved if we can identify features in the target structure that are unique to the target, i.e., "selectivity filters," and drug modifications that promote interactions with such nonconserved features. Thus, selectivity filters may serve as guidance to rational drug design [13–15, 27–29]. We advocate for this type of control of therapeutic impact to clean less selective drugs following an integral assessment of the diverse functional roles of targeted proteins.

10.2 Cleaning Dirty Drugs with Selectivity Filters: Basic Insights

As described in Chaps. 6 and 9, most protein drug targets have paralogs, that is, proteins that share a common ancestor with the target and have diverged away from it after speciation [30]. Thus, kinases belong to common ancestry groups (families) which typically share the same fold and basic structural features that often yield unexpected cross-reactivities [20, 21, 31].

In principle, much of this cross-reactivity could be removed by drug redesign guided by the identification of the common structural features that promote promiscuity and the nonconserved features that enable paralog discrimination. As shown in Chap. 8, this approach has been successfully attempted on the cross-reactive drug imatinib with promising results [13–15]. Its further utilization to clean truly promiscuous drugs requires further analysis, as described in the next section, and will enable us to propose its extended use to clean kinase inhibitors.

A number of paralog-discriminating nonconserved features different from wrapping have been identified and, although not amenable of systematic application, some can be exploited as selectivity filters [27, 28]. One approach uses high-resolution crystal structures of kinases in complex with non-ATP ligands to identify unique structural motifs in the purported targets. For example, a unique helical insert has been found in the activation loop of the NEK2 and MPSK1 kinases adjacent to the highly conserved DFG catalytic triad [27].

Specific inhibitors that bind to the inactive kinase conformations have also been developed targeting the "DFG-out" conformation [32]. In this unique conformation, the position of the phenylalanine (F) residue which is located at the start of the activation loop is flipped with respect to the active conformation, so that it points inward within the ATP pocket. This mode of association has been confirmed by the binding of imatinib to the inactive Abl kinase [33]. In addition to the DFG-out motif, nonconserved structural features within the inactive ensemble should perhaps be exploited to achieve paralog specificity. The inactive conformations of kinases make them more discernible, while the active conformation reveals fewer discriminatory features since it is constrained to be catalytically functional and hence more conserved. The substrate-discriminatory amino acid variations that define the different kinase functions are mostly located in loopy regions framing the ATP pocket, rather than in the pocket itself, making them less accessible targets. While targeting the inactive conformation may be a logical choice, there are also advantages in targeting the active conformation. The latter requires structure conservation for functional purposes, and hence it is less tolerant to drug-resistant mutations. Thus, a kinase that binds to a drug in the active conformation will seldom develop drug-resistant mutations since the latter can compromise the activity.

The selectivity filter advocated throughout this book is in principle of universal validity, as shown in Chap. 9, and may be used systematically. It arises from comparison of the wrapping patterns of protein targets [30]. The wrapping pattern may be turned into an operational selectivity filter for two reasons: (a) dehydrons may be targeted by drugs that further wrap them by bringing nonpolar groups to their proximity upon association and (b) dehydrons are *not* conserved across paralogs, as shown in Chap. 9. Thus, differences in dehydron distribution constitute promising selectivity filters to clean promiscuous inhibitors.

10.3 Cleaning Dirty Drugs by Exploiting the Wrapping Filter: Proof of Concept

In principle, most kinase inhibitors are susceptible of being turned into selective wrappers of packing defects through minor chemical modification and without altering their chemotype. Thus, clinically relevant compounds with considerable cross-reactivities such as sunitinib (inhibiting 55% of the 290 kinases screened [21]), dasatinib (30%), EKB-569 (20%), sorafenib (20%), erlotinib (15%), gefitinib (6%), and imatinib (4%) may be redesigned into drugs with enhanced specificity using the wrapping design concept. The generic strategy consists in modifying the parental compound to turn it into a wrapper of unique dehydrons while removing potential sources of cross-reactivity. The latter arise since chemical groups in the ligand often interact with conserved structural features in the protein such as backbone moieties or side chains that remain invariant within the target family. The wrapping strategy may in principle be applied to enhance the specificity of promiscuous drugs like sunitinib, dasatinib, EKB-569, sorafenib, and erlotinib toward their respective primary targets: KIT/VEGFR2, ABL/SRC, EGFR, VEGFR2, EGFR [21].

To illustrate the operational value of this approach, we focus on redesigning a truly promiscuous drug, EKB-569 (Wyeth-Ayerst, [34]), to significantly enhance its selectivity. This drug was launched as a major inhibitor of the epidermal growth factor receptor (EGFR) kinase (IC50 = 38.5 nM). It gained major therapeutic interest for the treatment of non-small cell lung cancer (NSCLC), colorectal neoplasia, and other EGFR-dependent solid tumors. Phase I and II trials for such therapeutic applications are currently in progress and closed to new patients [34, 35]. Preliminary high-throughput screening using a battery of 119 T7-phage-expressed kinases [20] revealed 25 sub-micromolar targets for EKB-569, making it a promiscuous drug with likely side effects. Other compounds such as *Iressa* (gefitinib) and *Tarceva* (erlotinib) share the same "4-anilinoquinoline" chemotype [36], yet they are more specific EGFR inhibitors [20]. The latter two gained FDA approval as anti-NSCLC agents.

EKB-569 promiscuity can be traced to its intermolecular interactions with highly conserved residues within the EGFR kinase family. As shown in Fig. 10.1a, the terminal acryl group in the ligand is a source of promiscuity because it reacts with the *conserved* residues Cys/Ser in the EGFR paralogs, acting as electrophile in an irreversible Michael adduct. Another source of EKB-569 promiscuity is the intermolecular electrostatic interaction between its cyanide group and the gatekeeper residue (Thr or Met), typically conserved within the family (Fig. 10.1a).

Thus, we may remove these sources of promiscuity by introducing the following modifications:

(a) Replace the double bond (the Michael acceptor) in the acryl group with a single bond

(b) Replace the cyanide group with a methyl to retain the chemotype while removing the electrostatic interaction with the conserved gatekeeper

To promote selectivity, we may introduce a wrapping modification to target a nonconserved dehydron in the primary target EGFR. A comparative structural analysis of the EGFR family reveals that the least-conserved EGFR dehydron is Asp831–Gly833. To wrap this dehydron, we may append a methyl group at position 3 of the terminal benzene ring (Fig. 10.1b). This wrapping modification is likely to possess inhibitory activity *exclusively* against the EGFR paralogs that retain the dehydron in its original position: EGFR, Aurora2 (STK6), CLK3, ERBB2, PIM1, PIM2, PTK2 (FAK), SLK, and STK10. The modified drug is expected to bind a subset of this target group since steric hindrances may arise as a result of the chemical medication. We thus provide a rational approach to clean the promiscuous inhibitor EKB-569 based on a selectivity filter, reducing its inhibitory spectrum from 25 confirmed targets [20] to only 9.

To validate this design strategy, we established a correlation between the affinities of EKB-569 for the 38 paralogs of EGFR reported in PDB and the extent of residue conservation at the Michael reaction site and at the gatekeeper position. To do so, we aligned each paralog structure with the EGFR structure and examined residues that align with Cys773 (Michael reactant) and Thr766 (gatekeeper). A statistical model was built to assess such correlation and revealed that the EKB-569 affinity is indeed dictated by these two sources of promiscuity (p-value $= 0.0003$). Thus, the terminal acryl group and the cyanide group in EKB-569 (Fig. 10.1a) are indeed the "dirty" moieties responsible for promiscuity.

The four dehydrons within the binding pocket of EGFR are Asp831–Gly833, Gly697–Phe699, Gly697–Gly700, and Gly772–Val821. By examining the conservation of these four dehydrons across the 38 EGFR paralogs, we found the least conserved is dehydron Asp831–Gly833, retained only in the nine paralogs mentioned above. Thus, we choose this dehydron as the nonconserved selectivity feature to be targeted. We then appended a methyl group at position 3 of the terminal benzene ring that acts as a wrapper or protector of such feature (Fig. 10.1b). This redesigned EKB-569 molecule was synthesized following a pathway that recapitulates the EKB-569 synthesis [37], as shown in Fig. 10.1c.

Based on the selectivity filter provided by the wrapping patterns, we predict the affinity profile of our prototype. The prediction is based on the conservation of the EGFR dehydron Asp831–Gly833 wrapped by the prototype (*but not by EKB-569*) and the existence of steric hindrances with nonpolar groups in the catalytic loop of the target structures. We predicted as "hits" only those targets with a conserved dehydron in such position and no steric hindrance. Only six hits are predicted (Table 10.1): CLK3, EGFR, ERBB2, PTK2 (FAK), SLK, and STK10. In the cases where the residues aligning with EGFR's Asp831-Gly833 are not engaged in a dehydron or a well-wrapped hydrogen bond, we examined whether such dehydron can be *induced* upon ligand binding with minimal structural adaptation (this issue is further discussed in Chap. 11). Only in BTK and LCK kinases this dehydron can be induced

---

Fig. 10.1 Structural features promoting promiscuity and selectivity in EGFR kinase guiding EKB-569 cleaning redesign. (**a**) Structural alignment of EGFR kinase (*blue ribbon* representation and atoms in licorice) and the paralog TNK2 kinase (*red ribbon* representation and atoms in balls and sticks), complexed with EKB-569 (licorice). Atoms are depicted following standard color convention (chlorine in *green* and fluorine in *light green*). The wrapping pattern of EGFR includes the poorly conserved Asp831–Gly833 dehydron that may be targeted to achieve selectivity since TNK2 lacks this dehydron. TNK2 contains the same two promiscuity-fostering features,

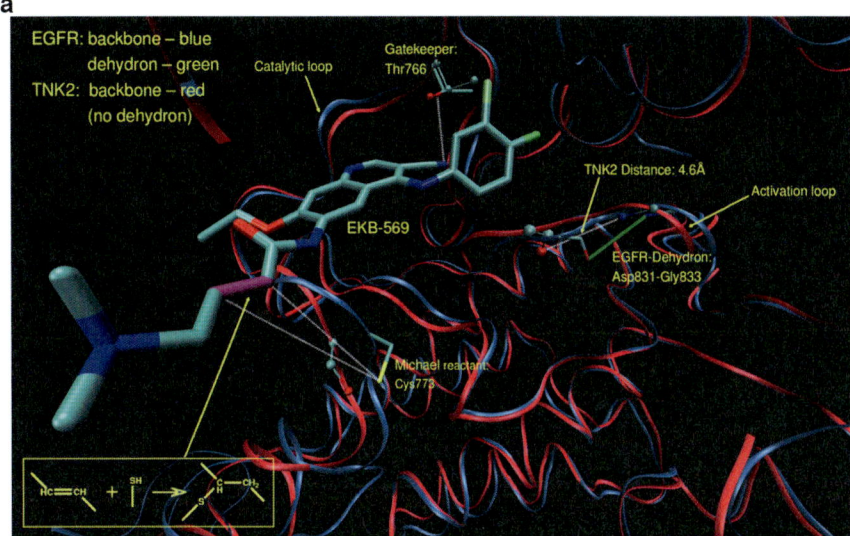

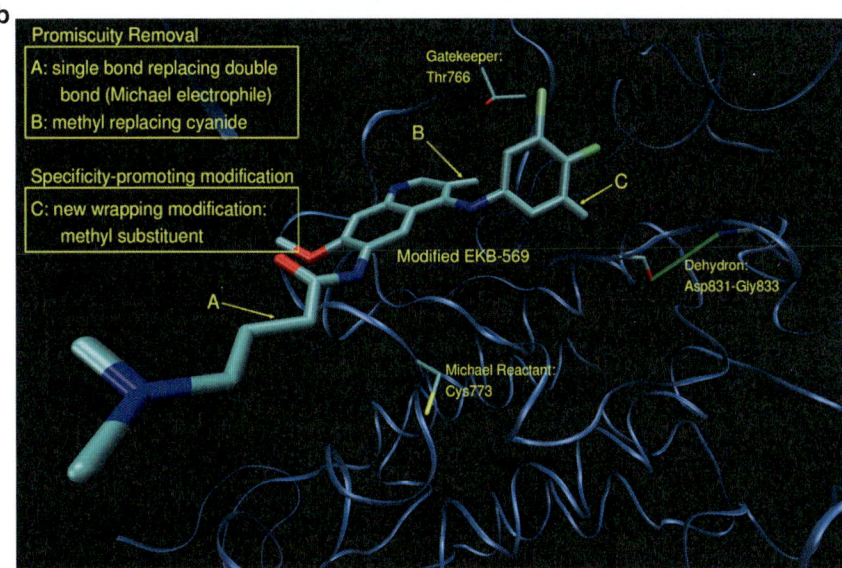

Fig. 10.1 (continued) while lacking the dehydron at the locus where EGFR contains the specificity-promoting feature. Thus, targeting the latter will ensure a discriminatory binding of EGFR without hitting TNK2. (**b**) EGFR kinase structure (same representation as above) complexed with the prototype EKB-569 redesigned inhibitor (licorice representation). To reduce EKB-569 promiscuity, the acrylic double bond (Michael electrophile) is replaced by a single bond and the gatekeeper-interacting cyanide is replaced by a methyl. To selectively target EGFR, a methyl group is added to the terminal benzene ring as a wrapper of the barely conserved Asp831–Gly833 dehydron. (**c**) Retro-synthetic pathway of the redesigned EKB-569 inhibitor. The synthesis recapitulates the EKB-569 pathway, albeit with different reactants [36]. Reprinted from (Trends in Biotechnology 26, [37]), copyright 2008 with permission from Elsevier

c

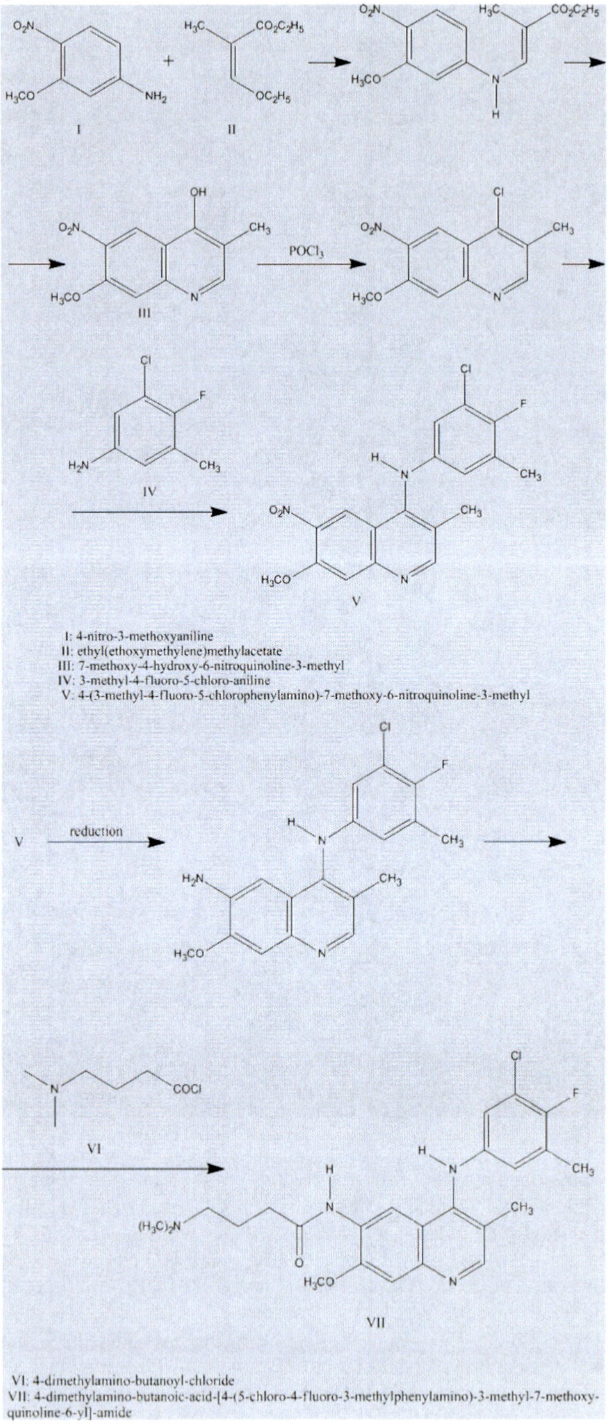
POCl$_3$
reduction
I
II
III
IV
V
VI
VII
I: 4-nitro-3-methoxyaniline
II: ethyl(ethoxymethylene)methylacetate
III: 7-methoxy-4-hydroxy-6-nitroquinoline-3-methyl
IV: 3-methyl-4-fluoro-5-chloro-aniline
V: 4-(3-methyl-4-fluoro-5-chlorophenylamino)-7-methoxy-6-nitroquinoline-3-methyl
VI: 4-dimethylamino-butanoyl-chloride
VII: 4-dimethylamino-butanoic-acid-[4-(5-chloro-4-fluoro-3-methylphenylamino)-3-methyl-7-methoxy-quinoline-6-yl]-amide

Fig. 10.1 (continued)

Table 10.1 Predicted and experimental affinity profiles for the "clean version" of kinase inhibitor EKB-569

Kinase	Dehydron conservation	Steric hindrance	Predicted affinity	Experimental affinity
ABL1	−		−	−
ACK1(TNK2)	−		−	−
Aurora2(AURKA)	+	−	−	−
Aurora3(AURKC)	−		−	−
BTK	Possibly induced	+	Possible hit	+
CDK2	−		−	−
CDK5	−		−	−
CLK1	−		−	−
CLK3	+	+	+	+
CSK	−		−	−
DAPK2	Possibly induced	−	−	−
DAPK3	−		−	−
EGFR	+	+	+	+
EPHA2	−		−	−
ERBB2	+	+	+	+
FGFR1	Possibly induced	−	−	−
FGFR2	−		−	−
HCK	−		−	−
INSR	−		−	−
JAK2	−		−	−
JNK1	−		−	−
JNK3	−		−	−
KIT	−		−	−
LCK	Possibly induced	+	Possible hit	+
MKNK2	−		−	−
NEK2	−		−	−
PAK1	−		−	−
PAK4	−		−	−
PAK6	−		−	−
PAK7(PAK5)	Possibly induced	−	−	−
PIM1	+	−	−	−
PIM2	+	−	−	−
PTK2(FAK)	+	+	+	+
RPS6KA5	−		−	−
SLK	+	+	+	+
SRC	−		−	−
STK10	+	+	+	+

upon drug binding with no steric hindrance with the catalytic loop, so they represent other "possible" hits (Table 10.1). The experimentally obtained affinity profile (Fig. 10.2) for the prototype agrees almost perfectly with our predicted profile, since the drug actually binds only to the six hits inferred with certainty and the two likely hits.

In this exercise, we have successfully cleaned the dirty inhibitor EKB-569 using an approach based on a selectivity filter. This was accomplished by first

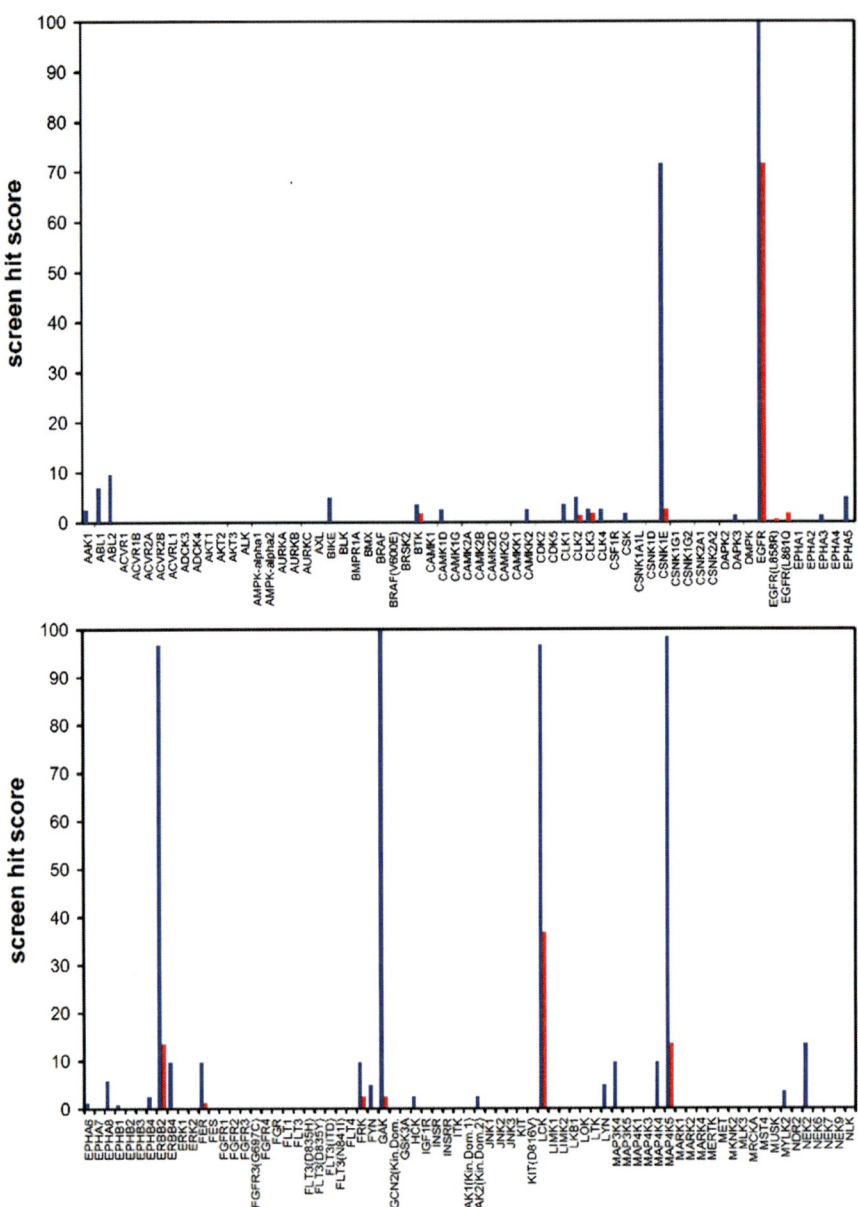

Fig. 10.2 Affinity profile of the redesigned EKB-569 prototype inhibitor. High-throughput screening at 10 mM EKB-569 (*blue*) and prototype (*red*) over a battery of 228 human kinases displayed in a T7-bacteriophage-expressing library. Hit values are reported as percentage-bound kinase. Reprinted from [37], copyright 2008 with permission from Elsevier

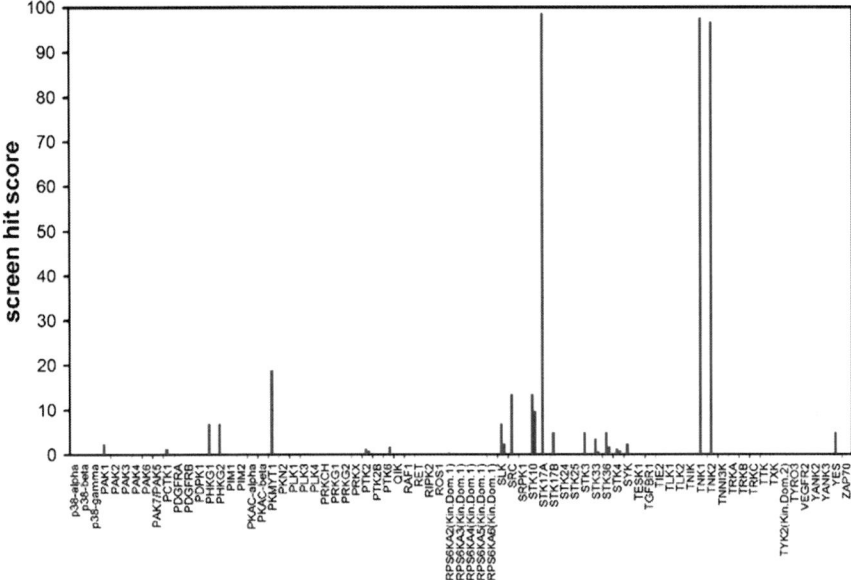

Fig. 10.2   (continued)

removing the chemical features that promote promiscuity. Subsequently, we introduced a wrapping modification to target a non conserved dehydron in the intended target and made the prototype more selective than the parental compound: Out of the 211 kinases experimentally screened [17, 25], our compound binds at nanomolar levels to 6 kinases only, whereas EKB-569 binds to 19.

10.4   Cleaning Staurosporine Through a Wrapping Modification: A Stringent Test

In a more stringent test, staurosporine, the most promiscuous kinase inhibitor known to man [21] was reengineered to elicit an inhibitory impact with controlled and enhanced specificity [38]. Staurosporine binds ($K_D < 3 \mu M$) to ~90% of the 290 human kinases recently screened [21]. To pursue a structure-based "cleaning" of staurosporine through a wrapping modification of the ligand, we compared all 28 PDB structures of kinases in complex with the ligand. A comparative wrapping analysis of their respective aligned ATP pockets prompted us to focus on the Src dehydron Q250–E267 which is the least conserved dehydron in this structural assortment. This dehydron is conserved in 7 out of 28 kinase structures, SRC, ABL, LCK, EGFR, ASK, TAO2, and GSK3, as indicated in Table 10.2 and Fig. 10.3a, b. The dehydron has been targeted by wrapping it through a specific methylation of staurosporine at the imide N6-position of the indole ring (Fig. 10.3a) [38]. Thus,

Table 10.2 Wrapping comparison of PDB-reported kinases in complex with staurosporine at position aligned with Src dehydron Q250–E267 (DH, dehydron; HB, well-wrapped hydrogen bond). The wrapping variant is predicted to bind *only* to the kinases that have a dehydron at the aligned position and the prediction is contrasted with the experimental screening of the wrapping variant of staurosporine

Group[a]	Protein	PDB	Wrapping classification	Hit in screening	Match prediction/experiment
TK	SRC	1BYG	DH	Yes	Yes
	ABL	2HZ4	DH	Yes	Yes
	LCK	1QPD	DH	Yes	Yes
	EGFR	2ITW (wt)	DH	Yes	Yes
		2ITQ (G719S)	HB	Not screened	
		2ITU (L858R)	HB	Not screened	
	FYN	2DQF	HB	No	Yes
	JAK3	1YVJ	HB	Not screened	
	SYK	1XBC	HB	No	Yes
	ZAP70	1U59	HB	No	Yes
	ITK	1SNU	HB	No	Yes
		1SM2	HB		
TKL	IRAQ4	2OIC	HB	Not screened	
		2NRY	HB		
AGC	PDK1	1OKY	HB	No	Yes
	PKAα	1STC	HB	No	Yes
	PKCθ	1XJD	HB	Not screened	
STE	ASK (MAP3K5)	2CLQ	DH	No	No
	TAO2	2GCD	DH	Not screened	
CAMK	CHK1	1NRV	HB	No	Yes
	MAPKAPK2	1NXK	HB	Not screened	
		2PZY	HB		
	MNK2	2HW7	HB	No	Yes
	DAPK1	1WVY	HB	Not screened	
	PIM1	1YHS	HB	No	Yes
CMCG	CDK2	1AQ1	HB	No	Yes
	GSK3β	1Q3D	DH	Not screened	
Unclassified	MPSK1 (STK16)	2BUJ	No HB	Not screened	

[a]Group names: AGC: containing PKA, PKG, PKC families; CAMK: calcium/calmodulin-dependent protein kinase; CK1: casein kinase 1; CMGC: containing CDK, MAPK, GSK3, CLK families; STE: homologs of yeast Sterile 7, Sterile 11, Sterile 20 kinases; TK: tyrosine kinase; TKL: tyrosine kinase-like

by redesigning staurosporine to turn it into a wrapper of the Src Q250–E267 dehydron, we can significantly restrict its inhibitory impact to the seven kinases that share the dehydron at the aligned position. The screening of the wrapping derivative of staurosporine is shown in Fig. 10.4a–c: only ∼11%, or 26 out of the 223 kinases assayed represent a hit, indicating a massive enhancement in specificity. *Most significantly, our structure-based prediction of affinity based on the presence or absence of the dehydron at the aligned position has only one false positive and not*

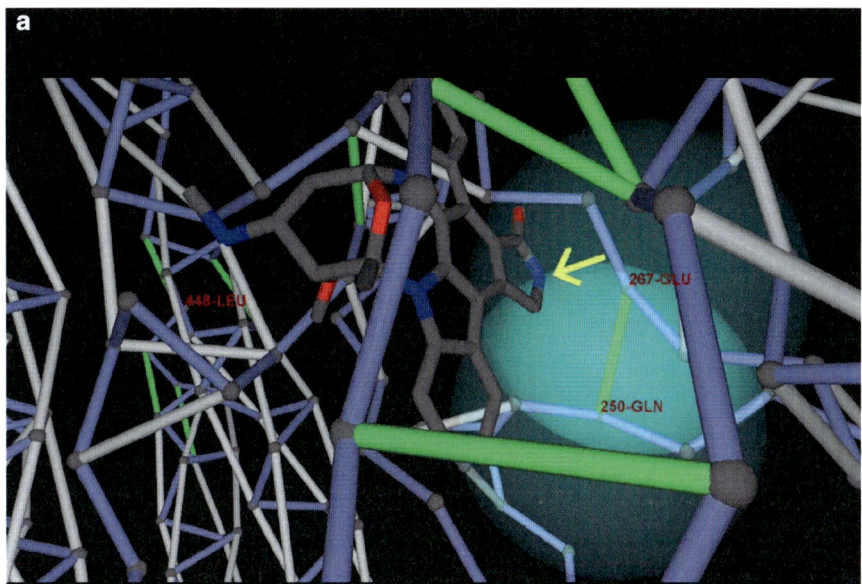

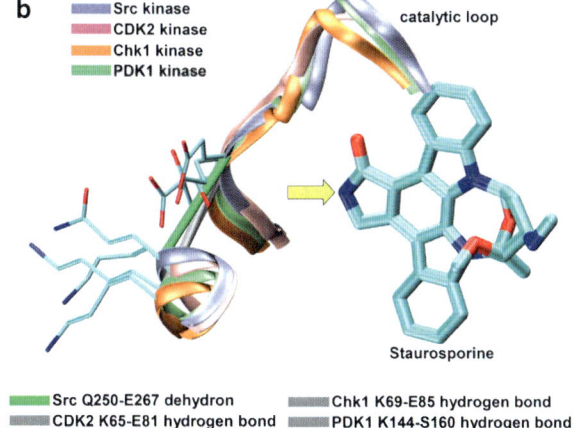

Fig. 10.3 (**a**) Relative position of Src kinase packing defects around the ligand indole region in a staurosporine–kinase complex (PDB.1BYG). Microenvironment of dehydron Q250–E267 in Src kinase framed by the desolvation spheres centered at the α-carbons of Q250 and E267. Methylation at the indole N5-position (indicated by the *yellow arrow*) would turn the ligand into a wrapper of the nonconserved packing defect. (**b**) Selected aligned backbones (*ribbon* representation) of Src (PDB.1BYG, *ice blue*), CDK2 (PDB.1AQ1, *pink*), Chk1 (PDB.1NVR, *orange*), and PDK1 (PDB.1OKY, *lime*) kinases complexed with staurosporine. The Src Q250–E267 dehydron (*green* virtual bond joining α-carbons) maps into well-wrapped backbone hydrogen bonds (*gray* virtual bond joining α-carbons): K65-E81 in CDK2, K69–E85 in Chk1 and K144–S160 in PDK1

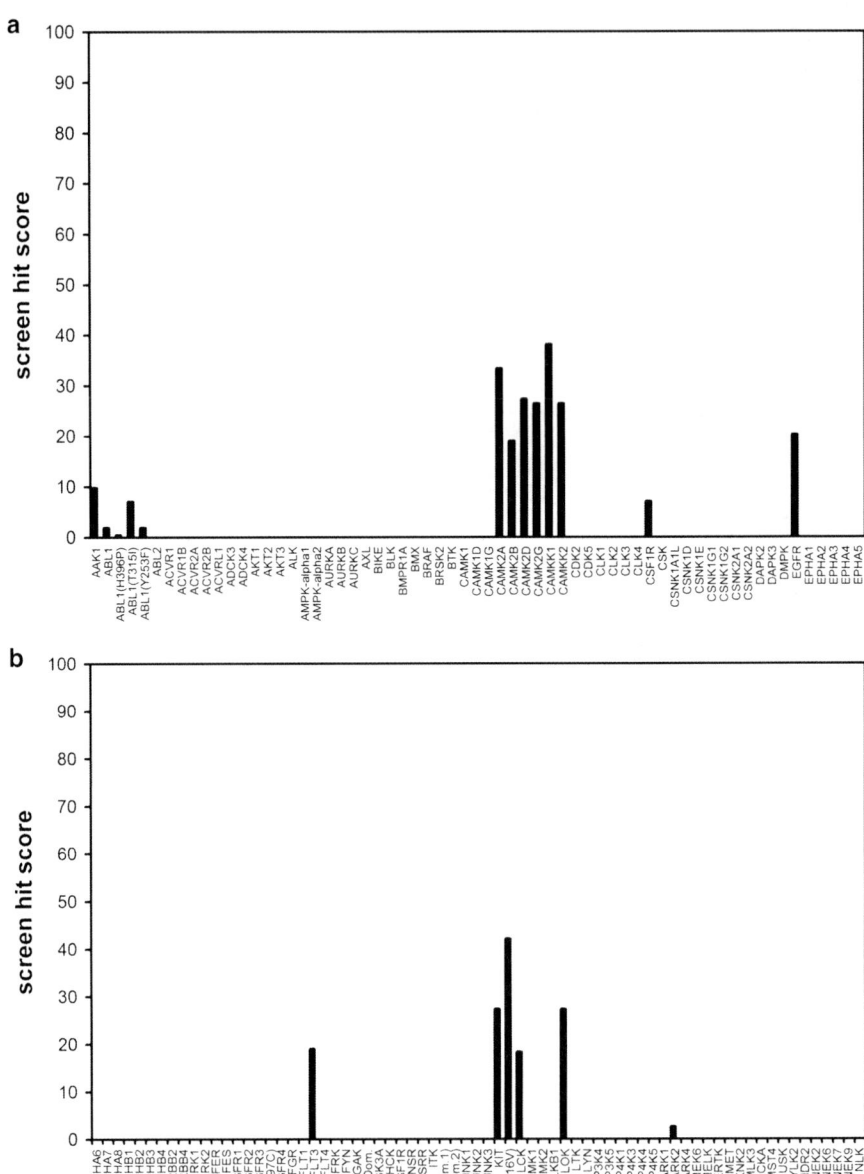

Fig. 10.4 (**a–c**) High-throughput screening at 10 mM for the wrapping variant of staurosporine over a battery of 223 human kinases displayed in a T7-bacteriophage library. Hit values are reported as percentage-bound kinase

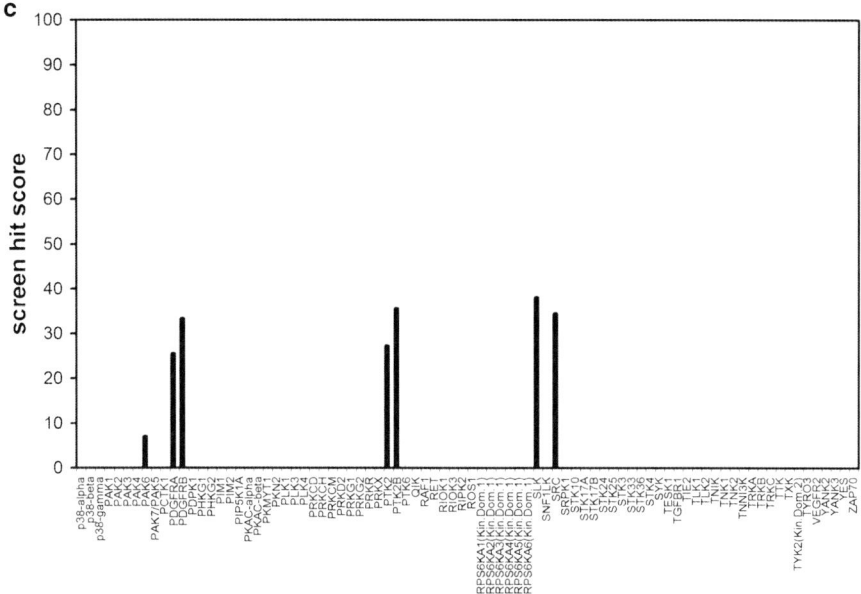

Fig. 10.4 (continued)

a single false negative over 15 instances (Table 10.2) where packing prediction can be contrasted with experiment! These results demonstrate that the packing differences across kinase homologs may guide molecular design to significantly enhance specificity in a highly controllable manner and even for the most promiscuous drugs.

10.5 Systems Biology Insights into Wrapping-Directed Design of Multi-target Kinase Inhibitors

Systems biology, that is, the multi-scale integration of the functional components of organismic complexity, is likely to broaden the bioinformatics platform of drug discovery [39, 40]. The modulation of clinical targets in living systems remains a central challenge and, for obvious reasons, systems biology is entrusted with providing meaningful approaches to face this challenge [41]. This emerging discipline is ultimately expected to bring meaningful insights to the issue of controlled promiscuity in anticancer therapy, yet its practical impact in this arena remains unclear and far from obvious. For all the interest in dirty or promiscuous therapeutic agents, drug discovery is still ingrained on the premise that successful therapeutic agents will result from targeting *single* disease-causing molecules, as the paradigmatic example of imatinib illustrated.

Tumors are perceived to be complex systems whose survival and propagation is a consequence of their resilience to external perturbations [41]. In the context of this discussion, these perturbations materialize as drug-based therapeutic interferences with the cell signaling apparatus. Thus, when facing a drug-induced node failure arising from the *specific* inhibition of a signal transducer, the tumor is likely to resort to basic robustness-enhancing possibilities that will ensure survival. These options arise especially from the resilient architecture of the signaling network and the genetic instability of the tumor [42], a feature that promotes sufficient target heterogeneity to render most selective drugs useless.

From a systems biology perspective, the features conferring robustness include the following [42–44]:

- Network modularity to hinder the propagation of effects arising from drug-induced node failure
- Network redundancy to maintain a biochemical flux in the face of a drug-induced node failure
- Feedback adaptive controls that ensure the habilitation of alternative pathways as one pathway is blocked by drug action
- Heterogeneity in the tumor cell population, promoted by genetic instability and ensuring the adaptive survival of a cell subpopulation which defines a drug-resistant trait

Thus, since the tumor is a masterful survivor, capable of displaying such a gamut of drug-resistant adaptations and of introducing manifold obstacles to molecular therapeutic intervention, it is hard to fathom what would be the advantage of a single-target therapeutic agent. It surely appears that a single drug-induced node failure can be readily overcome by merely resorting to the adaptive resilience of the network itself, unless the node happens to coincide with a point of network fragility. On the other hand, if the latter happens to be the case, the tumor can ultimately resort to its biological heterogeneity as an adaptive property. Thus, the genetic instability of the tumor is likely to promote a heterogeneous cellular population whereby the specific subpopulation that expresses drug-resistant kinase mutants or genetic variants is likely to survive drug treatment [42].

These considerations lead to the perception that an effective drug should target biochemical fluxes embedded in networks rather than specific molecules [45]. This goal may well be operationally unfeasible from the rational perspective of structure-based drug design. Furthermore, abandoning the single-target paradigm plunges us into the daunting challenges of controlling side effects described in the previous sections of this chapter.

The multiplicity of context-dependent roles for a target protein poses a danger of side effects *even in highly selective drugs*, such as the monoclonal antibodies *trastuzumab* (Herceptin, Genentech) [46] and *bevacizumab* (Avastin, Genentech/Roche) [47]. These therapeutic agents targeting, respectively, ERBB2 and VEGFA kinases are cardiotoxic and exhibit additional side effects like neutropenia, hemorrhage, hypertension, and thromboembolism.

As suggested by the systems biology assessment of robustness, a simultaneous modulation of multiple targets is probably necessary to alter a clinical phenotype as biological redundancies and alternative pathways can often bypass the inhibition of a single target. Thus, in spite of the issues pertaining to the control of side effects, a "magic shotgun" targeting multiple proteins may in some instances possess a higher therapeutic index than a more specific agent.

As noted previously in this chapter, multi-target drugs may be more resilient against drug-resistant mutations [48], providing another motivation to exploit promiscuity. This assertion readily follows from the fact that cross-reactive drugs typically make interactions with evolutionarily conserved residues and fewer interactions with residues that need to mutate to confer drug resistance.

Even if systems biology advocates the exploitation of multi-target kinase inhibitors, the latter probably entail a higher risk of promoting severe side effects than more specific compounds simply because they target more kinases. Yet, to counterbalance the argument, the highly specific anti-VEGFA bevacizumab introduces multiple and severe toxicities (hemorrhage, hypertension, cardiotoxicity) comparable to sunitinib [12]. In fact, nearly all drug-targeted therapies directed against a constitutively deregulated kinase or a target downstream of such a kinase present side effects. This is likely due to inhibition of "bystander" targets not essential for cancer cell killing or primary targets that may be essential to kill the cancer cell but play crucial roles associated with cell survival in other contexts.

10.6   Controlling the Cross-Reactivity of Sunitinib to Enhance Therapeutic Efficacy and Reduce Side Effects

From a systems biology perspective, the highly promiscuous but powerful anti-cancer agent sunitinib [49, 50] comes close to the ideal drug. Sunitinib action illustrates the power of multi-targeted kinase inhibitors with a dual pharmacological impact covering tumor angiogenesis targets VEGFR1-3 and PDGFR and tumor proliferation/survival targets KIT, CSF1R, FLT3 and RET. This level of promiscuity obviously entails a greater risk of cardiotoxicity due to the likelihood of inducing cardiomyocyte apoptosis through activation of pathways that compromise ATP replenishment or promote mitochondrial dysfunction [12, 51]. Intriguingly, treated patients with advanced GIST and metastatic renal carcinome often reveal significant reduction in left ventricular ejection fraction, a condition enhanced by other sunitinib-induced comorbidities, especially hypertension [51].

Thus, a wrapping-based redesign of sunitinib poses a far more serious challenge than that presented by imatinib (Chap. 8) because it is a far more promiscuous drug, because it targets major cancers (highly resilient solid tumors) and because of its confirmed severe side effects. We may regard the sunitinib redesign to substantially decrease its toxicity as the true proof of principle of the wrapping design concept.

A systems biology assessment of bystander sunitinib targets points to AMPK and RSK inhibition as culprit for cardiotoxicity [12]. Thus, RSK inhibition releases the pro-apoptotic factor BAD with activation of BAX and cytochrome *c* release, in turn promoting apoptosis and ATP depletion. The inhibition of AMPK promotes ATP depletion by preventing downstream inhibition of EEF2, mTOR, and ACC which are normally inhibited by AMPK in energy-compromising settings. In addition, AMPK inhibition affects cardiomyocyte survival if hypoxia occurs, since AMPK inhibits the energy-consuming processes of protein translation and lipid biosynthesis. Finally, the exacerbated activity of mTOR and EEF2 leads to cardiac hypertrophy [12, 51].

The differences in dehydron patterns across targets guide the redesign sunitinib into a discriminating therapeutic agent capable of retaining the inhibitory impact against the clinically relevant sunitinib targets (VEGFR, PDGFR, KIT, CSF1R, FLT3, and RET), while significantly reducing the activity toward the cardiotoxicity-related targets AMPK and RSK. The challenge from a structural biology standpoint is to identify a molecular feature that enables us to tell apart these two sets of targets and to harness on this difference to construct a therapeutically selective KI.

At first glance, a comparison of the wrapping patterns of primary sunitinib targets through structure alignment leads us to focus on a particular dehydron: the unburied backbone hydrogen bond N900–E917 in the VEGFR kinase, sunitinib's primary antiangiogenic target (Fig. 10.5a, b). This dehydron is conserved in all high-affinity targets relevant to angiogenesis, cancer-cell proliferation, and other clinically relevant processes, including PDGFR, CAMK1G, CAMK1D, CSF1R, FLT3, KIT, and RET. By contrast, the dehydron aligns with the well-packed hydrogen bond K78–E94 in AMPK2 (Fig. 10.6a, b) and with the well packed hydrogen bond T116–D132 in RSK. *That is, differences in the packing of the region that aligns*

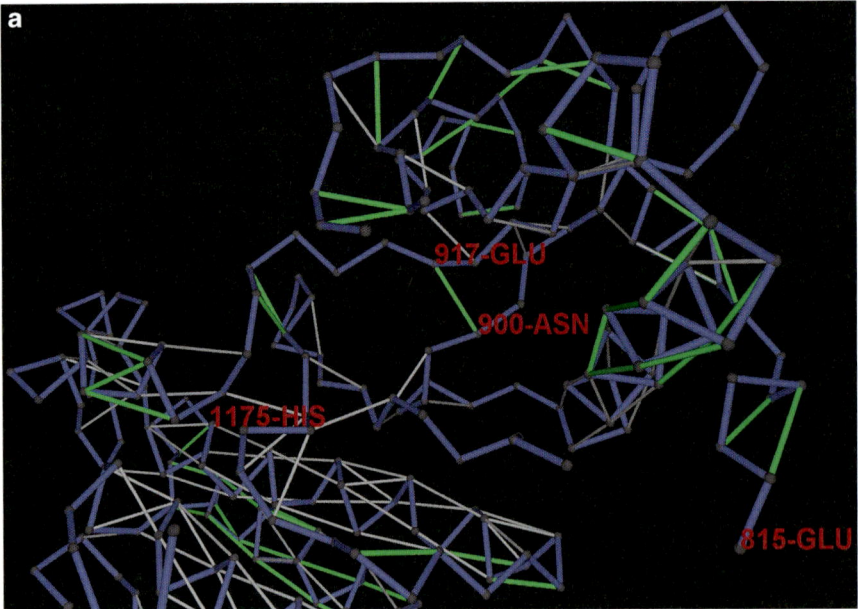

Fig. 10.5 (continued)

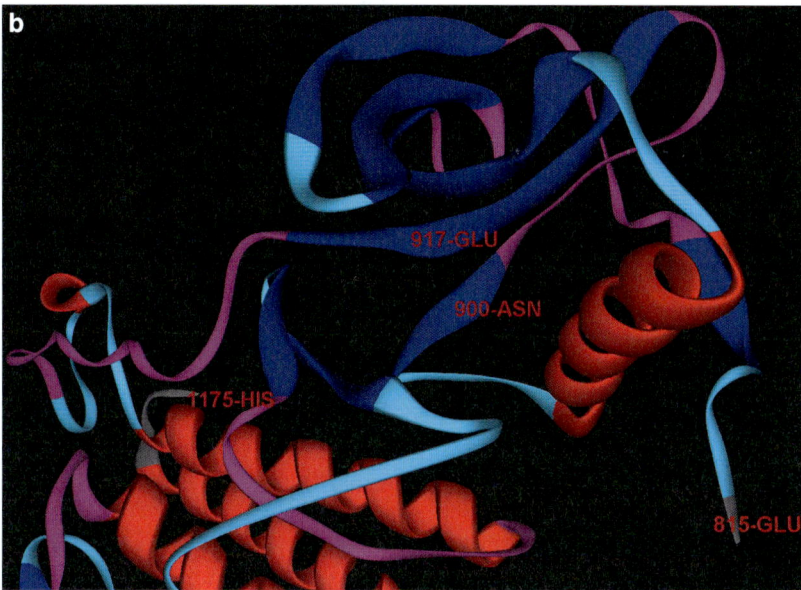

Fig. 10.5 (**a**) Wrapping of VEGFR kinase (PDB.2P2H), in the usual representation with backbone in *blue*, well-wrapped hydrogen bonds in *gray* and dehydrons in *green*. All bonds, including the backbone virtual bonds, are represented as *lines* joining the α-carbons of the paired residues. The N900–E917 dehydron is a *unique feature* conserved across all sunitinib's clinically relevant targets but aligned with well-wrapped hydrogen bonds in the cardiotoxicity-related targets AMPK and RSK. (**b**) Ribbon representation of VEGFR

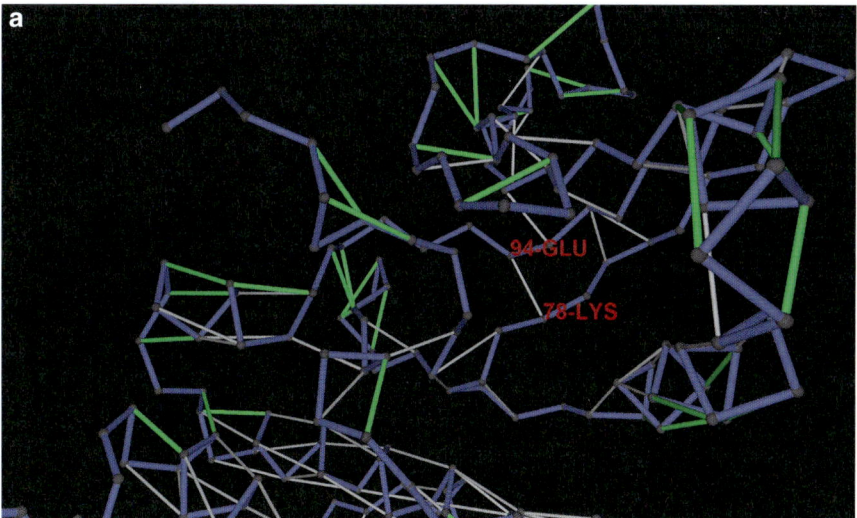

Fig. 10.6 (continued)

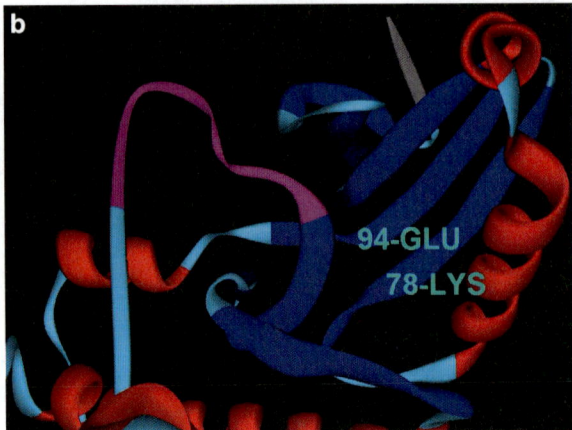

Fig. 10.6 (**a**) Wrapping of AMPK, highlighting the well-wrapped backbone hydrogen bond K78–E94. This bond aligns with a dehydron in all therapeutically relevant sunitinib targets. (**b**) Ribbon representation of AMPK structure

with dehydron N900–E917 in VEGFR may enable discrimination between clinically relevant and cardiotoxicity-promoting targets.

Thus, a promising wrapping difference between therapeutic and side-effect-related sunitinib targets is apparent as we contrast Figs. 10.5 and 10.6. This wrapping difference suggests possible modifications to sunitinib needed to remove

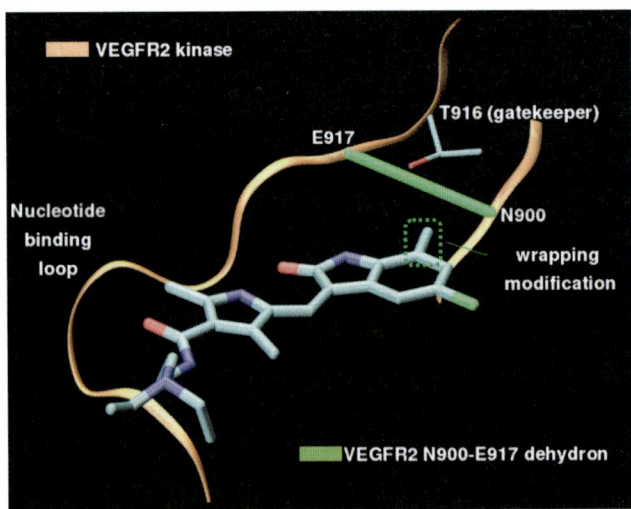

Fig. 10.7 Wrapping modification to sunitinib, making it selective for clinically relevant targets VEGFR1-3, PDGFR, KIT, CSF1R, FLT3, and RET, while avoiding AMPK and RSK

its cardiotoxicity and at the same time probe the potential of AMPK and RSK as cardiotoxicity biomarkers.

According to the premises of the wrapping technology, the methylation at the position indicated in Fig. 10.7 would turn the drug into a selective wrapper of dehydron N900–E917 in VEGFR. This drug is likely to bind selectively to all clinically relevant targets since the latter contain a dehydron at the position that aligns with N900–E917 in VEGFR. By contrast, this wrapping modification will produce a steric clash upon association with kinases RSK and AMPK2. Following this reengineering, a series of experimental studies will be necessary to demonstrate the target specificity, efficacy (in vitro and in vivo), and safety (with primary emphasis on cardiac safety using in vitro and in vivo models).

10.7 Is a Paradigm Shift in Drug Discovery Imminent?

Mounting evidence from the treatment of complex disorders and malignancies reveals that therapeutic efficacy does not correlate with drug specificity. This observation has motivated a reassessment of the therapeutic value of promiscuity and may well trigger a paradigm shift, from magic bullets to multi-target therapies. These conceptual leaps are supported by novel high-throughput screening technologies to assess cross-reactivity, hidden phenotypes, and side effects.

While a paradigm shift may be perceived as a necessity, it is far from becoming a reality. The major impediment arises because we do not have a clear picture of the therapeutic advantages of the novel concepts that would replace drug specificity. Knowing the realm of activity of "magic shotguns" is insufficient to warrant broad therapeutic application. The clinical uncertainties and risks they introduce are likely broader than the more limited side effects associated with more specific drugs.

Novel systems biology insights are leading to the belief that multi-target or even promiscuous drugs would be more effective to deal with complex diseases since dynamical objects like biochemical fluxes and not the molecules themselves are now perceived as the appropriate targets. However, the greater danger of side effects arising in multi-target therapies clearly counterbalances the recent faith bestowed upon them. Thus, we propose the use of selectivity filters to guide the rational control of specificity toward targets of clinical relevance within the systems biology frame of inference that dictated the use of multi-target drugs in the first place. It is clear that tumors may be more efficiently tackled by opening several therapeutic fronts but only a *selectively promiscuous drug* would be operative in such scenario.

At this juncture, multi-target molecular therapies could only be welcomed if their target selectivity can be controlled to curb side effects and clinical uncertainties. Thus, there is a niche in emerging biotechnologies for novel approaches to clean promiscuous drugs in order to achieve tighter specificity control. We have shown that such approaches are in principle feasible by redesigning highly cross-reactive kinase inhibitors, guided by novel selectivity filters. The drug redesign exercises and the proof of principle surveyed in this chapter suggest that cleaning a promiscuous drug guided by basic new concepts is in principle possible and will hopefully inspire further efforts in this regard. Promiscuity may become a welcomed feature in drug design and may well propel the needed paradigm shift, but only if cross-reactivity can be held under tight control through the effective implementation of selectivity filters.

References

1. Dancey J, Sausville EA (2003) Issues and progress with protein kinase inhibitors for cancer treatment. Nat Rev Drug Discov 2:296–313
2. Levitski A, Gazit A (1995) Tyrosine kinase inhibition: An approach to drug development. Science 267:1782–1788
3. Tibes R, Trent J, Kurzrock R (2005) Tyrosine kinase inhibitors and the dawn of molecular cancer therapeutics. Annu Rev Pharmacol Toxicol 45:357–384
4. Gibbs J, Oliff A (1994) Pharmaceutical research in molecular oncology. Cell 79:193–198
5. Donato NJ, Talpaz M (2000) Clinical use of tyrosine kinase inhibitors: Therapy for chronic myelogenous leukemia and other cancers. Clin Cancer Res 6:2965–2966
6. Roth BL, Sheffler DJ, Kroeze WK (2004) Magic shotguns versus magic bullets: Selectively non-selective drugs for mood disorders and schizophrenia. Nat Rev Drug Discov 3:353–359
7. Frantz S (2005) Drug discovery: Playing dirty. Nature 437:942–943
8. Keith CT, Borisy AA, Stockwell BR (2005) Multicomponent therapeutics for networked systems. Nat Rev Drug Discov 4:71–78
9. Mencher SK, Wang LG (2005) Promiscuous drugs compared to selective drugs (promiscuity can be a virtue). BMC Clin Pharmacol 5:3–9
10. McGovern SL, Helfand BT, Feng B, Shoichet BK (2003) A specific mechanism of nonspecific inhibition. J Med Chem 46:4265–4272
11. Feng BY, Shelat A, Doman TN et al (2005) High-throughput assays for promiscuous inhibitors. Nat Chem Biol 1:146–148
12. Force T, Krause D, van Etten RA (2007) Molecular mechanisms of cardiotoxicity of tyrosine kinase inhibition. Nat Rev Cancer 7:332–344
13. Fernández A, Sanguino A, Peng Z et al (2007) An anticancer C-kit kinase inhibitor is re-engineered to make it more active and less cardiotoxic. J Clin Invest 117:4044–4054
14. Demetri GD (2007) Structural reengineering of imatinib to decrease cardiac risk in cancer therapy. J Clin Invest 117:3650–3653
15. Crunkhorn S (2008) Anticancer drugs: Redesigning kinase inhibitors. Nat Rev Drug Discov 7:120–121
16. Hopkins AL, Mason J, Overington J (2006) Can we rationally design promiscuous drugs? Curr Opin Struct Biol 16:127–136
17. Hampton T (2004) 'Promiscuous' anticancer drugs that hit multiple targets may thwart resistance. J Am Med Assoc 292:419–422
18. Hopkins AL, Ren J, Milton J et al (2004) Design of non-nucleoside inhibitors of HIV-1 reverse transcriptase with improved drug resistance properties. J Med Chem 47:5912–5922
19. Owens J (2006) Screening: Dirty drugs' secrets uncovered. Nat Rev Drug Discov 5:542
20. Fabian MA, Biggs WH, Treiber DK et al (2005) A small molecule kinase interaction map for clinical kinase inhibitors. Nat Biotechnol 23:329–336
21. Karaman MW, Herrgard S, Treiber DK et al (2008) A quantitative analysis of kinase inhibitor selectivity. Nat Biotechnol 26:127–132
22. Fedorov O, Marsden B, Pogacic V et al (2007) A systematic interaction map of validated kinase inhibitors with Ser/Thr kinases. Proc Natl Acad Sci USA 104:20523–20528

23. MacDonald ML, Lamerdin J, Owens S et al (2006) Identifying off-targets effects and hidden phenotypes of drugs in human cells. Nat Chem Biol 2:329–337
24. Brunton LL, Lazo JS, Parker KL (2006) Goodman & Gilman's The Pharmacological Basis of Therapeutics. 11th ed. McGraw-Hill, New York
25. Rishton GM (2005) Failure and success in modern drug discovery: guiding principles in the establishment of high probability of success drug discovery organizations. Med Chem 1: 519–527
26. Kerkela R, Grazette L, Yacobi R et al (2006) Cardiotoxicity of the cancer therapeutic agent imatinib mesylate. Nat Med 12:908–916
27. Fedorov O, Sundstrom M, Marsden B, Knapp S (2007) Insights for the development of specific kinase inhibitors by targeted structural genomics. Drug Discov Today 12:365–372
28. Bogoyevitch MA, Fairlie DP (2007) A new paradigm for protein kinase inhibition: blocking phosphorylation without directly targeting ATP binding. Drug Discov Today 12:622–633
29. Crespo A, Fernández A (2007) Kinase packing defects as drug targets. Drug Discov Today 12:917–923
30. Chen J, Zhang X, Fernández A (2007) Molecular basis for specificity in the druggable kinome: Sequence-based analysis. Bioinformatics 23:563–572
31. Griffin JD (2005) Interaction maps for kinase inhibitors. Nat Biotechnol 23:308–309
32. Liu Y, Gray NS (2006) Rational design of inhibitors that bind to inactive kinase conformations. Nat Chem Biol 2:358–364
33. Noble ME, Endicott JA, Johnson LN (2004) Protein kinase inhibitors: Insights into drug design from structure. Science 303:1800–1805
34. Torrance CJ, Jackson PE, Montgomery E et al (2000) Combinatorial chemoprevention of intestinal neoplasia. Nat Med 6:1024–1028
35. Erlichman C, Hidalgo M, Boni JB et al (2006) Phase I study of EKB-569, an irreversible inhibitor of the epidermal growth factor receptor, in patients with advanced solid tumors. J Clin Oncol 24:2252–2260
36. Wissner A, Overbeek E, Reich MF et al (2003) Synthesis and structure-activity relationships of 6,7-disubstituted 4-anilinoquinoline-3-carbonitriles. The design of an orally active, irreversible inhibitor of the tyrosine kinase activity of the epidermal growth factor receptor (EGFR) and the human epidermal growth factor receptor-2 (HER-2). J Med Chem 46:49–63
37. Zhang X, Crespo A, Fernández A (2008) Turning promiscuous kinase inhibitors into safer drugs. Trends Biotechnol 26:295–301
38. Fernández A, Maddipati S (2006) The a-priori inference of cross reactivity for drug targeted kinases. J Med Chem 49:3092–3100
39. Butcher EC (2004) Systems biology in drug discovery. Nat Biotechnol 22:1253–1259
40. Butcher EC (2005) Can cell systems biology rescue drug discovery? Nat Rev Drug Discov 4:461–467
41. Kitano H (2007) A robustness-based approach to systems-oriented drug design. Nat Rev Drug Discov 6:202–209
42. Volgestein B, Kinzler KW (2004) Cancer genes and the pathway they control. Nat Med 10:789–799
43. Kitano H (2004) Biological robustness. Nat Rev Genet 5:826–837
44. Carlson JM, Doyle J (2002) Complexity and robustness. Proc Natl Acad Sci USA 99: 2538–2545
45. Hellerstein MK (2008) Exploiting complexity and the robustness of network architecture for drug discovery. J Pharmacol Exp Ther 325:1–9
46. Seidman A (2002) Cardiac dysfunction in the trastuzumab clinical trials experience. J Clin Oncol 20:1215–1221
47. Martel CL (2006) Bevacizumab-related toxicities: Association of hypertension and proteinuria. Commun Oncol 3:90–93
48. Hampton T (2004) 'Promiscuous' anticancer drugs that hit multiple targets may thwart resistance. J Am Med Assoc 292:419–422

49. Motzer RJ, Michaelson MD, Redman BG et al (2006) Activity of SU11248, a multitargeted inhibitor of vascular endothelial growth factor receptor and platelet-derived growth factor receptor, in patients with metastatic renal cell carcinoma. J Clin Oncol 24:16–24
50. Joensuu, H. (2007) Cardiac toxicity of sunitinib. Lancet 370:1978–1980
51. Chu TF, Rupnick MA, Kerkela R et al (2007) Cardiotoxicity associated with tyrosine kinase inhibitor sunitinib. Lancet 370:2011–2019

Chapter 11
Inducing Folding By Crating the Target

Kinases may be exploited as anticancer drug targets but their conformational plastic-ity often hinders the success of structure-based design. This is because of structural adaptation: target structures may change or proteins may adopt new conformations upon association with the ligand in unexpected and unpredictable ways. This may be the main reason for the modest interest in rational drug design when targeting regions with high disorder propensity such as the activation loop of a protein kinase. Yet this region presents the largest amino acid variability within the family and thus constitutes an attractive target to control specificity. In this chapter we advocate for a strategy to target flexible regions, offering a way to control the induced folding and turn it into a selectivity-promoting feature. Thus, drugs designed to wrap disordered regions in the target may be used to steer induced folding in very specific ways. The results surveyed herald the paradigmatic design concept of protective drugs for structurally adaptable targets.

11.1 Induced Folding: The Bête Noire of Drug Design

As described in Chaps. 7, 8, 9, and 10, kinases offer desirable therapeutic oppor-tunities because they transduce cell signals implicated in cancer progression and development and because they possess a natural binding pocket susceptible of being targeted by a man-made ligand [1–5]. These binding regions are framed by floppy regions (nucleotide-binding loop, activation loop, and catalytic loop) whose compo-sition determines the activation state of the kinase and its substrate specificity, hence conferring them their functional identity [5–10].

Precisely because of the conformational plasticity of the purported targets, structure-based design of kinase inhibitors has met with modest success. As an inhibitor binds to a kinase, it often promotes a largely unpredictable structural adaptation of the floppy regions. This adaptation is referred to as the induced fold-ing. This phenomenon poses design problems since the stability of the drug–target complex, and hence the drug affinity, cannot be meaningfully determined from the structures of the isolated binding partners [6–11]. The ab initio prediction of the induced fold remains forbiddingly difficult, often precluding a rational control of cross-reactivity for drugs that interact with loopy regions [6, 7]. In fact, the induced folding problem itself is comparable to the protein folding problem in terms of complexity and intractability [7–9].

However intractable, we are compelled to deal with this problem for a very simple reason: the highest variability in amino acid composition across kinases occurs precisely in the floppy regions, more specifically in the activation loop [12]. Thus, the most significant specificity-promoting feature becomes in practice the most difficult to target!

As illustrated in this chapter, the induced folding problem may be tractable in special contexts that are relevant to drug design: there are restrictive instances where structural adaptation may be steered in a controllable and predictable way, and selectivity control may be achieved through the identification of targets that undergo a similar induced fit upon association.

11.2 Wrapping the Target: A Tractable Case of Induced Folding

To deal with a tractable problem, we shall focus on small regions (four to eight residues) of the protein chain which are unstructured in the uncomplexed state and represent extreme cases of wrapping deficiency, that is, they are unable to hinder backbone hydration to any significant extent [11]. As discussed in Chap. 5, such regions are unequivocally characterized as being unable to contribute with at least seven nonpolar wrapping groups to the microenvironment of putative backbone hydrogen bonds. Since $\rho = 7$ is the threshold for existence of a backbone hydrogen bond, such regions are disordered with 100% certainty in the uncomplexed state of the protein. To distinguish the solvent shielding of induced hydrogen bonds from that of pre-formed hydrogen bonds, we shall introduce the term "crating" when referring to the former while retaining the usual term wrapping for the latter.

We shall focus on the problem of targeting regions that visit conformations where the backbone hydrogen bonds are too exposed to the solvent to be retained in soluble structure. The backbone amides and carbonyls in these regions are actually prone to become completely hydrated with concurrent loss of structure (Fig. 11.1). Thus, drug ligands may be specifically designed to "correct" the severe yet localized packing deficiency by removing surrounding water upon association, thereby inducing a hydrogen bond which is missing in the uncomplexed state of the protein. The design principle is illustrated in Fig. 11.1 and is based on ligand designs which are protective of specific conformations *visited but not retained* by the target protein in the unbound state.

As emphasized in previous chapters, in regions where kinase structure is retained upon binding, drug selectivity may be achieved or at least controlled by targeting dehydrons [12]. This special type of targeting is realized by engineering a drug that wraps the pre-formed intramolecular hydrogen bond [13–15], effectively contributing with nonpolar groups to the exclusion of water from the hydrogen bond microenvironment.

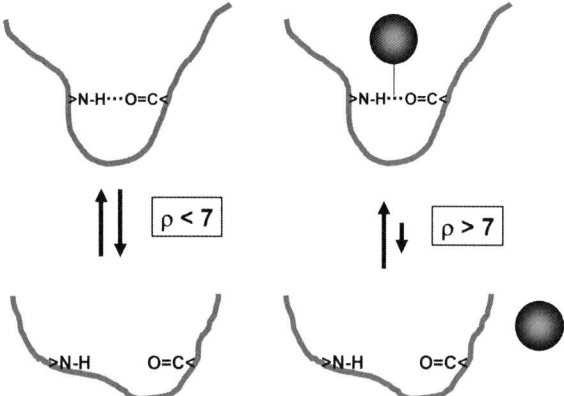

Fig. 11.1 Basic scheme of the folding-inducing drug. Due to insufficient crating in the unbound state ($\rho < 7$), the target protein *visits but does not retain* a hydrogen-bonded conformation that is in dynamic equilibrium with the fully hydrated amide and carbonyl. Since the drug (*gray* sphere) is able to provide sufficient crating ($\rho > 7$), the target protein in the drug-bound state is able to retain the hydrogen-bonded conformation. Since the proximity of a nonpolar moiety hinders hydration, the drug has no affinity for the alternative conformation of the protein target, with its solvent-exposed amide and carbonyl

Thus, a backbone hydrogen bond inferred from homolog structure alignment may be absent in a reported protein structure due to severe under-protection, i.e., when $\rho < 7$ [11]. Yet, it may be selectively induced if a drug is appropriately constructed to crate it, much like the drugs designed to "correct" wrapping defects in target proteins [12]. The crating design concept on a structurally adaptable target is illustrated in Fig. 11.2, which displays a complex between a protective drug and a kinase with a drug-induced folded region arising from the intermolecular crating of a hydrogen bond. The F811–A814 dehydron within the activation loop (in red) of the C-Kit kinase is being crated and hence induced by a purposely designed modification of the anticancer drug imatinib. The latter is known to target the C-Kit kinase, while the modified version, known as WBZ_7 [16], provides additional crating to the activation loop, enhancing affinity and selectivity beyond the levels achieved by the parental compound [16]. In other words, in the unbound state of the protein, the activation loop is likely to visit the conformation that contains the F811–A814 backbone hydrogen bond, but it is unlikely to retain it due to the local absence of tertiary structure. Yet, when the ligand is able to provide appropriate crating, the bond becomes inducible as the loop is able to adopt a specific conformation. Hence there is a crucial difference between wrapping and crating: while wrapping refers to an extra level of shielding or protection of a pre-formed hydrogen bond retained in the uncomplexed protein, crating provides exogenous shielding to a hydrogen bond that would otherwise not survive in the uncomplexed state.

The design concept of drug as crater of a specific local conformation adopted by a floppy region in the target has in fact been exploited in re-workings of imatinib [14]. As described in Chap. 8, this wrapping-based redesign not only targets a unique dehydron in the intended target C-Kit but also crates an otherwise floppy

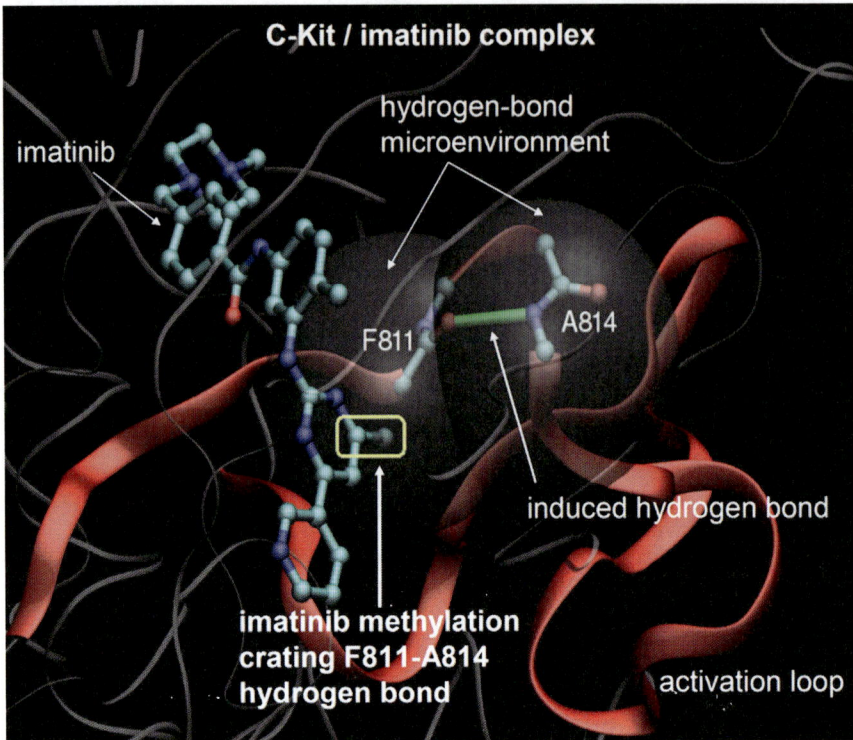

Fig. 11.2 *Design concept of protective drug illustrated by an imatinib modification.* A primary imatinib target, the C-Kit kinase (PDB.1T46), displayed in ribbon representation with activation loop in *red*, is shown in complex with a methylated version of imatinib (methylation highlighted by *yellow box*). The parental drug has been modified to crate the deficiently packed F811–A814 backbone hydrogen bond within the flexible activation loop. The protection materializes because the appended methyl in the drug penetrates the desolvation domain of the backbone hydrogen bond upon association. In consistence with Chap. 1, the bond microenvironment is defined by two intersecting balls centered at the α-carbons of residues F811 and A814. Thus, a specific local structure of the activation loop has been induced upon association with a drug purposely designed to crate a hydrogen bond. Reprinted from [21], copyright 2009 with permission from Elsevier

region in JNK, thus promoting an intended cross-reactivity needed to curb the side effects of the parent compound [14]. However, only a dynamic analysis of the target/drug complex can attest that the induced folding was indeed steered by the drug design. Thus, to fully assess this possibility, we revisit the clinically relevant imatinib redesign this time in light of a dynamic analysis of structural adaptation in the drug–target complex.

11.3 Kinase Inhibitors Designed to Crate Floppy Regions

The crating drug concept is evaluated by revisiting a recent redesign of imatinib [15], the anticancer drug adopted to treat chronic myeloid leukemia (CML) [3, 4]. As previously described in Chaps. 7 and 8, this inhibitor was originally designed to

target the deregulated kinase chimera from Bcr-Abl fusion, a result of chromosomal translocation causative of CML [4]. The redesign, named WBZ_4 [14, 15], was
engineered to funnel the drug specificity toward some clinically relevant targets (C-
Kit and PDGFR) [17] while reducing the cardiac risk attributed to Bcr-Abl kinase
targeting [18, 19]. One reinforcing way of averting this side effect is by ensuring
that the redesigned version of imatinib also inhibits JNK [12, 20], a robust downstream signal transducer in the ATP depletion and pro-apoptotic pathways triggered
by Bcr-Abl inhibition in cardiomyocytes [17, 19, 20]. *To avoid introducing new
cross-reactivities, the inhibition of JNK must involve the same crating modification
that enabled WBZ_4 to discriminate C-Kit (and its paralog kinase PDGFRK) from
the primary imatinib target Bcr-Abl. Thus, WBZ_4 acts as a wrapper of the unique
dehydron in C-Kit and as a crater of an inducible hydrogen bond in JNK.*

The JNK inhibitory activity of WBZ_4 ($K_D \approx 51$ nM) has being experimentally
confirmed and is not found in imatinib [14]. In fact, as our dynamic analysis shows,
this affinity results from a structure-inducing role of WBZ_4 that, in contrast with
imatinib, acts as a protector and hence a stabilizer of a specific JNK conformation.
As shown below, this conformation introduces an additional hydrogen bond in the
nucleotide-binding loop. This revision of the inhibitory impact of WBZ_4 on the
floppier JNK supports our proposed strategy to control the target-induced folding in
a selective manner.

As shown in Chap. 8, there is a dehydron present in C-Kit, the C673–G676
backbone hydrogen bond in the nucleotide-binding loop, that aligns with the well-
protected M318–G321 hydrogen bond in Bcr-Abl. Targeting this unique packing
defect by crating it with an added methylation in imatinib enhances drug selectivity
(Fig. 11.3), telling apart C-Kit, where the wrapping defect is present, from Bcr-Abl
kinase [14]. On the other hand, *the C-Kit C673–G676 packing defect does not align
with any backbone hydrogen bond in JNK* (Fig. 11.3). This is so since the purported
JNK M111–N114 hydrogen bond that would align with C673–G676 in C-Kit is
too severely under-protected to prevail (even as a packing defect) in the soluble
structure of JNK: a homology threading of the 109–118 nucleotide-binding loop
of JNK into the backbone conformation of C-Kit produces the backbone hydrogen
bond M111–N114 but with maximum possible extent of protection $\rho = 6$. This
value, obtained by maximizing over side chain rotamers in JNK 109–118 region, is
below the threshold protection needed to maintain the bond. In other words, JNK is
locally more flexible and less constrained than the primary imatinib targets and may
be inhibited precisely by targeting its unique region of structural plasticity. This is
precisely what the WBZ_4 compound accomplishes.

A molecular dynamics (MD) analysis enables a clear dissection of the protective
and folding-inducing role of WBZ_4 targeting the flexible region of JNK. This analysis provides a new framework to interpret the affinity of this ligand for the kinase
and supports the design concept of protective drug steering structural adaptation.

As MD computations corroborate [21], the M111–N114 hydrogen bond
(N–O distance < 3.4 Å) is unsustainable in the uncomplexed JNK or even in the
JNK/imatinib complex. The latter forms the bond intermittently and ephemerally
but its average lifetime, $\tau = 12$ ps over 20 runs spanning 730 ns, indicates metastability (Fig. 11.4a). This metastability results primarily from the lack of enough

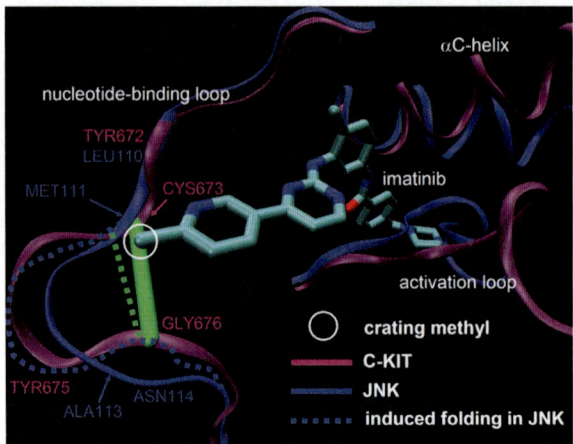

Fig. 11.3 *Protective drug as folding inducer. Structural alignment of targets JNK (isoform 1) and C-Kit for the imatinib redesigned prototype drug WBZ_4* [14]. The JNK backbone from PDB.1UKH is displayed as a *thicker line* than that of C-Kit (PDB.1T46). The C-Kit residues Y672 and Y675 contribute to crate the "crankshaft" C673–G676 hydrogen bond (*dashed line*) in C-Kit. This bond gets exogenous protection from the drug/ligand which has been purposely engineered to contribute to its crating. This crating capability represents the major difference between WBZ_4 and the parental drug imatinib. Protective interactions contributing to the dehydration of the C673–G676 hydrogen bond are indicated as *thin green lines*. The JNK residues aligning with C673 and G676 in C-Kit are M111 and N114. Lacking enough intramolecular protection (the protective Y675 in C-Kit is substituted for the poor protector A113 in JNK), they can only become paired by an induced backbone hydrogen bond (*arrow*) upon association with WBZ_4. This induced local folding does not prevail upon binding to the parental drug imatinib since the latter lacks the crating group at the terminal ring. Reprinted from [21], copyright 2009 with permission from Elsevier

intramolecular protection: the protective Y675 in C-Kit is substituted in JNK for the poor protector A113 (Fig. 11.2 and Table 11.1). By contrast, the hydrogen bond is induced upon association with WBZ_4 (Fig. 11.4a): the added methyl in WBZ_4 brings extra protection upon binding.

Fig. 11.4 *Folding upon binding dynamics of JNK in complex with a protective drug.* (**a**) Dynamics of the M111–N114 backbone hydrogen bond formation in JNK (isoform 1) in the uncomplexed form, within the complex with imatinib, and within the complex with the crating ligand WBZ_4. The induced folding is monitored by the M111–N114 carbonyl–amide O–N distance and materializes when this distance is below a hydrogen bond threshold (~3.4 Å, *blue line*). Twenty runs were generated for each case and the dynamics averaged over the trajectory ensemble is indicated by the *red line plot*. The *black line plot* corresponds to a single representative run. For the uncomplexed kinase, the runs covered an equilibration span of 500 ns (time extensions were deemed unnecessary due to equilibration) and represented the control results. The folding dynamics for the complexes covered 730 ns. (**b**) Induced hydrogen bond dynamics for the N114D JNK mutant in complex with the crating drug WBZ_4. The stochastic boundary molecular dynamics (SBMD) method [21] was employed to reduce simulation times while capturing the localized

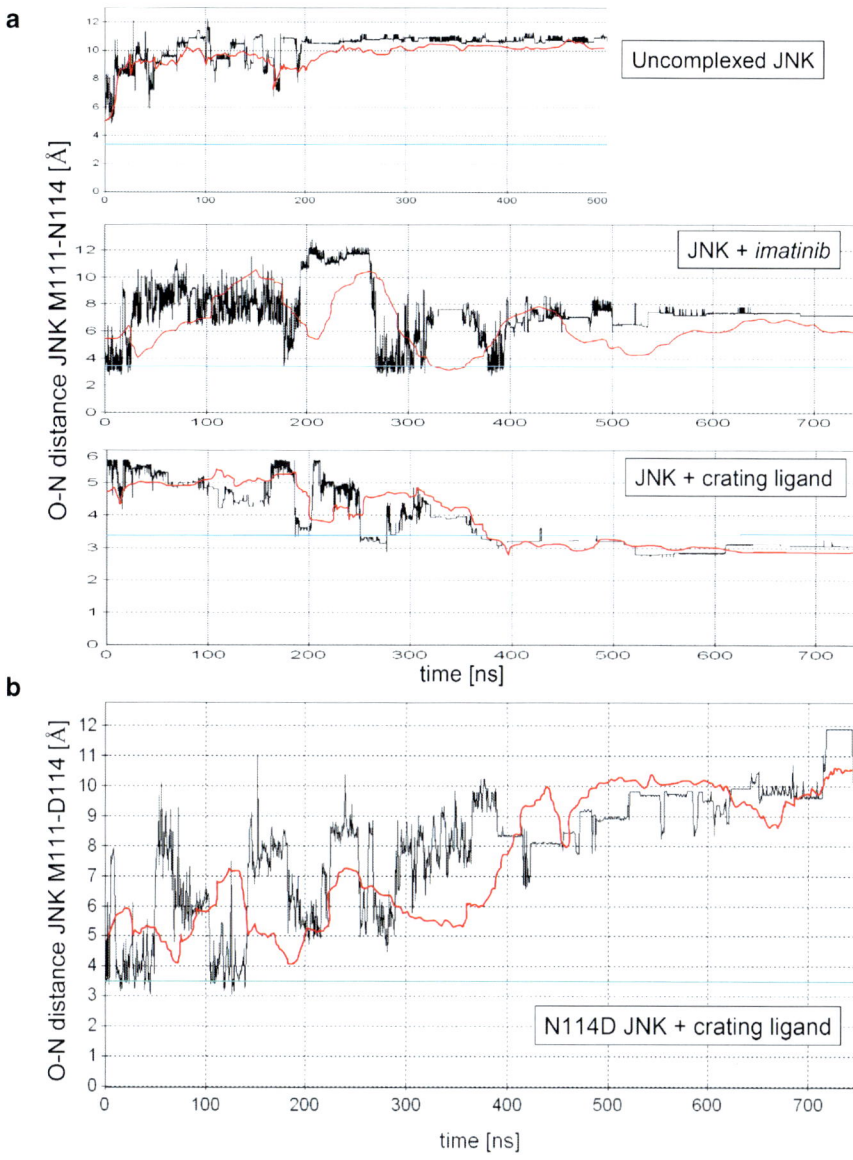

Fig. 11.4 (continued) binding-folding dynamics at the ATP-binding site of the JNK kinase. The system has a reaction zone of 42 residues and a reservoir region. The MD simulations were performed using the structural coordinates of free JNK (control), chimeric JNK/imatinib, and JNK/WBZ_4 complexes that resulted from alignment of the JNK1 crystal structure PDB.1UKH with that of the C-Kit/imatinib complex PDB.1T46, followed by in silico removal of the Kit domain. Hence, the simulations described the equilibration process of the JNK/imatinib and JNK/WBZ_4 complexes originally obtained from alignment and combination of the PDB files 1UKH and 1T46. Reprinted from [21], copyright 2009 with permission from Elsevier

The computational analysis supports and rationalizes the experimentally established affinity of WBZ_4 for JNK as being mainly due to its ability to crate and thus stabilize a particular JNK conformation. This conformation contains the M111–N114 backbone hydrogen bond that aligns with the packing defect C673–G676 in C-Kit and becomes an induced fit in JNK upon binding to WBZ_4.

To corroborate that the induction of the M111–N114 hydrogen bond is the dominant factor dictating the drug affinity for JNK, we examined the residues aligning with the C-Kit C673 and G676 across paralogs of JNK(1/3) with PDB-reported structure: MAPK1, MAPK6, p38α, and p38γ (Table 11.1). As shown elsewhere [14], WBZ_4 does not have any detectable affinity for such paralogs. Indeed, we expect that none of the paralogs is capable of maintaining the crankshaft hydrogen bond because one of the putatively paired residues is charged (Asp) rather than the polar N114 in JNK (Table 11.1). Furthermore, the extent of crating of the JNK paralogs is comparable to that of JNK, and less than that of C-Kit, mainly because of the replacement of the C-Kit Y675 for poorer protectors (A, T). The charged Asp aligning with position N114 in JNK has higher hydrophilicity and is therefore disruptive of the backbone hydrogen bond. This is corroborated not only by the absence of a crankshaft hydrogen bond in all reported structures of JNK(1/3) paralogs (Table 11.1) but also dynamically (Fig. 11.4b). In contrast with wild-type JNK, the crating drug WBZ_4 fails to induce the crankshaft hydrogen bond in the N114D in silico mutated JNK (Fig. 11.4b). Thus, the substitution of polar N114 for charged D114 introduces disruptive negative design, preventing side chain nonpolar groups from protectively clustering around the putative M111–D114 bond.

Table 11.1 Across-paralog alignment of residues that either form or crate the inducible backbone hydrogen bond M111–N114 in JNK

Kinase	Kit	Jnk1	Jnk3	MAPK6	MAPK1	p38α	p38γ
PDB code	1T46	1UKH	1PMN	2I6L	2OJI	1A9U	1CM8
Gatekeeper	THR670	MET108	MET146	GLN108	GLN103	THR106	MET109
HB crating	TYR672	LEU110	LEU148	TYR110	LEU105	LEU108	PHE111
Crankshaft HB	CYS673	MET111	MET149	MET111	MET106	MET109	MET112
HB crating	TYR675	ALA113	ALA151	THR113	THR108	ALA111	THR114
Crankshaft HB	GLY676	ASN114	ASN152	ASP114	ASP109	ASP112	ASP115
Crankshaft HB inducible?	Yes	Yes	Yes	No	No	No	No
Crating drug binding?	Yes	Yes	Yes	No	No	No	No

The alignment extends over all paralogs of JNK (isoforms 1 and 3) with PDB-reported structure and enables an inference of the feasibility of inducing the hydrogen bond. This feasibility depends on the purported extent of crating of the putative hydrogen bond and on the hydration propensity of the residues paired by the bond and becomes a necessary and sufficient condition for detectable affinity of the protective drug WBZ_4

11.4 Steering Induced Folding with High Specificity: The Emergence of the Crating Design Concept

Their therapeutic appeal notwithstanding, from a structural biology perspective, kinase targets pose daunting problems to the drug designer: their floppy regions can undergo considerable structural adaptation upon binding to the drug. Thus, due to the often unpredictable nature of induced folding, structure-based drug design has met limited success and is often relegated in favor of high-throughput screening methods and serendipitous engineering as the means to find or improve molecular leads. This leaves us with a conceptual conundrum since floppy regions, in particular the activation loop, are the most unique in the kinase targets, i.e., the ones that confer the kinase its functional and contextual identity.

A rational control of induced folding is nevertheless possible and of course highly desirable, as shown in this chapter. This control may be achieved by focusing on structural adaptations of deficiently wrapped regions of the protein chain. Thus, we propose a targeting approach suitable for floppy regions in kinases based on ligand/drug designs which crate specific conformations of the protein that would not be otherwise sustainable in the uncomplexed state. Such drugs contribute to improve the wrapping of the target upon association with the point where the latter is now capable of retaining a specific conformation. Thus, the target affinity of folding-inducing drugs is due in part to the fact that they stabilize the drug/target complex. Support for this concept is found through re-examination of a recently re-engineered version of imatinib in light of a dynamic analysis. Our dynamic assessment shows that the inhibitory activity of the compound results precisely from its ability to steer induced folding by protecting a conformation in a floppy kinase. On the other hand, since packing deficiencies are fairly unique to kinase targets and typically not conserved across proteins of common ancestry, the specificity of such protective drugs is basically guaranteed.

On the horizon we envision a new generation of protective kinase inhibitors stemming from known chemical scaffolds that may induce specific local structuring of the activation loop, the most extensive flexible region. Such drugs may exquisitely compete with ATP as they possess an extra anchoring site so far unexploited. This possibility may represent one of the few approaches to the problem of rationally exploiting induced folding in floppy targets to steer drug cross-reactivity in ways that provide clinical advantage.

References

1. Dancey J, Sausville EA (2003) Issues and progress with protein kinase inhibitors for cancer treatment. Nat Rev Drug Discov 2:296–313
2. Tibes R, Trent J, Kurzrock R (2005) Tyrosine kinase inhibitors and the dawn of molecular cancer therapeutics. Annu Rev Pharmacol Toxicol 45:357–384
3. Schindler T, Bornmann W, Pellicena P et al (2000) Structural mechanism for STI-571 inhibition of Abelson tyrosine kinase. Science 289:1938–1942

4. Schiffer CA (2007) BCR-ABL tyrosine kinase inhibitors for chronic myelogenous leukemia. N Engl J Med 357:258–265
5. Crespo A, Fernández A (2007) Kinase packing defects as drug targets. Drug Discov Today 12:917–923
6. Teague S (2003) Implications of protein flexibility for drug discovery. Nat Rev Drug Discov 2:527–541
7. Damm KL, Carlson HA (2007) Exploring experimental sources of multiple protein conformations in structure-based drug design. J Am Chem Soc 129:8225–8235
8. Hornak V, Simmerling C (2007) Targeting structural flexibility in HIV-1 protease inhibitor binding. Drug Discov Today 12:132–138
9. Erickson J (2004) Lessons in molecular recognition: The effects of ligand and protein flexibility on molecular docking accuracy. J Med Chem 47:45–55
10 Noble ME, Endicott JA, Johnson LN (2004) Protein kinase inhibitors: Insights into drug design from structure. Science 303:1800–1805
11. Pietrosemoli N, Crespo A, Fernández A (2007) Dehydration propensity of order-disorder intermediate regions in soluble proteins. J Proteome Res 6:3519–3526
12. Chen J, Zhang X, Fernández A (2007) Molecular basis for specificity in the druggable kinome: Sequence-based analysis. Bioinformatics 23:563–572
13. Fernández A (2004) Keeping dry and crossing membranes. Nat Biotechnol 22:1081–1084
14. Fernández A, Sanguino A, Peng Z et al (2007) An anticancer C-kit kinase inhibitor is re-engineered to make it more active and less cardiotoxic. J Clin Invest 117:4044–4054
15. Crunkhorn S (2008) Anticancer drugs: Redesigning kinase inhibitors. Nat Rev Drug Discov 7:120–121
16. Fernández A, Sanguino A, Peng Z et al (2007) Rational drug redesign to overcome drug resistance in cancer therapy: Imatinib moving target. Cancer Res 67:4028–4033
17. Demetri GD, von Mehren M, Blanke CD et al (2002) Efficacy and safety of imatinib mesylate in advanced gastrointestinal stromal tumors. N Engl J Med 347:472–480
18. Kerkela R, Grazette L, Yacobi R et al (2006) Cardiotoxicity of the cancer therapeutic agent imatinib mesylate. Nat Med 12:908–916
19. Force T, Krause D, van Etten RA (2007) Molecular mechanisms of cardiotoxicity of tyrosine kinase inhibition. Nat Rev Cancer 7:332–344
20. Demetri GD (2007) Structural reengineering of imatinib to decrease cardiac risk in cancer therapy. J Clin Invest 117:3650–3653
21. Fernández A, Bazan S, Chen J (2009) Taming the induced folding of drug-targeted kinases. Trends Pharmacol Sci 30: 66–71

Chapter 12
Wrapper Drugs as Therapeutic Editors of Side Effects

Wrapping designs have some limitations. While they enable control of specificity, they may not be able to exclude a toxicity-related target if the latter shares the same wrapping pattern with the therapeutically relevant targets. In such circumstances, we may need to resort to a multicomponent therapy where one drug acts synergistically with the other while selectively antagonizing it in the specific context where the action of the first drug promotes toxicity. This chapter explores these a priori therapeutic possibilities.

As previously described, due to their ability to interfere with signal transduction events controlling cell proliferation and fate, kinase inhibitors hold promise as anticancer agents. Nevertheless, the functional role of a kinase depends on the cellular context and hence kinase inhibition in an off-target cell may lead to side effects. Motivated by these observations, we explore a mode of "therapeutic editing" where one drug – the editor – suppresses the side effect promoted by the primary drug as it impacts off-target cells. Editor and primary drug have overlapping therapeutic impact, while the editor suppresses the downstream propagation of toxicity-related signaling.

12.1 The Editor Concept

Side effects in molecular targeted anticancer therapy are common [1–5], hard to avoid [6], often unexpected [7], and potentially health-threatening [8], thereby introducing daunting difficulties to drug development. At least two major problems must be dealt with when addressing side effects [4–10]: (a) target proteins have paralogs with similar 3D structure [9, 10] and (b) the target protein plays a role in off-target cells that is different from the role it plays in target (cancer) cells [5–7]. Accordingly, undesired cross-reactivities are expected due to (a), while (b) implies that target inhibition may be therapeutically efficacious in cancer cells, but toxic in off-target cells. This chapter will be limited to kinase inhibitors (KIs), the drugs that interfere with cell-signaling pathways, because of their promising prospects for anticancer therapy [9–12] and because such treatments require addressing the problems outlined above [5, 11–13].

A. Fernández, *Transformative Concepts for Drug Design: Target Wrapping*,
DOI 10.1007/978-3-642-11792-3_12, © Springer-Verlag Berlin Heidelberg 2010

Problem (a) motivated the exploitation of selectivity filters for drug design [9, 10], leading to an efficacious control of specificity through paralog discrimination [13]. This problem has been dealt with extensively in Chaps. 7, 8, and 9. where we showed that dehydrons in proteins can be used to distinguish family members, and that such packing defects can be targeted using ligands engineered to "wrap" or shield them from interactions with water [10, 13], resulting in high specificity.

Dealing with problem (b) requires advanced network annotation [5, 14]. The problem can be dissected by comparing protein interaction networks across tissues and placing the target protein in different interacting contexts represented in various cell types [14]. In this chapter we focus on problem (b) and advocate a strategy to avert the resulting side effects.

The context-dependence problem has been effectively dealt with through formulations for tissue-specific delivery [15–17]. The strategies to enhance tissue specificity include incorporating the drug molecules into nanosystems such as liposomes, coupling drugs to antibodies or ligands that target cell receptors, exploiting permeation enhancers or passive or active translocators. These approaches often fail to be completely specific and introduce additional problems like limited delivery to the site of action.

In light of this situation, we explore an alternative systemic approach to deal with the context-dependence problem: a mode of "therapeutic editing" using two-component free-drug treatments. One drug – the editor – suppresses the side effect promoted by the primary drug, while both drugs overlap in their impact on therapeutically relevant targets. Thus, the editor selectively suppresses downstream propagation of toxicity-related signaling while circumventing the delivery issues that plague the conventional approaches to the context-dependence problem.

12.2 Editing Drugs to Curb Side Effects

Context-dependent functionalities introduce a risk of toxicity in molecular target therapy [4–7]. For example, the constitutively active chimera Bcr-ABL kinase is a clinical target for treating chronic myeloid leukemia (CML), as evidenced by the success of the KI imatinib (Gleevec) [18], but ABL inhibition in cardiomyocytes (off-target cells in CML treatment) may introduce a cardiotoxicity risk [5–7], as shown in Fig. 12.1. This risk is more pronounced for promiscuous KIs like sunitinib [5, 8].

In general, prevention of side effects arising from context dependence requires tissue-specific delivery [15–17] to selectively avoid inhibiting a target kinase in off-target cells. In practice, this level of selectivity is difficult to achieve. In view of this problem, we propose a two-component treatment in which one drug modulates a downstream component of the toxicity-signaling pathway recruited by the other.

Two-component treatments have been exploited to improve therapeutic impact [19–21]. In this chapter we advocate an alternative application with the aim of

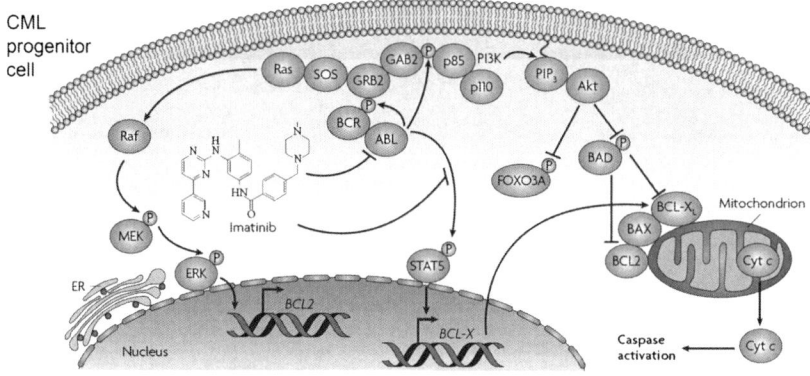

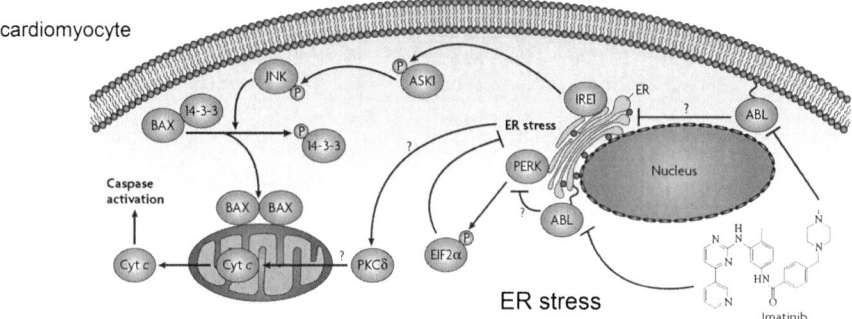

Fig. 12.1 Molecular dissection of imatinib side effects due to the context-dependent functionality of one of its primary targets. On display is the constitutive signaling in chronic myeloid leukemia (CML) progenitor cell through the cytoplasmic BCR–ABL tyrosine kinase, which leads to activation of Ras–ERK (extracellular signal-regulated kinase), phosphatidylinositol 3-kinase (PI3K)–Akt, and signal transducer and activator of transcription 5 (STAT5) pathways. Several of these pathways converge on antiapoptotic mechanisms: the Ras–ERK pathway stimulates the expression of *BCL2*, STAT5 activates *BCL-X*, and Akt inhibits BCL2 antagonist of cell death (BAD) and forkhead box O3A (FOXO3A). Imatinib blocks all BCR–ABL-dependent phosphorylation and signaling events, leading to reversal of pro-survival effects and activation of apoptosis. On the other hand, ABL (localized to the endoplasmic reticulum, ER) seems to maintain ER homeostasis in cardiomyocytes. The ABL kinase inhibitor, imatinib, induces ER stress, leading to activation of the PKR-like ER kinase (PERK) and IRE1 pathways, and to overexpression of protein kinase Cδ (PKCδ). PERK phosphorylates the eukaryotic translation initiation factor 2α (EIF2α) as part of a protective response, and on sustained ER stress IRE1 activates Jun N-terminal kinases (JNKs), leading to phosphorylation of 14-3-3 and release of BAX followed by mitochondrial depolarization, ATP depletion, cytochrome *c* (Cyt *c*) release, and features of necrotic and apoptotic cell death. Notation: 14-3-3, 14-3-3 protein; ASK1, apoptosis signal-regulating kinase 1; BAX, BCL2-associated X protein; GAB2, GRB2-associated binding protein 1; GRB2, growth factor receptor-bound protein 2; MEK, mitogen-activated ERK kinase; PIP_3, phosphatidylinositol trisphosphate; SOS, son of sevenless. Reprinted by permission from MacMillan Publishers Ltd: [5], copyright 2007

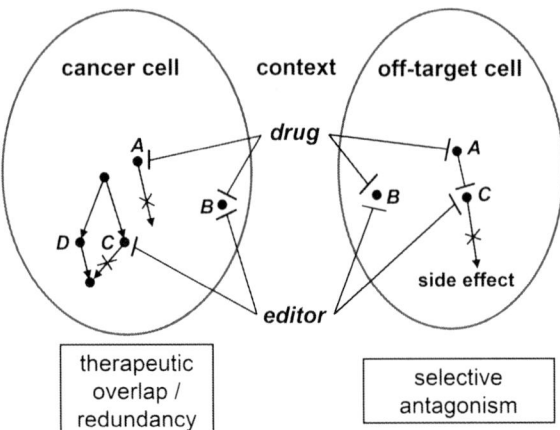

Fig. 12.2 Scheme of editing strategy. Drug and editor overlap therapeutically in on-target cancer cells, while the editor interferes selectively and antagonistically with the primary drug in off-target cells. Drug and editor have a common therapeutically relevant target B, while the drug inhibits additional target A. In addition, the editor inhibits target C which is redundant in cancer cells (biochemical flow may be restored through alternative signal transducer D), but becomes indispensable in the side-effect-related pathway recruited by the action of the primary drug on target A in normal (off-target) cells. Thus, phosphorylation of C is promoted by the inhibitory action of the drug on target A (A normally inhibits C). In this way, the editor interferes to suppress a side effect resulting from the action of the primary drug in off-target cells, while this interference becomes innocuous in cancer cells due to a built-in redundancy of the latter (D acts as an alternative transducer to C). Reprinted from [32], copyright 2009 with permission from Elsevier

curbing the side effects of one drug by "editing" its impact in off-target cells with another drug. This requires that we introduce a dual-role editor, so that the two drugs overlap therapeutically in target tumor cells, but behave antagonistically in off-target cells (Fig. 12.2).

Two premises must hold for this editing therapy to apply: (I) the side effect of the primary drug is attributable to an alternative role of the target kinase in off-target cells and (II) target inhibition in the off-target cell activates a pathway that compromises cell survival. Consequently, the editor must fulfill the following conditions (Fig. 12.2): (i) in cancer cells, drug and editor have overlapping impact; (ii) inhibition of a primary target activates a detrimental pathway in off-target cells; (iii) in off-target cells, the editor has an antagonistic effect through downstream inhibition, impairing recruitment of the deleterious pathway triggered by the action of primary drug; and (iv) this antagonistic effect is not carried over to cancer cells, where the biochemical flow is restored through alternative (redundant) pathways that warrant robustness vis-à-vis drug-induced perturbations [14].

We focus on editing cardiotoxicities of KIs, which are significant side effects [5–8] and fulfill the applicability premises described above (cf. Fig. 12.1). In general, KI cardiotoxicity remains a controversial subject [5–8, 22, 23]. However, in vivo and animal model studies [13] and clinical studies [8] corroborate the original characterization of KI cardiotoxicities [5–7]. The discrepancy with alternative reports that focus on imatinib [22, 23] is mainly due to the fact that the latter are not prospective studies, as patients did not have baseline functional studies and clinical endpoints were not delineated.

The therapeutic concept is illustrated by considering possible editors for sorafenib, a major KI successfully exploited to treat renal cell carcinoma (RCC) [24]. Sorafenib entails a cardiotoxicity risk traced to RAF1 inhibition in cardiomyocytes (Fig. 12.3 [5]). The drug binding to RAF1 hinders the MST2-inhibitory RAF1–MST2 association [5, 25], as suggested by the results described below. Thus, RAF1 inhibition activates pro-apoptotic pathways mediated by MST2 and further downstream by JNK, ultimately leading to cytochrome c release from mitochondria [5]. As shown in Fig. 12.3, two compounds are potentially capable of interfering with this pathway, acting as sorafenib editors: the highly specific imatinib-derived compound WBZ_4, a JNK inhibitor with $K_D = 34nM$ [13], already introduced in Chap. 8, and SUDE, a sunitinib-derived editor described below that inhibits MST2 with $K_D = 11nM$ [9]. Both compounds overlap therapeutically with sorafenib while they can suppress its potential cardiotoxicity by blocking the sorafenib-recruited pro-apoptotic pathway (Fig. 12.3a, b).

As shown in Chap. 8, WBZ_4 is a re-engineered version of imatinib with an enhanced specificity needed to curb imatinib's potential cardiotoxicity. The inhibitory impact of WBZ_4 is restricted to KIT, PDGFR, and JNK, all with nanomolar affinity, as corroborated by screening against a T7-bacteriophage library of kinase displays (Ambit Biosciences, CA) [13].

JNK is an indispensable downstream transducer of pro-apoptotic signals in cardiomyocytes [5, 25, 26] (Fig. 12.3), thus, its inhibition by WBZ_4 should antagonize sorafenib's inhibition of RAF1. In cancer cells, both drugs overlap on therapeutic targets PDGFR and KIT [5, 13], while JNK inhibition by WBZ_4 interferes with a redundant pro-apoptotic pathway that may be restored through p38 mediation [26]. The surrogate role of p38 may not be universal, but holds in the MAPK cancer-related contexts of proliferation, differentiation, and apoptosis [26].

SUDE was designed using the wrapping-based selectivity filter [27] with two purposes in mind: (A) decrease cross-reactivity and (B) target MST2. As a sunitinib derivative [9], SUDE overlaps with sorafenib on therapeutic targets VEGFR, KIT, PDGFR, FLT1 and FLT3 [24, 28]. With MST2, these kinases constitute the complete target set for SUDE as corroborated by kinase phage-display screening, while SUDE becomes an MST2 inhibitor more potent than the parent compound ($K_D = 11nM$ *versus* $56nM$). Thus, SUDE should edit sorafenib, counteracting its deleterious pathways in cardiomyocytes through its anti-MST2 activity (Fig. 12.3b). This editing role was experimentally validated for RCC, as shown below, but may not hold for other cancers like ovarian carcinoma. In the latter, MST2 inhibition may actually promote cancer survival since the tumor suppressor LATS1 becomes activated by MST2 [29]. This fact introduces a limitation in the editing applicability of SUDE.

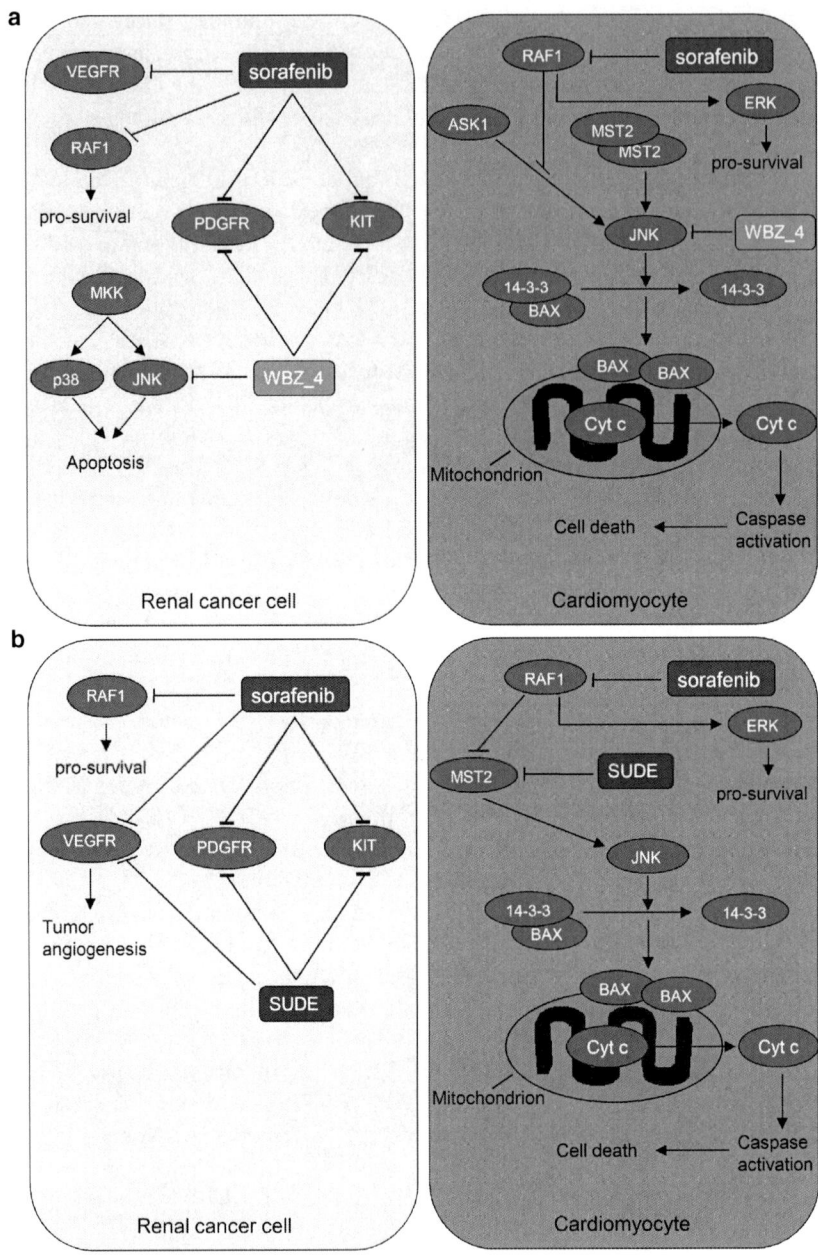

Fig. 12.3 (continued)

12.3 Designing a Therapeutic Editor Using the Wrapping Selectivity Filter

SUDE was primarily engineered to edit sorafenib by inhibiting the kinase MST2, a downstream transducer in the pro-apoptotic pathway recruited by sorafenib in cardiomyocytes (Fig. 12.3). Additionally, it was necessary for SUDE to overlap therapeutically with sorafenib in cancer cells, to avoid disrupting anticancer activity. Furthermore, SUDE needed to be specific enough to reduce the cardiotoxicity risks of its parent drug sunitinib, adopted as the scaffold (see Chap. 10 for details). Sunitinib overlaps with sorafenib on the clinical targets VEGFR, KIT, PDGFR, FLT1, and FLT3 [5, 24, 28]. Sunitinib was redesigned guided by the wrapping selectivity filter defined in Chap. 9. The first goal is to make it selective and less cardiotoxic, while the second is to turn it also into an MST2 inhibitor. To achieve the first goal, we exploit the wrapping differences across sunitinib's reported targets [30]. As described in Chap. 10, alignment of wrapping patterns revealed a targetable dehydron in sunitinib's clinical targets that is absent in the cardiotoxicity-related targets AMPK and RSK (Fig. 12.4, [5, 8]). The distinct and targetable feature is the backbone hydrogen bond pairing residues N900 and E917 in VEGFR, a primary target of sunitinib (Fig. 12.4b), a bond significantly exposed to solvent in all primary targets, but well wrapped in AMPK and RSK (Fig. 12.4a). This targeted dehydron is also present in MST2 and involves residues K69 and E85 at the aligned position (Fig. 12.4a). Thus, a methyl group was incorporated to sunitinib to provide a "wrapping" protection to the MST2 dehydron, also present in all clinically relevant targets VEGFR, KIT, PDGFR, FLT1, and FLT3, but absent in AMPK and RSK. Thus, we redesigned sunitinib into SUDE (Fig. 12.4c), an MST2 inhibitor with controlled specificity.

The predicted SUDE target set {MST2, VEGFR, KIT, PDGFR, FLT1, FLT3} was confirmed using the AMBIT high-throughput assay that uses a phage-display kinase library [30]. With $K_D = 11nM$, SUDE proved to be an MST2 inhibitor more

---

Fig. 12.3 Network level analysis of editing therapy, illustrated by sorafenib/WBZ_4 and sorafenib/SUDE combined activity. (**a**) Sorafenib edited by WBZ_4. In renal cancer cells sorafenib and WBZ_4 overlap therapeutically by inhibiting common clinically relevant targets (PDGFR and KIT). The inhibitory action of WBZ_4 on JNK confers editing power to this drug. This inhibitory activity is likely inconsequential for cancer cells, as the biochemical flow is restored through p38, while in cardiomyocytes, the inhibitory action of WBZ_4 on JNK antagonizes a pro-apoptotic pathway activated by upstream sorafenib-promoted inhibition of RAF1. This pathway leads to a BAX-mediated cytochrome c release to the cytosol triggering the caspase cascade activation. (**b**) Sorafenib edited by SUDE. SUDE overlaps therapeutically with sorafenib on VEGFR, PDGFR, and KIT in renal cancer cells, whereas in cardiomyocytes, the SUDE inhibition of MST2 impairs the recruitment of pro-apoptotic pathways activated by the upstream sorafenib-promoted inhibition of RAF1. Reprinted from [32], copyright 2009 with permission from Elsevier

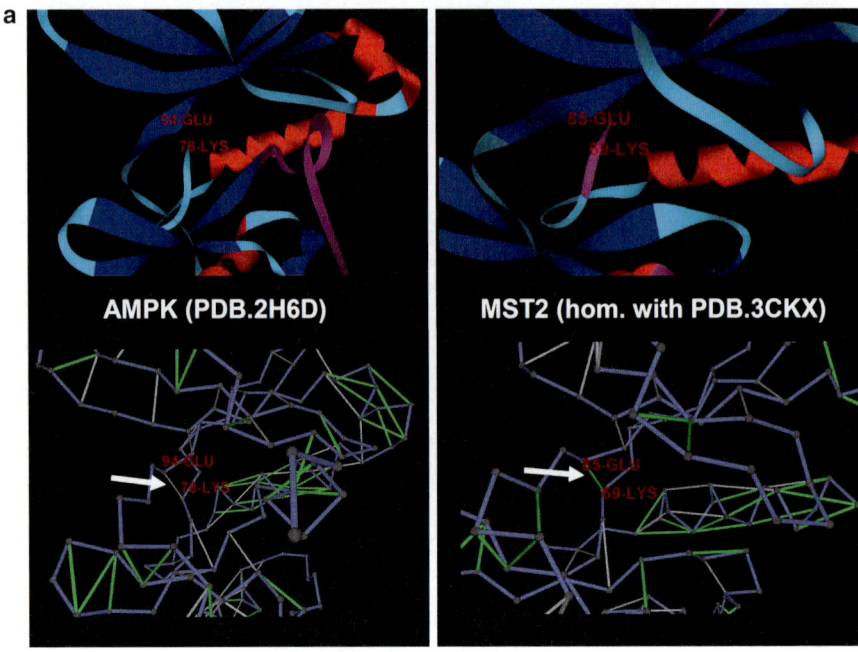

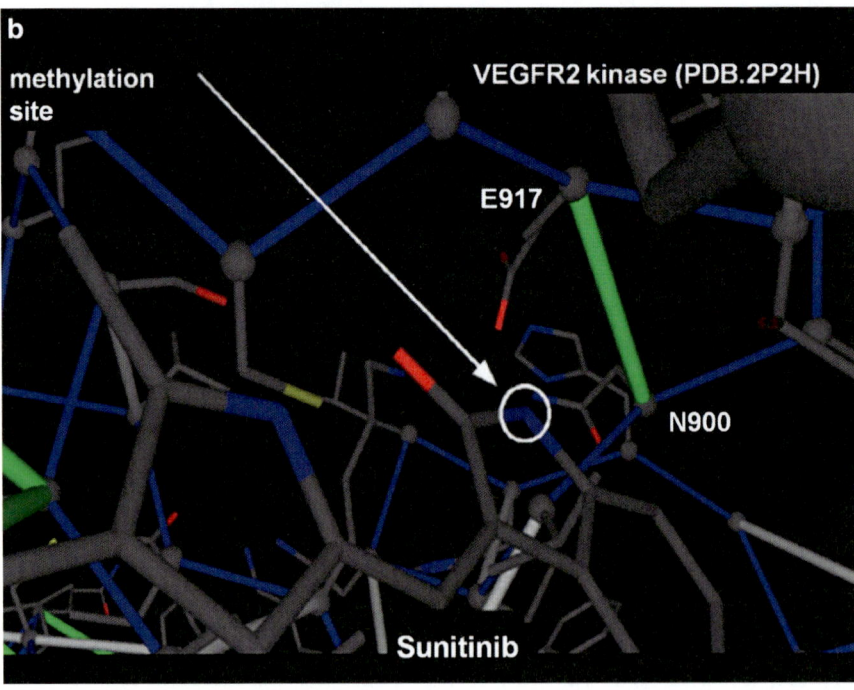

Fig. 12.4 (continued)

Fig. 12.4 Molecular design of selective sunitinib-derived editor (SUDE) guided by structural/wrapping comparison of sunitinib targets, by the need to remove cardiotoxicity-related anti-AMPK and anti-RSK activity and by the need to incorporate the new target MST2 for editing purposes. Protein structures are also rendered in ribbon representation for visual aid. (**a**) Structures and dehydron patterns of sunitinib target AMPK (PDB.2H6D) and MST2 (homologous to PDB.3CKX) kinase in complex with sunitinib. These targets need to be discriminated by SUDE, which should have affinity for MST2 but not for AMPK (or RSK). The K69–E85 dehydron in MST2 is conserved in the high-affinity targets (KIT, PDGFR, VEGFR, FLT1, and FLT3) relevant to angiogenesis and cancer cell proliferation. Crucially, it aligns with the well-wrapped hydrogen bond K78–E94 in AMPK2 (and with a well-wrapped hydrogen bond in RSK, not shown). *This important packing difference suggests the modification to sunitinib needed to selectively target MST2: methylation at the N1 indol position turns the drug into SUDE, a wrapper of the MST2 dehydron K69–E85.* Thus, by inhibiting MST2, a signal transducer in the pro-apoptotic pathways triggered by sorafenib, SUDE overlaps antagonistically with sorafenib. (**b**) SUDE retains activity against clinically relevant sunitinib targets KIT, PDGFR, VEGFR, FLT1, and FLT3. Structure and packing pattern of VEGFR2 (PDB.2P2H) kinase in complex with sunitinib. The N900–E917 dehydron in VEGFR2 is conserved in other clinically relevant targets of sunitinib and hence the modification to sunitinib serves as a wrapper to this common dehydron. (**c**) Chemical structure of SUDE. The predicted SUDE cross-reactivities were confirmed using the AMBIT high-throughput assay that uses a phage-display kinase library. Reprinted from [32], copyright 2009 with permission from Elsevier

potent than sunitinib ($K_D = 56nM$, [30]). The higher affinity of SUDE is due to the additional wrapping protection on MST2 dehydron K69–E85.

12.4 Therapeutic Editing: Toward a Proof of Principle

At a cellular level, the purported sorafenib editors should be assayed using the renal cancer cell line RCC-786-O and NRVMs (neonatal rat ventricular monocytes). In RCCs, the sorafenib/WBZ_4 combination yields agonistic synergy, marked by an increased inhibition of cell proliferation when compared to sorafenib-alone levels at equivalent bulk concentrations (Fig. 12.5a): The dose-dependent inhibition is greater than the Loewe-additive values [20, 21] (Fig. 12.5b). By contrast, the sorafenib-induced recruitment of the pro-apoptotic pathway in cardiomyocytes should be impaired by the downstream interference of WBZ_4 through JNK inhibition (Fig. 12.3a). Consequently, cytosolic release of

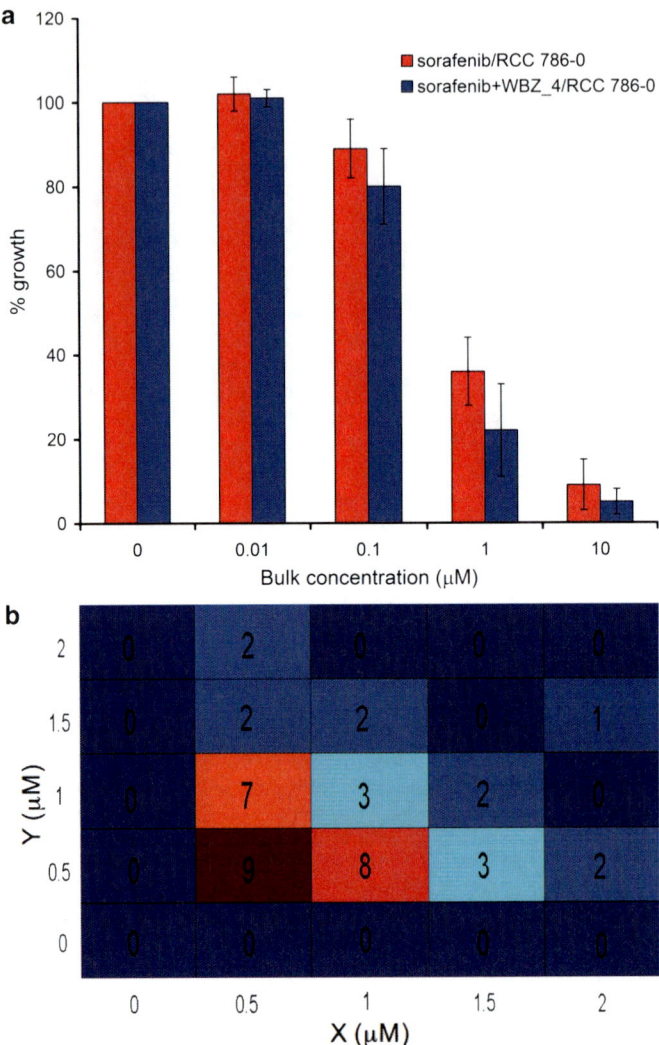

Fig. 12.5 (**a**) Cell proliferation assays for renal cancer cell line RCC 786-O for sorafenib/WBZ_4 combination treatments. The cells were seeded in 96-well plates (4×10^4 cells per well), cultured for 24 h and treated for 48 h with sorafenib alone and in 50–50% combinations with WBZ_4 at overall 0, 0.01, 0.1, 1, and 10 μM bulk concentrations. Cell proliferation was determined by Alamar Blue assay (Bio Source International, Camarillo, CA). Plates were read at dual wavelength (570–595 nm) in an Elisa plate reader (Molecular Devices Corporation, Sunnyvale, CA). Cell proliferation is expressed as the percentage of proliferating cells relative to untreated cells. The sorafenib/WBZ_4 combination treatment led to an increased inhibition of cell proliferation when compared to the sorafenib-alone levels at equivalent bulk concentrations. Also, the dose-dependent inhibition surpassed the Loewe-additive values indicating agonistic synergy. The sorafenib/SUDE combination treatment (not shown) suppressed RCC proliferation to sorafenib-alone levels at identical overall concentrations. In this case, the dose-dependent inhibition fits the Loewe-additive scheme indicating therapeutic reinforcement without synergy. (**b**) Agonistic synergy matrix for sorafenib/WBZ_4 combinations assayed in cell proliferation assays. Matrix elements are defined

cytochrome c should also be impaired. A dose-dependent study revealed antagonistic synergy: If Y_{50} denotes the WBZ_4 concentration yielding 50% reduction in sorafenib-induced cytochrome c release, we consistently obtained $Y_{50} = Y_{50}(X)$ values $(X = sorafenib\,concentration, 1\mu M \leq X \leq 10\mu M) \sim 10\%$ below the additive (linear) (X, Y_{50}) plot. Furthermore, the cardioprotective effect of the editor should become apparent as NRVM sarcolemmal integrity and ATP levels are restored to normal values.

Alternatively, the sorafenib/SUDE combination suppressed RCC proliferation to sorafenib-alone levels at identical overall concentrations. This dose-dependent inhibition fits the Loewe-additive scheme [20, 21] indicating therapeutic reinforcement without synergy. In NRVMs, SUDE inhibits MST2, hence it should block the pro-apoptotic pathway triggered by sorafenib (Fig. 12.3b). The detailed mechanism of the sorafenib-induced activation of MST2 is not known precisely, but it is likely that drug binding to RAF1 impairs its association with MST2 known to inhibit the latter [25]. Furthermore, in the combined sorafenib/SUDE treatment we expect reduction of cytosolic release of cytochrome c, restored sarcolemmal integrity, and restored ATP levels all confirming SUDE's role as sorafenib editor.

The clinical relevance of the editing therapy remains to be established. The potential sorafenib-induced cardiotoxicity [5] and the possibility of antagonist interference exploiting WBZ_4 or SUDE hints at the therapeutic value of this mode of side-effect removal. We already established the cardioprotective role of WBZ_4 [13], while that of SUDE is expected to be established on the same premises: antagonistic interference on a pro-apoptotic pathway recruited by an upstream kinase inhibition. A decisive test on the restoration of cardiac function through editing therapy requires determining whether the left ventricular ejection fraction (LVEF) decreases significantly upon sorafenib treatment and is restored to normal levels as sorafenib cardiotoxicity is antagonized downstream by WBZ_4 or SUDE.

The editors impact a set of therapeutic targets that are more limited than that of the parent compounds [13], probably precluding the editors from serving as therapeutic agents by themselves. This is evident in RCC786-O-proliferation assays. Thus, in contrast with sorafenib effects (Fig. 12.5a), cell growth decreases by a

---

Fig. 12.5 (continued) as $M(X,Y) = I(X,Y) - I'(X,Y)$, where $X =$ sorafenib concentration in μM, $Y =$ WBZ_4 concentration in μM, $I(X, Y) =$ observed inhibition for the (X, Y) combination, $I'(X,Y) =$ predicted inhibition according to the Loewe-additive scheme [20, 21] which requires the (X, Y) concentration pair to satisfy the linear equation $X/X_{I'} + Y/Y_{I'} = 1$, where $X_{I'}$ sorafenib concentration yielding inhibition I' and $Y_{I'} =$ WBZ_4 concentration yielding inhibition I'. For clarification, a two-component treatment involving two drugs at respective concentrations X and Y is said to obey Loewe additivity if the concentration-dependent inhibition $I(X,Y)$ is equal to $I'(X, Y)$, the predicted inhibition for the concentration pair (X, Y) that satisfies the linear equation $X/X_{I'} + Y/Y_{I'} = 1$, where $X_{I'}$ and $Y_{I'}$ are the single-agent effective concentrations that produce inhibition I'. Reprinted from [32], copyright 2009 with permission from Elsevier

meager 9.5 and 4.5%, respectively, when these cells are treated solely with SUDE or WBZ_4 at 10 μM.

12.5 Future Perspectives for the Editing Therapy

To assess the applicability breadth of this editing strategy we surveyed signaling pathways presumably causative of side effects and recruited through the action of other anticancer KIs. In an effort to curb side effects through therapeutic editing, we explored the editing potential of WBZ_4, exploiting its selective activity against tyrosine kinases and its additional impact on a downstream kinase (JNK). We predict that WBZ_4 could also serve as editor of primary tyrosine KIs like imatinib (Fig. 12.6) and dasatinib [31]. WBZ_4 is a potential editor because it inhibits JNK, a downstream non-redundant transducer of pro-apoptotic pathways recruited by imatinib or dasatinib inhibition of ABL in cardiomyocytes (Figs. 12.1 and 12.6). Moreover, WBZ_4 overlaps therapeutically with imatinib and dasatinib in cancer

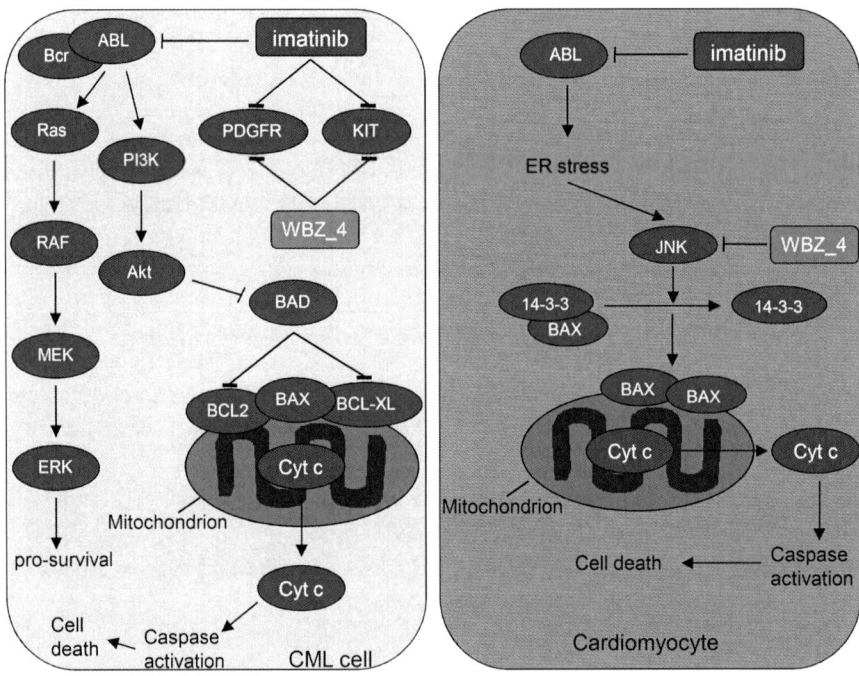

Fig. 12.6 Network level analysis of the proposed editing therapy involving imatinib/WBZ_4 combination treatment. In the CML cells imatinib and WBZ_4 overlap therapeutically by inhibiting common clinically relevant targets PDGFR and KIT, whereas in cardiomyocytes, the inhibitory action of WBZ_4 on JNK antagonizes the pro-apoptotic pathways activated by upstream inhibition of ABL by imatinib. Reprinted from [32], copyright 2009 with permission from Elsevier

cells on the clinical targets PDGFR and KIT [13]. The synergies of a possible imatinib/WBZ_4 treatment are described in Fig. 12.6.

To summarize, therapeutic editors may become a realistic engineering possibility as we harness the molecular factors controlling drug specificity (cf. Chap. 9). Thus, the success of the editing examples described depends pivotally on the fact that the high-throughput screening of the engineered kinase inhibitors fit the modeling predictions and the drug synergies found experimentally.

The therapeutic editor is a systemic concept whose clinical impact will ultimately be unraveled as the in vitro and in vivo assays described pave the way to clinical trials. This development may encounter difficulties arising, for instance, from differences in bioavailability, half-life, and pharmacodynamics of drug and editor. Thus, the engineering of adequate delivery vehicles and possibly even drug redesign to tune these parameters will become an essential component of the editing strategy in the advanced stages of development.

References

1. Widakowich C, de Castro G, de Azambuja E, Dinh P, Awada A (2007) Review: Side effects of approved molecular targeted therapies in solid cancers. Oncologist 12: 1443–1455
2. Schmidinger M, Zielinski CC, Vogl UM et al (2008) Cardiac toxicity of sunitinib and sorafenib in patients with metastatic renal cell carcinoma. J Clin Oncol 26:5204–5212
3. Lacouture ME (2006) Mechanisms of cutaneous toxicities to EGFR inhibitors. Nat Rev Cancer 6:803–812
4. Verheul HM, Pinedo HM (2007) Possible molecular mechanisms involved in the toxicity of angiogenesis inhibition. Nat Rev Cancer 7:475–485
5. Force T, Krause D, van Etten RA (2007) Molecular mechanisms of cardiotoxicity of tyrosine kinase inhibition. Nat Rev Cancer 7:332–344
6. Force T, Kerkela R (2008) Cardiotoxicity of the new cancer therapeutics – mechanisms of, and approaches to, the problem. Drug Discov Today 13:778–784
7. Kerkela R, Grazette L, Yacobi R et al (2006) Cardiotoxicity of the cancer therapeutic agent imatinib mesylate. Nat Med 12:908–916
8. Chu TF, Rupnick MA, Kerkela R et al (2007) Cardiotoxicity associated with tyrosine kinase inhibitor sunitnib. Lancet 370:2011–2019
9. Fernández A, Crespo A, Tiwari A (2009) Is there a case for selectively promiscuous anticancer drugs? Drug Discov Today 14:1–5
10. Zhang X, Crespo A, Fernández A (2008) Turning promiscuous kinase inhibitors into safer drugs. Trends Biotechnol 26:295–301
11. Dancey J, Sausville EA (2003) Issues and progress with protein kinase inhibitors for cancer treatment. Nat Rev Drug Discov 2:296–313
12. Levitzki A, Gazit A (1995) Tyrosine kinase inhibition: An approach to drug development. Science 267:1782–1788
13. Fernández A, Sanguino A, Peng Z et al (2007) An anticancer C-kit kinase inhibitor is re-engineered to make it more active and less cardiotoxic. J Clin Invest 117: 4044–4054
14. Kitano H (2007) A robustness-based approach to systems-oriented drug design. Nat Rev Drug Discov 6:202–209
15. Torchilin VP (2006) Multifunctional nanocarriers. Adv Drug Del Rev 58:1532–1555
16. Langer R (2001) Drug delivery: Drugs on target. Science 293:58–59

17. Allen TM, Cullis PR (2004) Drug delivery systems: Entering the mainstream. Science 303:1818–1822

18. Deninger M, Buchdunger E, Druker BJ (2005) The development of imatinib as a therapeutic agent for chronic myeloid leukemia. Blood 105:2640–2653

19. Dancey JE, Chen HX (2006) Strategies for optimizing combinations of molecularly targeted anticancer drugs. Nat Rev Drug Discov 5:649–659

20. Keith CT, Borisky AA, Stockwell BR (2005) Multicomponent therapeutics for networked systems. Nat Rev Drug Discov 4:71–78

21. Zimmermann GR, Lehar J, Keith CT (2007) Multi-target therapeutics: When the whole is greater than the sum of the parts. Drug Discov Today 12:34–42

22. Verweij J, Casali P, Kotasek D et al (2007) Imatinib does not induce cardiac left ventricular failure in gastrointestinal stromal tumours patients: Analysis of EORTC-ISG-AGITG study 62005. Eur J Cancer 43:974–978

23. Atallah E, Durand JB, Kantarjian H, Cortes J (2007) Congestive heart failure is a rare event in patients receiving imatinib. Blood 110:1233–1237

24. Escudier B, Eisen T, Stadler WM et al (2007) Sorafenib in advanced clear-cell renal-cell carcinoma. N Engl J Med 356:125–134

25. O'Neill E, Rushworth L, Baccarini M, Kolch W (2004) Role of the kinase MST2 in suppression of apoptosis by the Proto-Oncogene product Raf-1. Science 306:2267–2270

26. Dhillon AS, Hagan S, Rath O, Kolch W (2007) MAP kinase signalling pathways in cancer. Oncogene 26:3279–3290

27. Chen J, Zhang X, Fernández A (2007) Molecular basis for specificity in the druggable kinome: Sequence-based analysis. Bioinformatics 23:563–572

28. Faivre S, Demetri G, Sargent W, Raymond E (2007) Molecular basis for sunitinib efficacy and future clinical development. Nat Rev Drug Discov 6:734–745

29. O'Neill EE, Matallanas D, Kolch W (2005) Mammalian sterile 20-like kinases in tumor suppression: An emerging pathway. Cancer Res 65:5485–5487

30. Karaman MW, Herrgard S, Treiber DK et al (2008) A quantitative analysis of kinase inhibitor selectivity. Nat Biotechnol 26:127–132

31. Kantarjian H, Jabbour E, Grimley J, Kirkpatrick P (2006) Dasatinib. Nat Rev Drug Discov 5:717–718

32. Fernández A, Sessel S (2009) Selective antagonism of anticancer drugs for side-effect removal. Trends Pharmacol Sci 30: 403–410

Chapter 13
Wrapper Drugs for Personalized Medicine

Personalized molecular therapy ("the right drug for the right person") is regarded as a major imperative of post-genomic medicine. This perception is reinforced almost daily as promising therapeutic agents are recalled because of idiosyncratic side effects detected in small subpopulations of patients. However pressing the need, rational approaches to personalized drug therapy will ultimately and pivotally depend on our ability to translate genomic individualities and variations into molecular biomarkers that can guide a patient-tailored design. This chapter addresses this issue and describes how the wrapping design concept can be brought to fruition in this area. The chapter introduces plausible scenarios in which genomic idiosyncrasies and oncogenic variations may promote targetable differences in the wrapping patterns of the gene products. Ultimately, the chapter extends an invitation to adopt and exploit protein wrapping as a molecular biomarker for personalized medicine and the wrapping technology as the enabling platform to tailor drugs to patient idiosyncrasies.

13.1 Wrapping as a Biomarker in Personalized Drug Therapy

The therapeutic imperative to make more efficacious and safer drugs by exploiting the enormous output of genomic forays has proven to be a daunting challenge, far more difficult than originally expected. No reliable rational approach has been identified so far to meet such a challenge. In addressing this imperative, the dearth of ideas of general applicability with the potential to become transformative and broaden the technological base of the pharmaceutical industry is quite evident [1, 2]. From the perspective of a rational drug designer, one can only speculate that a personalized approach would require an effective translation of genomic idiosyncrasies and disease-related variations into structural differences that may guide the molecular engineering of the therapeutic agent. While this conclusion may be inevitable, it has not been pursued beyond a limited number of anecdotal instances [3].

It is seldom the case that idiosyncratic or disease-related genomic differences would translate as significant structural changes in the gene products. The functional demands of the wild-type locus cannot generally be compromised without a

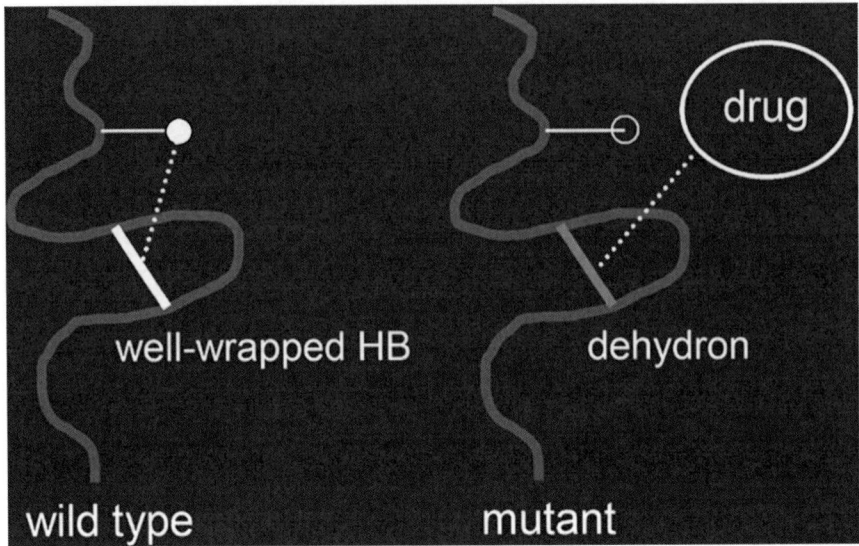

Fig. 13.1 An idiosyncratic variation substituting a good hydrogen bond wrapper (*solid circle*) for a bad wrapper (*empty circle*) can produce a new dehydron in the protein structure. This "idiosyncratic dehydron" may in turn be targeted or wrapped intermolecularly by a purposely designed drug

serious fitness cost and misfolded proteins are generally degraded unless they can aberrantly aggregate (as in poly-glutamine or poly-asparagine in-frame insertions). On the other hand, subtler variations, leading, for instance, to an oncogenic deregulation of structural elements may still be exploitable from the vantage point of the drug designer. Thus, while such accidents are unlikely to alter the functionally competent topology of the protein, they may affect its wrapping, as shown below. We may thus speculate that wrapping patterns may well be on their way to become molecular biomarkers for drug personalization in the particular context where a genetic accident or idiosyncrasy impacts a coding region and produces a change in the microevironment of a pre-formed hydrogen bond (Fig. 13.1).

In other words, a single amino acid substitution can turn a wrapping residue into a bad wrapper (that is, one of the following amino acids: G, A, N, D, S, T, or P) and hence generate a dehydron in lieu of the well-wrapped hydrogen bond in the wild type. Thus, a mutation can produce a new dehydron, actually an "idiosyncratic dehydron," which in turn may be wrapped intermolecularly by a purposely designed drug, as shown schematically in Fig. 13.1.

Conversely, an amino acid substitution that generates a well-wrapped hydrogen bond by increasing the number of nonpolar groups in the bond microenvironment may not yield a productive opportunity from the perspective of the wrapping designer: if the bond is well wrapped, the wrapping drug would sterically clash with the intramolecular wrappers and has nothing to add in terms of hydrogen bond stabilization (dehydration). Yet, there is a conspicuous exception arising in the case where the augmentation in the number of wrapping side chain groups triggered by the mutation *induces* a dehydron that was absent in the wild type due to a local scarcity of wrapping residues (the $\rho < 7 \rightarrow \rho \geq 7$ transition upon drug binding, described in Chap. 11).

To illustrate this case, we focus on the Asp816→Val substitution in the activation loop that confers resistance to the C-Kit kinase against the powerful anticancer drug imatinib [3]. This substitution renders imatinib ineffective to treat the gastrointestinal stromal tumor (GIST), which has C-Kit as the primary target. Yet, Val is a good wrapper residue while Asp is not. Hence, the Val816-Kit mutant is able to sustain the dehydron Phe811–Ala814, which is absent or unstable in the wild-type Kit kinase due to insufficient wrapping (see Chap. 11). In turn, the induced dehydron may be exploited to design a wrapping variant of imatinib with an extra methyl group, as needed to target the idiosyncratic dehydron Phe811–Ala814

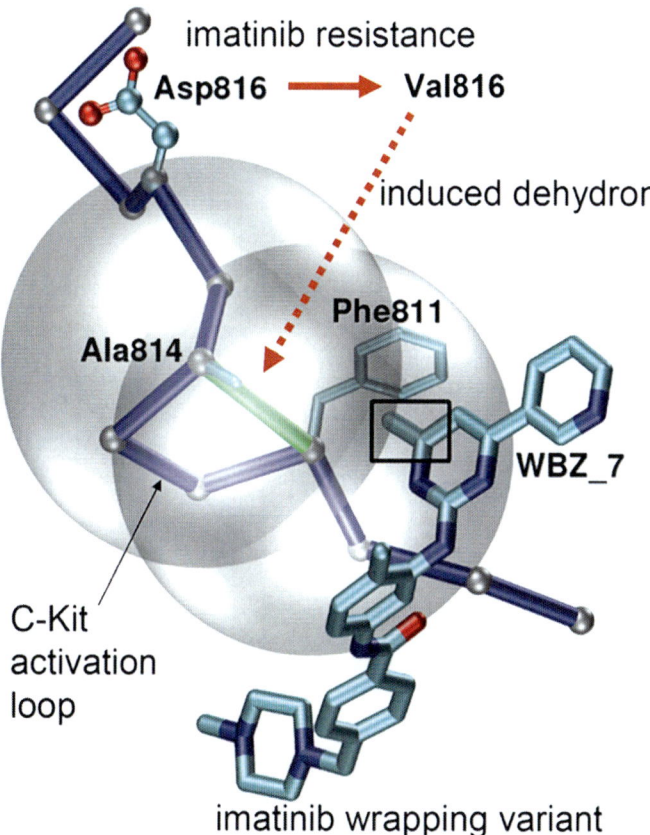

Fig. 13.2 Imatinib wrapping modification engineered to overcome drug resistance in the C-Kit kinase. This resistance is promoted by the somatic mutation Asp816Val. The mutation induces the dehydron Phe811–Ala814 in the activation loop which in turn can be targeted or wrapped by suitably modifying imatinib at the position highlighted by the *rectangle* [3]

by contributing to its dehydration upon association (Fig. 13.2). This wrapping modification has been named WBZ_7 and has been shown in cellular assays to be highly effective against imatinib-resistant GISTs [3]. Thus, the subpopulation of GIST patients with the Asp816Val substitution in the C-Kit kinase domain may potentially be treated with a drug that targets the induced dehydron arising from the single-site mutation.

13.2 Targeting Oncogenic Mutations with Wrapper Drugs

The impact on protein structure of amino acid substitutions stemming from onco-genic mutations has not been systematically studied. In the particular context of cell signaling related to mitosis or to the onset of anti-apoptotic pathways, it is likely that many such mutations will promote deregulation, yielding constitutively active kinases. Under normal conditions, these enzymes typically become acti-vated through phosphorylation of specific residues in the activation loop. This is a reversible event that fundamentally alters the conformation of the loop, making it more soluble and less obstructive, thereby promoting ATP binding. Thus, a sin-gle amino acid substitution within the activation loop turning a nonpolar residue

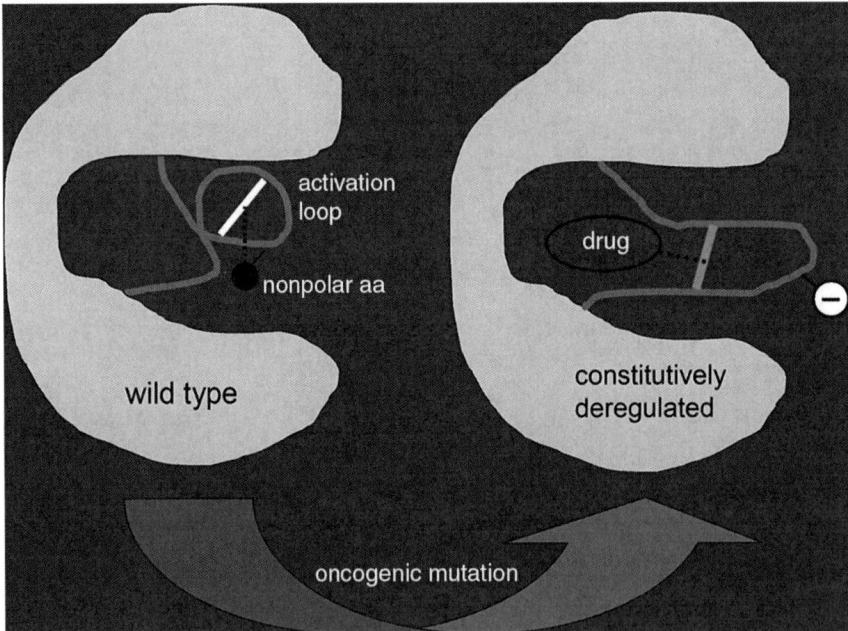

Fig. 13.3 An oncogenic mutation in a kinase domain produces a constitutively active signal trans-ducer by permanently solubilizing the activation loop. Being less structured, this deregulated kinase is richer in dehydrons than the wild type in its autoinhibited state and hence may be targeted by a wrapping drug

into a negatively charged one can promote the deregulation of the kinase, turning it into a constitutively active signal transducer. This accident can of course promote tumor development and cancer progression [4], depending on whether the transduced signal recruits pathways leading to mitosis, cell immortalization, focal adhesions, etc.

On the other hand, since the oncogenic substitution solubilizes the activation loop, it should promote a loss of structure. This process is likely to yield a loop conformation far worse wrapped than that of the autoinhibited state and hence, far richer in dehydrons (Fig. 13.3). Ergo, a constitutively deregulated kinase, should be more susceptible to be targeted by wrapping/crating drugs than the wild type. This operational hypothesis begs a validation in the light of the ideas described in this book.

13.3 Closing Remarks

By interfering with cell signaling pathways, single-molecule kinase inhibitors hold great promise as anticancer agents and probably hold potential for the molecular targeted treatment of other diseases. Yet, their therapeutic efficacy and clinical benefit are typically limited to a fraction of the treated patients. Genomic variations across patients and idiosyncratic patterns of drug resistance are likely to contribute significantly to such differences in clinical outcome. The acknowledgment of these facts has triggered the need for a personalized drug design. Perhaps this challenge is not so daunting in so far as we can translate idiosyncratic or disease-related variations into molecular differences that can be systematically targeted by molecular engineers. The wrapping pattern may prove to be a transformative concept that may be brought to fruition in this key area of molecular medicine.

References

1. Gazdar A (2009) Personalized medicine and inhibition of EGFR signaling in lung cancer. New Engl J Med 361:1018–1020
2. Janne PA, Gray N, Settleman J (2009) Factors underlying sensitivity of cancers to small molecule kinase inhibitors. Nat Rev Drug Discov 8:709–723
3. Fernández A, Sanguino A, Peng Z et al (2007) Rational drug design to overcome drug resistance in cancer therapy: Imatinib moving target. Cancer Res 67:4028–4033
4. Hubbard SR (2004) Oncogenic mutations in B-Raf: Some losses yield gains. Cell 116, 764–766

Chapter 14
Last Frontier and Back to the Drawing Board: Protein–Water Interfacial Tension in Drug Design

As we have advocated throughout this book, the protein–water interface is at the crux of the drug design problem. Understanding this interface is essential when addressing the frontier problem of disrupting a protein–protein interaction with a competing ligand (that is, as man tries to beat nature at its own recognition game). One lesson to be drawn from the preceding pages is that it is quite unproductive to think of a drug and a target in a context where the solvent is merely regarded as background or as a statistical bath. Yet, little has been said about the thermodynamic nature of the protein–water interface and about the future exploitation of this knowledge to the advantage of molecular targeted design. This chapter squarely addresses this problem, dealing with questions like What is the nature of the interfacial tension between protein and water? What is the thermodynamic cost of creating a protein–water interface? How can we identify regions that beget large interfacial tension to guide targeting designs? In plain terms, which regions on the surface of the protein generate unstable interfaces that may be removed upon association with a drug? The results presented open up an avenue to implement a nanoscale thermodynamics approach to drug design, where the *dehydron* concept will play again a pivotal role. This chapter heralds the advent of drugs adapted to reduce the tension of protein–water interfaces, hence stabilizing the structure of the target. This strategy takes us back to the wrapping concept: Dehydron patches that generate significant interfacial tension between target and water may be favorably wrapped by the drug.

14.1 Interfacial Tension Between Protein and Water: A Missing Chapter in Drug Design

The physical nature of the protein–water interface remains elusive and yet, as advocated in the preceding chapters, this interface is truly at the crux of rational drug design. This shift in focus from protein structure to protein–water interface will surely become instrumental to fulfill our expectations across the next tier of Frontier problems in drug design, including interfering with a protein–protein association by disrupting it with a small ligand. Chemical heterogeneities of the protein surface and a variety of local morphologies for the nanoscale confinement of interfacial water

A. Fernández, *Transformative Concepts for Drug Design: Target Wrapping*,
DOI 10.1007/978-3-642-11792-3_14, © Springer-Verlag Berlin Heidelberg 2010

make it difficult to treat the protein–water interface using classical thermodynamic approaches. Thus, we cannot readily introduce statistical thermodynamic concepts like interfacial surface tension without extending the validity of the physical picture to the nanoscale realm. Concepts like interfacial tension may be most useful when dealing with the separation of homogeneous phases with smooth solution of continuity but are far harder to define and apply when dealing with the complex morphologies and heterogeneous surfaces of proteins. Thus, while the length-scale dependence of nonpolar hydration and hydrophobic interactions is relatively well established [1], nontrivial complications arise mainly from the presence of nearby hydrophilic groups that modulate the interfacial tension in a curvature-dependent manner.

This chapter squarely addresses these problems and does so by introducing a nanoscale thermodynamic treatment based on first principles. Thus, we deal with questions like What is the thermodynamic cost of creating a protein–water interface? Which regions on the surface of the protein generate unstable interfacial patches with significant surface tension that may be removed upon drug association? The treatment lays the foundations of a nanoscale thermodynamics approach to drug design.

An intuitive picture emerges in the midst of the apparently murky problem: Given the heterogeneous chemical composition of the protein surface and its complicated morphology, the protein–water interfacial tension is likely to be *patchy*. Thus, we envision hot spots defined by solvent cavities whose creation entails a positive increment in interfacial free energy and regions of high solvent polarizability yielding negative free-energy increments (no tension spots). The positive free-energy patches (begetting tension) do not necessarily interface with hydrophobic moieties on the protein surface. *An analysis of the dielectric response under nanoscale confinement shows that polar groups with low-density hydration may also generate interfacial tension*, as shown below.

Drawing inspiration from Chaps. 3 and 5, we note that a tractable problem emerges when the extent of interfacial water confinement is described by a position ($\mathbf{r}$)-dependent parameter $\Gamma = \Gamma(\mathbf{r})$, the expected number of hydrogen bond partnerships of a water molecule at position $\mathbf{r}$ ($\Gamma = 4$ for bulk water). Thus, the solvent-accessible envelope of the protein surface may be covered by a minimal set W of water-confining osculating (first-order contact) spheres $D_j s, j \in W$, each with its average value $\Gamma = \Gamma_j$ which depends on the radius $\theta = \theta_j$ of D_j (the curvature radius of the protein envelope) and the physicochemical nature of protein group(s) in the contact region (Fig. 14.1).

Complementing the information conveyed by Γ_j, a fugacity f_j is introduced and defined as $f_j = N_j/N_j(b)$, where N_j is the expected number of water molecules in D_j at equilibrium and $N_j(b)$, the number associated with the same volume in bulk solvent. Thus, the chemical potential μ_j of water in D_j becomes $\mu_j = k_B T \ln [N_j/N_j(b)]$ ($k_B =$ Boltzmann constant). The θ-dependence of Γ, f-values (Fig. 14.1) is obtained at equilibrium determined from classical trajectories generated by molecular dynamics (MD) simulations within an NPT ensemble of the type described in Chap. 4. The computations start with the protein

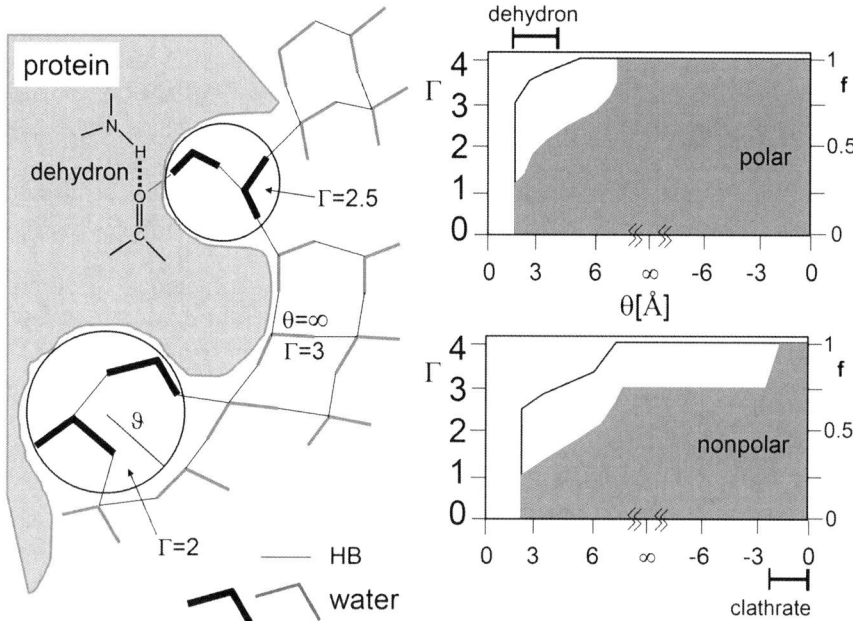

Fig. 14.1 Scheme and plots illustrating the dependence of nanoscale-confinement parameters Γ and f on the radius θ of the osculating sphere at the protein–water interface. First-order contacts with a polar or nonpolar patch on the protein surface are treated individually. Values were determined at equilibrium obtained by integrating Newton's equations of motion in an NPT ensemble with box size 10^3 nm^3, starting with the PDB structure embedded in a pre-equilibrated cell of water molecules. The box size was calibrated so that the solvation shell extended at least 10 Å from the protein surface at all times. Simulations were performed as described in Chap. 4

structure embedded in a pre-equilibrated cell of explicitly represented water molecules. A nonredundant set of 206 structures of monomeric proteins reported in PDB (protein data bank) with chain lengths ranging from $N= 34$ (PDB.1QUZ) to 223 (PDB.1AZ8) was used to generate the data shown in Fig. 14.1 (see Chap. 4 for computational details).

Nanoscale confinement significantly affects water polarizability $\mathbf{P} = \mathbf{P(r)}$, whose dependence on Γ may be obtained rigorously from first principles, starting from the Debye relation [2]: $\nabla \cdot (\varepsilon_0 \mathbf{E} + \mathbf{P})(\mathbf{r}) = \rho(\mathbf{r})$ ($\epsilon_0=$ vacuum permittivity, $\rho(\mathbf{r})$ = charge density, $\mathbf{E}=$ electrostatic field). In Fourier-conjugate frequency space ($\mathbf{v}$-space) we get [2]

$$F(\mathbf{P})(\mathbf{v}) = K_p(\mathbf{v})F(\mathbf{E})(\mathbf{v}) \tag{14.1}$$

where F denotes 3D-Fourier transform $F(\mathbf{g})(\mathbf{v}) = (2\pi)^{-3/2} \int e^{i\mathbf{v} \cdot \mathbf{r}} \mathbf{g(r)} d\mathbf{r}$ and the kernel $K_p(\mathbf{v})$ is the frequency Lorentzian $K_p(\mathbf{v}) = (\varepsilon_b - \varepsilon_0)/(1 + \tau^2 |\mathbf{v}|^2)$, with $\tau=$ dielectric relaxation time ($\tau = \tau_b \approx 100$ ps in bulk) and $\varepsilon_b =$ bulk permittivity. Equation (14.1) yields the following equation in r-space:

$$\nabla.\left[\int F^{-1}(K)(\mathbf{r}-\mathbf{r}')\mathbf{E}(\mathbf{r}')d\mathbf{r}'\right] = \rho(\mathbf{r}), \tag{14.2}$$

with $K(\mathbf{v}) = \varepsilon_0 + K_p(\mathbf{v}), F^{-1}(K)(\mathbf{r}) = (2\pi)^{-3/2}\int e^{-i\mathbf{v}.\mathbf{r}}K(\mathbf{v})d\mathbf{v}$ The reader should note that (14.2) *is not the commonly used Poisson–Boltzmann equation, which requires an arbitrary proportionality between the fields* **E** *and* **P** *under the incorrect assumption $K(\mathbf{v}) \equiv constant$.* Rather, (14.2) includes the convolution between $F^{-1}(K)(\mathbf{r})$ and $\mathbf{E}(\mathbf{r})$ that accurately translates in **r**-space the **v**-dependent proportionality between **E** and **P** that rigorously holds in frequency space.

Upon water confinement, the dielectric relaxation undergoes a frequency redshift (a shift to lower frequencies or to larger relaxation times) arising from the reduction in hydrogen bond partnerships that translates in a reduction in orientational possibilities. Thus, $\tau = \tau_b \exp(B(\Gamma)/k_BT)$, where the kinetic barrier $B(\Gamma) = -k_BT\ln(\Gamma/4)$ is obviously entropic, yielding $\tau = \tau_b(\Gamma/4)^{-1}$. In order to determine the Γ-dependence of the polarization **P**, we introduce the electrostatic potential $\Phi = \Phi(\mathbf{r})$, yielding $\mathbf{E} = \nabla\Phi$. Thus, (14. 2) may be solved in frequency space:

$$F(\Phi)(\mathbf{v}) = F(\rho)(\mathbf{v})/\left[|\mathbf{v}|^2 K(\mathbf{v})\right]. \tag{14.3}$$

Taking the inverse Fourier transformation we obtain the electrostatic filed under water confinement in explicit form:

$$\begin{aligned}\mathbf{E}(\mathbf{r}) &= \nabla F^{-1}\left\{F(\rho)(\mathbf{v})/\left[|\mathbf{v}|^2 K(\mathbf{v})\right]\right\}(\mathbf{r}) \\ &= (2\pi)^{-3/2}\nabla\int d\mathbf{v}e^{-i\mathbf{v}.\mathbf{r}}F(\rho)(\mathbf{v})/[|\mathbf{v}|^2 K(\mathbf{v})]\end{aligned} \tag{14.4}$$

This equation may be solved explicitly as we assume a net charge density distribution:

$$\rho(\mathbf{r}) = \sum_{j\in W}\sum_{n\in M(j)} 4\pi q_n\delta(\mathbf{r}-\mathbf{r}_n), \tag{14.5}$$

where $M(j)$ denotes the set of charges labeled by dummy index n, localized on the protein surface and interfacing with D_j, and $\mathbf{r}_n =$ position vector of charge n. Thus, from (14.3) and noting that

$$F(\rho)(\mathbf{v}) = (2\pi)^{-3/2}\sum_{j\in W}\sum_{n\in M(j)} 4\pi q_n\exp(i\mathbf{v}.\mathbf{r}_n), \tag{14.6}$$

we obtain the explicit form of the electrostatic potential:

$$\begin{aligned}\Phi(\mathbf{r}) &= \sum_{j\in W}\Phi_j(\mathbf{r}) \\ &= (2\pi)^{-3}\sum_{j\in W}\sum_{n\in M(j)}\int d\mathbf{v}e^{-i\mathbf{v}.(\mathbf{r}-\mathbf{r}_n)}4\pi q_n/\left[|\mathbf{v}|^2 K_{(j)}(\mathbf{v})\right],\end{aligned} \tag{14.7}$$

with $K_{(j)}(\mathbf{v}) = \varepsilon_0 + K_{p,(j)}(\mathbf{v}) = \varepsilon_0 + (\varepsilon_0 - \varepsilon_0) / \left(1 + \tau_j^2 |\mathbf{v}|^2\right)$ and $\tau_j = \tau_b(\Gamma_j/4)^{-1}$. Combining (14.1), (14.4), and (14.6), we get explicitly the polarization of water under nanoscale confinement:

$$\mathbf{P}(\mathbf{r}) = \int F^{-1}(K_p)(\mathbf{r} - \mathbf{r}')\mathbf{E}(\mathbf{r}')d\mathbf{r}' = \sum_{j \in W} \mathbf{P}_{(j)}(\mathbf{r})$$

$$= (2\pi)^{-3} \sum_{j \in W} \sum_{n \in M(j)} \int d\mathbf{r}' F^{-1}\left(K_{p,(j)}\right)(\mathbf{r} - \mathbf{r}') \qquad (14.8)$$

$$\nabla_{r'} \int d\mathbf{v} e^{-i\mathbf{v}.(\mathbf{r}' - \mathbf{r}_n)} 4\pi q_n / \left[|\mathbf{v}|^2 K_{(j)}(\mathbf{v})\right],$$

with $F^{-1}\left(K_{p,(j)}\right)(\mathbf{r}) = \int d\mathbf{v} e^{-i\mathbf{v}.\mathbf{r}} (\varepsilon_b - \varepsilon_0) / \left(1 + \tau_j^2 |\mathbf{v}|^2\right)$.

Thus, the overall free energy change ΔG associated with creating the protein–water interface by transferring the water molecules from the bulk to the interface is

$$\Delta G = \sum_{j \in W} \left[(-1/2) \int |\mathbf{P}_{(j)}(\mathbf{r})|^2 d\mathbf{r} - T\Delta S_j\right]$$

$$= -(1/2) \sum_{j \in W} \int d\mathbf{r} \left|(2\pi)^{-3} \sum_{n \in M(j)} \int d\mathbf{r}' F^{-1}\left(K_{p,(j)}\right)(\mathbf{r} - \mathbf{r}')\right.$$

$$\left.\nabla_{r'} \int d\mathbf{v} e^{-i\mathbf{v}.\left(\mathbf{r}' - \mathbf{r}_n\right)} 4\pi q_n / \left[|\mathbf{v}|^2 K_{(j)}(\mathbf{v})\right]\right|^2 \qquad . \quad (14.9)$$

$$- \sum_{j \in W} k_B T c_b N f_j (4/3)\pi \theta_j^3 \ln(\Gamma_j/4)$$

Since $N_j = c_b N f_j (4/3)\pi \theta_j^3$ ($c_b = 55.5$ mol/(10^{-3} m^3) = bulk water concentration and $N = 6.023 \times 10^{23}$/mol = Avogadro's number), the last term in (14.9) represents the sum of entropic contributions $\left(-\sum_{j \in W} T\Delta S_j\right)$, where the jth term in the sum is associated with transferring N_j water molecules from bulk solvent ($\Gamma = 4$) to the interfacial osculating sphere D_j (with $\Gamma = \Gamma_j$). Thus, we get the relation $-T\Delta S_j = -k_B T c_b N f_j (4/3)\pi \theta_j^3 \ln(\Gamma_j/4) \geq 0$.

Interfacial tension arises in an osculating sphere D_j when $\Delta G_j > 0$, that is, when generating the interface region entails a thermodynamic cost translated in reversible work performed on the system. As shown below, the most common patches generating interfacial tension involve backbone hydrogen bonds of the protein that are partially exposed to water, so that the amide or carbonyl is hydrated by low-density water. *In other words, protein dehydrons generate interfacial tension.* Thus, the most common osculating spheres contact *polar* patches arising from incomplete shielding of the backbone. These spheres constitute 66% of the interfacial spheres that fulfill $\Delta G_j > 0$ in the database. They are described by the ranges $\theta_j = 2.3 \pm 0.4\text{Å}$, $\Gamma_j = 1.2 \pm 0.2$, and $f_j = 0.75 \pm 0.1$ (Fig. 14.1), yielding a redshifted relaxation $\tau_j = (3.42 \pm 0.58)\tau_b$, with $-T\Delta S_j = (2.48 \pm 0.45)k_B T$ and $\Delta G_j = (1.54 \pm 0.40) k_B T$. *Thus, dehydrons yield interfacial tension through insufficient polar hydration.*

Since the Γ, f-parameters remain invariant for $\theta \geq 7\text{Å}$, interfacial polar regions with curvature radius $\theta > 7\text{Å}$ are covered with disjoint spheres with $\theta_j = 7\text{Å}$ (Fig. 14.1). At $\theta_j = 7\text{Å}$, we get $\Gamma_j = 4 (\Delta S_j = 0)$, hence these osculating spheres generate no interfacial tension. By contrast, nonpolar interfacial regions with $\theta \geq 7\text{Å}$, may also be covered with disjoint spheres of radius $\theta_j = 7\text{Å}$, (the corresponding values $\Gamma=3$, $f=1$ remain invariant for $\theta_j \geq 7\text{Å}$) *but in this case, these regions yield the highest interfacial tension* at $\Delta G_j \approx -T\Delta S_j = -k_B T c_b N f (4/3)\pi 7^3 10^{-30} m^3 \ln(3/4) = 13.17 k_B T$. The same contribution is associated with each disjoint $\theta = 7\text{Å}$ sphere covering convex regions $(\theta < -2.7 \text{ Å})$. By contrast, the "clathrate" range $0 \geq \theta \geq -2.7$ Å, where nonpolar moieties can be hydrated while preserving the tetrahedral hydrogen bond lattice ($\Gamma=4$) do not generate tension: as expected this leads to $\Delta S = 0$.

To summarize, three conclusions transpire from the nanoscale thermodynamics results: (a) The interfacial tension between protein and water is patchy and the result of both nanoscale confinement of interfacial water and local redshifts in dielectric relaxation; (b) the poor hydration of polar groups (a curvature-dependent phenomenon) generates interfacial tension, a property previously attributed only to hydrophobic patches; and (c) because of its higher occurrence at protein–water interfaces, the poorly hydrated dehydrons become collectively bigger contributors to the interfacial tension than the rarer nonpolar patches on the protein surface.

These conclusions open up an avenue to actually implement a *nanoscale thermodynamics approach to drug design*, where no doubt the dehydron concept will play a pivotal role. This rigorous treatment heralds the advent of drugs adapted to reduce the tension of protein–water interfaces, hence stabilizing the structure of the protein target. This strategy takes us back to the wrapping concept: patches that generate significant interfacial tension between target and water may be favorably wrapped by the drug, and we already know how to do that.

14.2 Disrupting Protein–Protein Interfaces with Small Molecules

Protein–protein associations, be they ephemeral or obligatory, constitute core biological events that enable basic cellular operations like signal transfer, catalytic activity, oligomeric functionalization. Understandably, their disruption using man-made ligands that have the capacity to bind competitively has become an important goal in molecular therapeutics [3]. Rational approaches to this drug design problem are based on the identification of "hot spots" of protein–protein interfaces [3]. These hot spots consist of interfacial residues that make the most significant contribution to the association free energy of the protein. In practice, they are operationally identified through "alanine scanning," a thermodynamic assessment of the effect of site-specific substitution on the complexation free energy. Thus, specific interfacial residues are mutated to alanine (essentially a truncation of the side chain) and the effect is quantified by $\Delta \Delta G$, the differential free energy of association. This quantity is in turn obtained calorimetrically by comparing the association affinity of mutant and wild-type protein. Alanine scanning has been used in conjunction with drug design with moderate success to selectively disrupt hormone-receptor associations using competitive man-made ligands [3].

By definition, patches that significantly contribute to protein–water interfacial tension must translate into hot spots for protein–protein associations. Hence the laborious alanine scanning may be effectively replaced by a direct computational identification of patches of interfacial tension in the free subunits. This computational advance will no doubt simplify the engineering of therapeutic interference with protein-protein associations.

References

1. Chandler D (2005) Interfaces and the driving force of hydrophobic assembly. Nature 437: 640–647
2. Debye P (1929) Polar Molecules. Dover, New York
3. Wells JA, McClendon CL (2007) Reaching for high-hanging fruit in drug discovery at protein–protein interfaces. Nature 450:1001–1009

Epilogue

This book advocates a rational translational approach to molecular-targeted therapy based on the wrapping concept. As such, it stands at the antipodes of drug discovery endeavors based on high-throughput screening and trial-and-error approaches. Before incarnating as a molecular design concept in Chaps. 7, 8, 9, 10, 11, 12, 13, and 14, wrapping is explored from architectural, biophysical, bioinformatics, and evolutionary perspectives in preparatory chapters 1, 2, 3, 4, 5, and 6. Wrapping is a novel category in structural biology, in which the 3D structure of the target protein is examined in relation to the surrounding solvent. Thus, wrapping becomes particularly insightful as we attempt to examine the protein–water interface. The emerging drug discovery platform is rooted in fundamental principles that shaped our current understanding of biological water. This "wrapping technology" is articulated with the aid of a bioinformatics toolbox that enables us to explore the biomolecular and evolutionary basis of drug specificity and encode the necessary information in a digitalized matrix format. In this regard, the book advocates that controlled specificity is not only the safest and therapeutically most efficacious alternative but also a plausible one within the realm of rational design.

While the book portends to introduce transformative concepts, capable of broadening the technological base of the pharmaceutical industry, we are acutely aware that a sobering note is in order. Wrapping designs are physically sound and seem to withstand the test of academic scrutiny and, yet, ultimately, only clinical trials can fully assess their therapeutic value and safety. In this regard, and from a practical perspective, current business models in the pharmaceutical R&D may not be sufficiently resilient to accommodate this type of core innovation.

Current business models predicated around prevalent discovery paradigms appear – at least to some of us – to be getting progressively obsolete. Thus, more than ever, the lead in the pharmaceutical industry will depend on alternative models with the suppleness to incorporate innovative research and harvest its fruits. As far as we can see, this type of innovation can come from only one source: *fundamental knowledge*. On the other hand, the business context required for its effective exploitation remains elusive and perhaps requires a new breed of entrepreneurship.

Standing at the cross-roads of drug discovery and academic pursuit, this book squarely addresses basic translational and integrative problems in the pharmaceutical industry. It does so by introducing fundamental concepts in basic

A. Fernández, *Transformative Concepts for Drug Design: Target Wrapping*,
DOI 10.1007/978-0-387-88630-5, © Springer-Verlag Berlin Heidelberg 2010

biomolecular research that hold potential to transform molecular engineering, with the overarching goal of designing safer drugs with reduced or at least controllable side effects. The actual implementation of the purported innovations within current or future business models lies outside the scope of this monograph, yet it would be desirable – if not imperative – that both sides of the endeavor evolve harmoniously together.

Index